BIOSENSORS
in
FOOD PROCESSING,
SAFETY,
and
QUALITY CONTROL

Contemporary Food Engineering

Series Editor

Professor Da-Wen Sun, Director

Food Refrigeration & Computerized Food Technology
National University of Ireland, Dublin
(University College Dublin)
Dublin, Ireland
http://www.ucd.ie/sun/

Biosensors in Food Processing, Safety, and Quality Control, *edited by Mehmet Mutlu* (2011)

Physicochemical Aspects of Food Engineering and Processing, *edited by Sakamon Devahastin* (2010)

Infrared Heating for Food and Agricultural Processing, *edited by Zhongli Pan and Griffiths Gregory Atungulu* (2010)

Mathematical Modeling of Food Processing, *edited by Mohammed M. Farid* (2009)

Engineering Aspects of Milk and Dairy Products, *edited by Jane Sélia dos Reis Coimbra and José A. Teixeira* (2009)

Innovation in Food Engineering: New Techniques and Products, *edited by Maria Laura Passos and Claudio P. Ribeiro* (2009)

Processing Effects on Safety and Quality of Foods, *edited by Enrique Ortega-Rivas* (2009)

Engineering Aspects of Thermal Food Processing, *edited by Ricardo Simpson* (2009)

Ultraviolet Light in Food Technology: Principles and Applications, *Tatiana N. Koutchma, Larry J. Forney, and Carmen I. Moraru* (2009)

Advances in Deep-Fat Frying of Foods, *edited by Serpil Sahin and Servet Gülüm Sumnu* (2009)

Extracting Bioactive Compounds for Food Products: Theory and Applications, *edited by M. Angela A. Meireles* (2009)

Advances in Food Dehydration, *edited by Cristina Ratti* (2009)

Optimization in Food Engineering, *edited by Ferruh Erdoğdu* (2009)

Optical Monitoring of Fresh and Processed Agricultural Crops, *edited by Manuela Zude* (2009)

Food Engineering Aspects of Baking Sweet Goods, *edited by Servet Gülüm Sumnu and Serpil Sahin* (2008)

Computational Fluid Dynamics in Food Processing, *edited by Da-Wen Sun* (2007)

Contemporary Food
Engineering Series
Da-Wen Sun, Series Editor

BIOSENSORS
in
FOOD PROCESSING, SAFETY, and QUALITY CONTROL

EDITED BY

Mehmet Mutlu

CRC Press
Taylor & Francis Group
Boca Raton London New York

CRC Press is an imprint of the
Taylor & Francis Group, an **informa** business

CRC Press
Taylor & Francis Group
6000 Broken Sound Parkway NW, Suite 300
Boca Raton, FL 33487-2742

First issued in paperback 2017

© 2011 by Taylor and Francis Group, LLC
CRC Press is an imprint of Taylor & Francis Group, an Informa business

No claim to original U.S. Government works

ISBN-13: 978-1-4398-1985-2 (hbk)
ISBN-13: 978-1-138-11600-9 (pbk)

Library of Congress Cataloging-in-Publication Data

Biosensors in food processing, safety, and quality control / edited by Mehmet Mutlu.
     p. ; cm. -- (Contemporary food engineering)
  Includes bibliographical references and index.
  Summary: "This book details the latest developments in sensing technology and its application in the food industry. It explores the opportunities created by chemical and biosensensing technology and improvements performed in recent years for better food quality, better food safety, better food processing and control, and better input for the food industry. The chapters in this book have been divided into three sections: basic principles of chemical and biosensing technology, biosensors for food processing and control, and biosensors for food safety."--Provided by publisher.
  ISBN 978-1-4398-1985-2 (hardcover : alkaline paper)
  1. Food--Safety measures. 2. Food--Food--Quality. 3. Biosensors. I. Mutlu, Mehmet, editor. II. Title. III. Series: Contemporary food engineering (Unnumbered)
  [DNLM: 1. Biosensing Techniques. 2. Food Handling. 3. Food Contamination--prevention & control. 4. Food Technology. WA 695]

TX531.B56 2011
363.19'26--dc22   2010043720

Visit the Taylor & Francis Web site at
http://www.taylorandfrancis.com

and the CRC Press Web site at
http://www.crcpress.com

Contents

Series Preface

CONTEMPORARY FOOD ENGINEERING

Food engineering is the multidisciplinary field of applied physical sciences combined with the knowledge of product properties. Food engineers provide the technological knowledge transfer essential to the cost-effective production and commercialization of food products and services. In particular, food engineers develop and design processes and equipment in order to convert raw agricultural materials and ingredients into safe, convenient, and nutritious consumer food products. However, food engineering topics are continuously undergoing changes to meet diverse consumer demands, and the subject is being rapidly developed to reflect market needs.

In the development of food engineering, one of the many challenges is to employ modern tools and knowledge, such as computational materials science and nanotechnology, to develop new products and processes. Simultaneously, improving food quality, safety, and security continue to be critical issues in food engineering study. New packaging materials and techniques are being developed to provide more protection to foods, and novel preservation technologies are emerging to enhance food security and defense. Additionally, process control and automation regularly appear among the top priorities identified in food engineering. Advanced monitoring and control systems are developed to facilitate automation and flexible food manufacturing. Furthermore, energy saving and minimization of environmental problems continue to be important food engineering issues, and significant progress is being made in waste management, the efficient utilization of energy, and the reduction of effluents and emissions in food production.

The *Contemporary Food Engineering Series*, consisting of edited books, attempts to address some of the recent developments in food engineering. Advances in classical unit operations in engineering applied to food manufacturing are covered as well as such topics as progress in the transport and storage of liquid and solid foods; heating, chilling, and freezing of foods; mass transfer in foods; chemical and biochemical aspects of food engineering and the use of kinetic analysis; dehydration, thermal processing, nonthermal processing, extrusion, liquid food concentration, membrane processes, and applications of membranes in food processing; shelf life, electronic indicators in inventory management; sustainable technologies in food processing; and packaging, cleaning, and sanitation. The books are aimed at professional food scientists, academics researching food engineering problems, and graduate-level students.

The books' editors are leading engineers and scientists from many parts of the world. All the editors were asked to present their books to address the market need and pinpoint the cutting-edge technologies in food engineering.

All contributions are written by internationally renowned experts who have both academic and professional credentials. All authors have attempted to provide critical, comprehensive, and readily accessible information on the art and science of a relevant topic in each chapter, with reference lists for further information. Therefore, each book can serve as an essential reference source to students and researchers in universities and research institutions.

Da-Wen Sun
Series Editor

Preface

A healthy life, a suitable environment, sustainable high-quality food, and inexpensive energy are inevitable components of a better life for human beings. With respect to food, the utmost level of health standards in the process "from field to fork" is vital. Due to this fact, along with other fields, food engineering and technology are also being transformed by continually increasing levels of automation. While the objective in other sectors of industry is simply to increase efficiency in food technology due to system theory and safety considerations, a high level of automation is required. The processes are complex; generally multifunctional control with feedback is employed, safety requirements allow for only a small degree of tolerance in the measurements, and human error as a risk factor needs to be eliminated.

During the last two decades, a rapid technological evolution has occurred in the field of chemical sensors in general, and biorecognition element–based sensors, or *biosensors*, in particular. It is fueled by an ever-growing need for improved sensors for early detection, which would allow remedial steps in a shortened time period for biomedical, industrial, environmental, and military applications. The success in biosensors is owed as much to the fundamental research in finding novel biorecognition mechanisms as to a number of rapidly evolving technologies, such as micro/nanofabrication of sensors and the production and immobilization of enhanced biorecognition elements.

A biosensor consists of two main parts: *biorecognition agent*(s) and *physical transducer*(s). The biological part of a biosensor is the unique part of the "instrument" that separates it from other sensors. Enzyme–substrate, antibody–antigen, DNA–DNA, and aptamer–target interactions are the most well known interactions used in biosensor design. The transducers, ranked in order of importance, include: electrochemical, optical, mass (piezoelectric), electrochemical/optical combination, and calorimetric (enzyme thermistor).

This book gives a brief summary about the past, present, and future of biosensors with an emphasis on food technology. Although we will see more advances in biosensors in the future, I believe this comprehensive and authoritative text will continue to serve the intended users for many years.

Food and chemical engineers, food technologists, and biochemists will find this book useful, as well as graduate students working in biosensor-related fields. It might also serve as a reference textbook for schools offering graduate courses in food technology and biosensors. With the help of leading scientists, I am most pleased to bring this book to the readership.

MATLAB® is a registered trademark of The MathWorks, Inc. For product information, please contact:

The MathWorks, Inc.
3 Apple Hill Drive
Natick, MA 01760-2098 USA
Tel: 508 647 7000
Fax: 508-647-7001
E-mail: info@mathworks.com
Web: www.mathworks.com

Mehmet Mutlu

Series Editor

Professor Da-Wen Sun, Ph.D., was born in Southern China and is a world authority on food engineering research and education. His main research activities include cooling, drying, and refrigeration processes and systems; quality and safety of food products; bioprocess simulation and optimization; and computer vision technology. His innovative studies on vacuum cooling of cooked meats, pizza quality inspection by computer vision, and edible films for shelf-life extension of fruits and vegetables have been widely reported in the national and international media.

Dr. Sun received first-class B.Sc. honors and an M.Sc. in mechanical engineering, and a Ph.D. in chemical engineering in China before working at various universities in Europe. He became the first Chinese national to be permanently employed in an Irish university when he was appointed college lecturer at the National University of Ireland, Dublin (University College Dublin) in 1995, and was then continuously promoted in the shortest possible time to senior lecturer, associate professor, and full professor. Dr. Sun is now professor of food and biosystems engineering and director of the Food Refrigeration and Computerized Food Technology Research Group at the University College Dublin.

As a leading educator in food engineering, Dr. Sun has contributed significantly to the field of food engineering. He has trained many Ph.D. students who have made their own contributions to the industry and academia. He has also, on a regular basis, given lectures on the advances in food engineering at academic institutions internationally and delivered keynote speeches at international conferences. As a recognized authority in food engineering, Dr. Sun has been conferred adjunct/visiting/consulting professorships from 10 top universities in China including Zhejiang University, Shanghai Jiaotong University, Harbin Institute of Technology, China Agricultural University, South China University of Technology, and Jiangnan University. In recognition of his significant contribution to food engineering worldwide and for his outstanding leadership in the field, the International Commission of Agricultural and Biosystems Engineering (CIGR) awarded him the CIGR Merit Award in 2000 and again in 2006; the Institution of Mechanical Engineers based in the United Kingdom named him Food Engineer of the Year 2004; in 2008 he was awarded the CIGR Recognition Award in recognition of his distinguished achievements as the top 1% of agricultural engineering scientists around the world; in 2007, Dr. Sun was presented with the AFST(I) Fellow Award by the Association of Food Scientists and Technologists (India); and in 2010, he was presented with the CIGR Fellow Award, the title of "Fellow" is the highest honor in CIGR, and is conferred to individuals who have made sustained, outstanding contributions worldwide.

Dr. Sun is a fellow of the Institution of Agricultural Engineers and a fellow of the Institution of Engineers of Ireland. He has also received numerous awards for teaching and research excellence, including the President's Research Fellowship, and has received the President's Research Award from the University College Dublin on two occasions. He is editor-in-chief of *Food and Bioprocess Technology—An International Journal* (Springer); series editor of the *Contemporary Food Engineering Series* (CRC Press/Taylor & Francis); former editor of the *Journal of Food Engineering* (Elsevier); and an editorial board member for the *Journal of Food Engineering* (Elsevier), the *Journal of Food Process Engineering* (Blackwell), *Sensing and Instrumentation for Food Quality and Safety* (Springer), and the *Czech Journal of Food Sciences*. Dr. Sun is also a chartered engineer.

On May 28, 2010, Dr. Sun was awarded membership to the Royal Irish Academy (RIA), which is the highest honor that can be attained by scholars and scientists working in Ireland. At the 51st CIGR General Assembly held during the CIGR World Congress in Quebec City, Canada, in June 2010, he was elected as incoming president of CIGR, and will become CIGR president in 2013–2014, the term of the presidency is six years, two years each for serving as incoming president, president, and past president.

Acknowledgments

I gratefully acknowledge the encouragement and kind collaboration of Professor Da-Wen Sun, the series editor of the *Contemporary Food Engineering Series*, for editing this book. I wish to express my deepest appreciation to my youngest but extraordinarily talented student, Nurşen Ziğal, for her great effort and patience to complete this book. I am sincerely thankful to my students Ebru Akdoğan, Eren Tur, Yasin Şen, Başak Beyhan Güdüllüoğlu, Nesrin Şir, Beyhan Günaydın, and Demet Ataman for their efforts in working on the preparation of this book.

Mehmet Mutlu

Contributors

Ebru Akdoğan
Plasma Aided Bioengineering and
   Biotechnology Research Group
Hacettepe University, Beytepe Campus
Ankara, Turkey

Salvador Alegret
Grup de Sensors i Biosensors
Departament de Química
Universitat Autònoma de Barcelona
Barcelona, Spain

İsmail Hakki Boyaci
Department of Food Engineering
Faculty of Engineering
Hacettepe University, Beytepe Campus
Ankara, Turkey

Carole Calas-Blanchard
Université de Perpignan
Perpignan Cedex, France

Mònica Campàs
Institut de Recerca: Tecnologia
   Agroalimentaries (IRTA)
Ctra. Poble Nou
Sant Carles de la Ràpita (Tarragona),
   Spain

Raghuraj S. Chouhan
Fermentation Technology and
   Bioengineering Department
Central Food Technological Research
   Institute
Mysore, India

Montserrat Cortina-Puig
Université de Perpignan
Perpignan Cedex, France

Frank Davis
Cranfield Health
Cranfield University
Bedford, United Kingdom

Séamus P. J. Higson
Cranfield Health
Cranfield University
Bedford, United Kingdom

Pınar Kara
Department of Analytical Chemistry
Faculty of Pharmacy
Ege University
Bornova, Izmir, Turkey

Ozan Kılıçkaya
Department of Analytical Chemistry
Faculty of Pharmacy
Department of Biotechnology
Ege University
Bornova, Izmir, Turkey

Jean-Louis Marty
Université de Perpignan
Perpignan Cedex, France

Selma Mutlu
Department of Chemical Engineering
Faculty of Engineering
Hacettepe University, Beytepe Campus
Ankara, Turkey

Thierry Noguer
Université de Perpignan
Perpignan Cedex, France

Mehmet Şengün Özsöz
Department of Analytical Chemistry
Faculty of Pharmacy
Ege University
Bornova, Izmir, Turkey

María Isabel Pividori
Grup de Sensors i Biosensors
Departament de Química
Universitat Autònoma de
    Barcelona
Barcelona, Spain

Beatriz Prieto-Simón
Université de Perpignan
Perpignan Cedex, France

P. Narender Raju
National Dairy Research Institute
Karnal (Haryana), India
and
National Academy of Agricultural
    Research Management
Hyderabad, India

K. Hanumantha Rao
National Academy of Agricultural
    Research Management
Hyderabad, India

Munna S. Thakur
Fermentation Technology and
    Bioengineering Department
Central Food Technological Research
    Institute
Mysore, India

José S. Torrecilla
Department of Chemical Engineering
Universidad Complutense de Madrid
Madrid, Spain

Aaydha C. Vinayaka
Fermentation Technology and
    Bioengineering Department
Central Food Technological Research
    Institute
Mysore, India

Liju Yang
Biomanufacturing Research Institute
    and Technology Enterprise (BRITE)
Department of Pharmaceutical Sciences
North Carolina Central University
Durham, North Carolina

1 Amperometric Biosensors in Food Processing, Safety, and Quality Control

İsmail Hakki Boyaci and Mehmet Mutlu

CONTENTS

1.1   INTRODUCTION

The area of electrochemical biosensors, particularly amperometric biosensors, has greatly benefited from the remarkable advances made in micro- and nanoelectronics during the last decade to give powerful and reliable instrumentation.

Although various types of transducers may be combined with different types of biorecognition elements to build a certain type of a biosensor, *amperometric biosensors* have a distinct advantage over the second most important and popular transducers and biosensors, *optic biosensors*. The advantage of amperometric biosensors (mainly enzyme electrodes and microbial sensors) over optic biosensors can be exemplified with the *fermentation* processes, which have an important role in biotechnological processes (Karube et al. 1991). Precise control of these systems is essential if complex and changeable reaction mixtures and broths are to be used on the industrial scale. Without such control substrates, products, catalysts, pH, and other variables cannot be kept at optimal levels. If food processing is considered, most of these materials in broth can be determined by spectrophotometric methods. When natural materials are used as raw materials in fermentations, however, the broths are not optically clear and may therefore be unsuitable for online measurement. Rapid and sensitive online monitoring and control of the variable factors already mentioned calls for sensors that are not only specific to the substrates and products of fermentation but can also measure the numbers of viable whole cells present in the fermentation broth.

Biosensor design and technology play a prominent role in fulfilling these criteria because they have distinct advantages: For example, test samples do not need to be optically clear and can be measured over a wide range of concentrations without pretreatment. Biosensors also offer the possibility of real-time analysis, which is particularly important for the rapid measurement of analytes in industry, for example, in process monitoring and control, in which there is a demand for the in-situ determination of flow rates, levels of contaminants, and so on.

This chapter deals with the basic mechanism and principles of amperometric biosensors and their potential usage in food processing, safety, and quality control.

1.2   AMPEROMETRIC BIOSENSORS

A *biosensor* is a device incorporating a biological sensing element either intimately connected to or integrated within a transducer (Turner et al. 1987). This device responds selectively and reversibly to the concentration or activity of chemical species in biological samples. No mention is made here of a biologically active material involved in the device; thus, any sensor physically or chemically operated in biological samples can be considered as a biosensor. Although all types of sensing elements and transducers are attempted to be used, developed, or commercialized as biosensors, amperometric biosensors have a special importance due to their historical background. Furthermore, commercially, the amperometric type is by far the most important.

1.2.1 PRINCIPLES OF AMPEROMETRIC TRANSDUCTION

The measure of relation between electric current and potential (I-V) in an electrochemical cell is called *voltammetry*, and it is the main output of the amperometric transducer. If an amperometric enzyme electrode is considered, the main function of the enzyme is to generate (or to consume) an electroactive species in a stoichiometric relationship with its substrate or target analyte. In amperometry, the electrode is held at a constant potential to detect the relevant electroactive species by oxidation or reduction. The optimum potential for detection in amperometry is chosen after obtaining the current response of the analyte as a function of electrode potential (Wijayawardhana and Heineman 2002). It means that the amperometric transducer allows the electrochemical reaction (oxidation or reduction) to proceed at the electrode surface, giving rise to a current. This current is directly related to the bulk substrate concentration. It should also be noted that in detecting an analyte in a complex solution like blood, serum, industrial wastewater, juice, broth, and so forth, the presence of interferents is inevitable. The potential interferents in the amperometric detection of food, food products, and food processing are discussed in another section of this chapter.

In amperometry, electrochemical processes are generally complex and may be considered a succession of electron transfers and chemical events. Those mechanisms are well defined in the literature. The overall sensor current is thus dependent on many factors, including charge transfer, adsorption, chemical kinetics, diffusion, convection, and substrate mass transport. Numerous relationships, each giving the generalized equation of current as a function of these parameters, have been elaborated (Turner et al. 1987; Wilson 1987; Rogers and Mascini 1998; Mutlu et al. 1998; Thevenot et al. 2001). Many types of amperometric biosensors have been developed keeping the above-mentioned factors in regard.

There are several books that deal with biosensors in general, and some also have a significant part dedicated to electrochemical biosensors (Eggins 1996; Turner et al. 1997; Cunningham 1998; Diamond 1998; Ramsay 1998; Law et al. 2002).

1.2.2 AMPEROMETRIC ENZYME ELECTRODE

An enzyme-based electrochemical biosensor is the most popular combination of biosensor investigated in the laboratories and placed in the market. An amperometric electrode is generally preferred in electrochemical sensor due to its easy and cheap setup. A typical amperometric sensor consists of a three-electrode (working, reference, and auxiliary electrodes) cell, a voltage source, and a device to measure voltage or current change. First, a biochemical reaction is catalyzed by an enzyme in the bioactive layer of the sensor. Then products of the reaction are oxidized or reduced on the working electrode. Current or voltage change on the electrochemical system is proportional to the concentration of the electroactive product and also the target molecule catalyzed by the enzyme. This is the basic reaction that takes place in the biochemical detection using an enzyme-based amperometric biosensor. While it seems that the amperometric biosensor can be built easily, its construction varies depending upon the properties of the target molecules, matrix structures of the sample from which the target molecule is analyzed, and the sensor structure that is used in biosensor design.

Clark's *oxygen electrode* is historically the first transducer associated with soluble glucose oxidase for glucose monitoring (Clark and Lyons 1962). Clark built this biosensor for monitoring of blood glucose level by following glucose oxidase (GOD) enzyme activity in the presence of glucose and oxygen. The sensor consisted of a platinum or gold disk cathode polarized at a potential of about –700 mV versus a silver/silver chloride anode having a ring shape. The set was covered with a polymeric film, such as polytetrafloroethylene (PTFE), polypropylene (PP), polyvinylchloride (PVC), and so forth, with a high permeability toward oxygen.

This arrangement allows oxygen reduction at the cathode to proceed. Different steps are involved in this electrochemical process, leading finally to water as shown in the overall reaction:

$$\frac{1}{2}O_2 + 2H^+ + 2e \xrightarrow{\text{Pt}} H_2O \qquad (1.1)$$

The measured cathodic current resulting from this electrochemical reduction is directly proportional to the oxygen level in the solution. Thus the depletion of oxygen at the biosensor tip due to the oxidase reaction can be easily correlated with the substrate concentration, and a linear relationship between the current variation and the analyte concentration can be established in a definite range.

Due to the importance and the nature of *oxidases*-based reactions in biological systems and food processing, which is given in detail in Section 1.2.3 of this chapter, the detection of the by-product *hydrogen peroxide* can also be performed using a platinum disc anode polarized at about +700 mV versus a silver/silver chloride reference. At this potential, hydrogen peroxide oxidation occurs at the platinum surface, according to the reaction

$$H_2O_2 \xrightarrow{\text{Pt}} O_2 + 2H^+ + 2e^- \qquad (1.2)$$

The enzymic membrane is positioned on the sensor tip confining a buffered and chlorinated solution. The sensor responds linearly to the hydrogen peroxide generated by the enzyme reaction, the output current being correlated with the analyte concentration.

The amperometric enzyme electrodes operating with the reactions given above are called *first-generation biosensors* (Figure 1.1a). The main drawback of these sensors is that the reaction is limited to the ambient partial pressure of oxygen (pO_2), which may vary during the measurement. The system needs to be calibrated frequently or the sample should be diluted to eliminate the kinetically insufficient oxygen supply.

Limitations of oxygen-dependent measuring techniques lead the researchers to develop sensors that are insensitive to oxygen. This topic is discussed in Section 1.2.3 of this chapter.

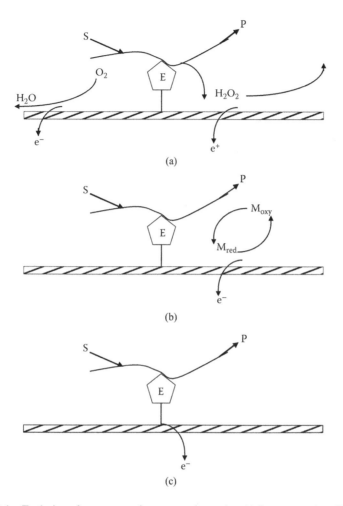

FIGURE 1.1 Evolution of amperometric enzyme electrodes: (a) first-generation, (b) second-generation, and (c) third-generation electrodes.

1.2.3 MEDIATED AMPEROMETRIC ENZYME ELECTRODES

Mediated amperometric enzyme electrodes are also classified as "second-generation biosensors" (Figure 1.1b). Due to the absence of oxygen (anaerobic) or insufficient oxygen concentration in the media, mediators were developed as an oxygen substitute (Cass et al. 1984).

The reaction sequence for glucose monitoring utilizing flavine adenine dinucleotide (FAD) is given below:

Enzyme Reaction

$$\text{GOD-FAD} + \text{Glucose} \longleftrightarrow \text{GOD-FADH}_2 + \text{Gluconic Acid}$$

$$\text{GOD-FADH}_2 + \text{Med}_{(oxidized)} \longleftrightarrow \text{GOD-FAD} + \text{Med}_{(reduced)}$$

TABLE 1.1

Oxidases Applied in Amperometric Biosensors

Enzyme	Substrate
Glucose oxidase	Glucose
Lactate oxidase	Lactate
Cholin oxidase	Choline
Alcohol oxidase	Ethanol
	Methanol
	Formaldehyde
Glutamate oxidase	Glutamate
Tryptophan-2-monooxygenase	Tryptophan
Lysine oxidase	Lysine
Xantine oxidase	Hypoxantine

Sources: Dzyadevych et al. 2008; Prodromidis and Karayannis 2002; Mehrvar and Abdi 2004; Alaejos and Montelongo 2004; Mello and Kubota 2002.

Electrode Reaction

$$Med_{(reduced)} \longleftrightarrow Med_{(oxidized)} + e^-$$

Hill, Higgins, and coworkers (Davis et al. 1984) have shown that various ferrocene ferrocinium couples are efficient mediators.

The list of oxidases mainly used in amperometric enzyme electrodes and their products are given in Table 1.1.

From a food technology point of view, another very useful mediator in biosensors is nicotinamide adenine dinucleotide (NAD^+), a cofactor found in more than 250 reactions involving dehydrogenases (DH) (Wijayawardhana and Heineman 2002). The basic detection scheme can be represented as follows:

Enzyme Reaction

$$NAD^+ + SH_2 \xleftarrow{\text{dehydrogenase}} S + NADH + H^+$$

Electrode Reaction

$$NADH \longleftrightarrow NAD^+ + H^+ + 2e^-$$

Although such a detection scheme can be used in principal to detect any one of the numerous dehydrogenases or their substrates, the high cost of NAD^+ limits its use. Furthermore, unlike ferrocene, NAD^+ is highly soluble and thus not easily immobilized for making solid-state sensors. Those drawbacks led the researchers to develop a new generation of biosensors, which is the topic of the next section. The list of dehydrogenases mainly used in amperometric biosensors is presented in Table 1.2.

TABLE 1.2

Dehydrogenases Applied in Amperometric Biosensors

Enzyme	Substrate
Aldehyde dehydrogenase	Aldehydes
Alcohol dehydrogenase	Ethanol
Lactate dehydrogenase	Lactate
Glutamate dehydrogenase	Glutamate
Glucose dehydrogenase	Glucose
Glycerol dehydrogenase	Glycerol

Sources: Dzyadevych et al. 2008; Prodromidis and Karayannis 2002; Mehrvar and Abdi 2004; Alaejos and Montelongo 2004; Mello and Kubota 2002.

1.2.4 AMPEROMETRIC ENZYME ELECTRODES WITH NONDIFFUSING MEDIATORS

The *third-generation biosensors* are based on redox polymers and associated redox monomers (Figure 1.1c). As a matter of fact, the need for reagentless biosensors arises in order to avoid leaching, because the ultimate purpose is to prepare an implantable biosensor to be used in humans. Leaching would cause degradation of the sensor and, moreover, release potentially toxic substances into the body.

Enzymes generally tend to denature at surfaces of electrodes. If the general mechanism of the first- and second-generation biosensors is considered, direct electron transfer between the enzyme, which can be considered as a natural insulator, and the electrode occurs extremely slowly. Unfortunately, the ideal situation of perfectly immobilized enzymes with direct electrical contact between the redox centers of the enzyme and electrode has been difficult to realize. Therefore an electron acceptor mediator is required to make the electrochemical reaction rapid and effective. The search for such a mediator has indeed become an important and active area of research, and some successful results were obtained by employing conducting organic salts (Albery et al. 1987a,b) and redox polymers (Karan et al. 1994; Akmal and Usmani 2002).

1.2.5 MULTIENZYME ELECTRODES

Today, *immobilized enzymes* have become buzzwords for people developing new enzyme applications. One of the most important pieces of information about an enzyme is its specific activity. This describes the enzymatic strength or the concentration of enzyme activity. Enzyme activity is derived from an experimental measurement of its ability to catalyze a reaction under set conditions. It is a measure of the amount of product formed or substrate reacted in a unit of time per weight of enzyme preparation (Hartmeier 1988).

Application areas where biosensors are set to make a significant impact reach well beyond the established needs of medicine and veterinary science, including food processing and bioprocessing, environmental monitoring and control, and pharmaceuticals where efficient access to biochemical information has always been at a premium (Vadgama and Crump 1992).

The glucose oxidase enzyme electrode has received the greatest attention as described in previous sections. However, the substrate to be detected, D-glucose, is supposed to be in [β] form. Therefore, the measurement of the β-D-glucose can be managed by employing a single enzyme combined with an amperometric transducer. Of course, this method is easy and consequently the ultimate way to reach the main goal, that is, the precise determination of the concentration of substrate with a single enzyme electrode.

In some special cases of analyte tracing, enzme activity measurement, or some other process, two or more enzymes may be needed for tracking (Krysteva and Yotova 1992; Loechel et al. 1998; Mello et al. 2003; Rotariu and Bala 2003; Li et al. 2006; Morales et al. 2007). A typical example in food processing engineering is the determination of enzymatic activity of invertase by an amperometric glucose oxidase (EC 1.1.3.4 from *Aspergillus niger*) electrode (Mutlu et al. 1997). The recognition system of the sensor needs to be extended by a second enzyme, mutarotase (EC 5.1.3.3 from porcine kidney), which converts the α-D-glucose to β-D-glucose. The enzymatic and amperometric reaction sequences of the process are given below:

$$\text{Sucrose} + H_2O \xrightarrow{\text{invertase}} \alpha\text{-D-glucose} + \text{D-fructose}$$

$$\alpha\text{-D-glucose} \xrightarrow{\text{mutarotase}} \beta\text{-D-glucose}$$

$$\beta\text{-D-glucose} + O_2 \xrightarrow{\text{glucose oxidase}} \text{gluconic acid} + H_2O_2$$

$$H_2O_2 \xrightarrow[700\,mV]{\text{Pt electrode}} 2H^+ + O_2 + 2e^-$$

The last three steps of the reactions take place in the electrode. So, this would enable us to determine the immobilized invertase (EC 3.12.1.26. from bakers yeast) activity in terms of electrode response where the enzyme is immobilized on spherical particules of ethyleneglycoldimetacrilate (EGDMA) and acrylamide (AAm) copolymer by the method described elsewhere (Okubo and Takahashi 1994).

A similar method for determination of lactose needs to be proposed due to a medical condition called lactose intolerance, which is the inability to metabolize lactose, caused by a lack of lactase enzyme in the digestive system. The usage of amperometric lactose sensors for dairy products is discussed in Chapter 9. The enzymatic and anodic reaction sequences are given below (Boyaci and Mutlu 2002):

$$\text{Lactose} + H_2O \xrightarrow{\beta\text{-galaktosidase}} \alpha\text{-D-glucose} + \text{D-fructose}$$

$$\alpha\text{-D-glucose} \xrightarrow{\text{mutarotase}} \beta\text{-D-glucose}$$

$$\beta\text{-D-glucose} + O_2 \xrightarrow{\text{glucose oxidase}} \text{gluconic acid} + H_2O_2$$

$$H_2O_2 \xrightarrow[700\,mV]{\text{Pt electrode}} 2H^+ + O_2 + 2e^-$$

TABLE 1.3

Enzyme Combinations for Multienzyme Electrodes in Food Technology

Enzymes	Substrate
Cytrate lyase–oxaloacetatecarboxilase–piruvate oxidase	Cytrate
Invertase–mutarotase–glucose oxidase	Sucrose
Amyloglucosidase–glucose oxidase	Maltose
β-Galactosidase–glucose oxidase	Lactose
Acetylcholinesterase–choline oxidase	Acetyl choline
Alcaline phosphatase–glucose oxidase	Phosphates
Phospholipase–choline oxidase	Lecithin

From a food processing engineering point of view, another typical example is the determination of the concentration of starch.

$$\text{Starch} + H_2O \xrightarrow{\text{amyloglucosidase}} \alpha\text{-D-glucose}$$

$$\alpha\text{-D-glucose} \xrightarrow{\text{mutarotase}} \beta\text{-D-glucose}$$

$$\beta\text{-D-glucose} + O_2 \xrightarrow{\text{glucose oxidase}} \text{gluconic acid} + H_2O_2$$

$$H_2O_2 \xrightarrow[\text{700 mV}]{\text{Pt electrode}} 2H^+ + O_2 + 2e^-$$

The list of possible enzyme combinations to determine specific substrates mainly in food processing engineering and food technology is given in Table 1.3.

In all multiple enzyme electrode researches, it should be noted that the kinetic and dynamic behavior of the electrode should be on the basis of mass transfer and diffusion properties of intermediate products, apparent kinetic constants, and mole ratio of enzymes leading to minimize the rate restricting steps that determine the overall performance of the device. The mathematical modeling of such reaction sequence including enzymatic and amperometric steps was intensively studied (Albery and Knowles 1976; Albery and Craston 1987; Mutlu et al. 1998).

1.3   BASIC CONSTRUCTION AND MEASUREMENT PRINCIPLES

Since the development of Clark's oxygen electrode in 1962, the measuring principles of the amperometric electrode have not changed drastically. However, regarding to the basic needs for an amperometric biosensor, such as (1) linearity, (2) sensitivity, (3) response time, (4) signal-to-noise ratio, and (5) selectivity, the design of the biosensor has shown a variety from a sandwich type to a single layer, from a self-assembled monolayer to carbon nanotubes, due to findings in novel biorecognition mechanisms and remarkable advances in polymer science and technology, microelectronics and micro- and nanotechnologies in the last two decades.

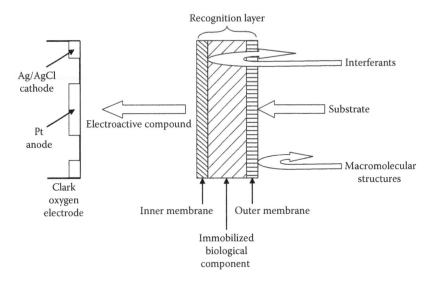

FIGURE 1.2   Functions of membrane layers in a sandwich-type enzyme electrode.

First, in order to construct an amperometric biosensor, we must understand the mechanism of the sensing system and consider the factors affecting the performance. The operation of a biosensor integrated with an amperometric transducer can be seen as a five-step process: (1) the substrate transfers to the surface of the recognition layer; (2) depending on the recognition layer structure, the substrate diffuses in to the matrix where the biological component is immobilized; in some biosensor designs, the substrate directly meets with the biological component; (3) the reaction occurs and the electroactive component and side products diffuse in three dimensions; (4) the electroactive components transport through the perm-selective membrane covering the electrode surface; that membrane mainly keeps away the possible interferants from the electrode surface; and (5) the electroactive component is measured at the electrode surface. These steps are exemplified on a sandwich-type enzyme electrode in Figure 1.2 (Mutlu et al. 1998).

There are many options for the recognition layer in the first step, such as the "old-fashioned" sandwich type or the recent nanoparticules, and even though the sandwich type is the oldest approach in the literature, most of the commercialized sensors are still this type.

Numerous research articles and review papers in the literature exemplify the usage of sandwich-type recognition layers for food analysis, processing, and quality control. Those studies are concerning different biological components, such as enzymes (Moody and Thomas 1991; Moody 1992; Maines et al. 1996; Cokeliler et al. 2002; Boyacı and Mutlu 2002; Gerard et al. 2002; Mello and Kubota 2002; Pal and Sarkar 2002; Prodromidis and Karayannis 2002; Mehrvar and Abdi 2004; Mak et al. 2005; Ricci and Palleschi 2005; Amine et al. 2006; Andreescu and Marty 2006; Ahuja et al. 2007; Nikolaus and Strehlitz 2008; Singh et al. 2008, 2009; Zeravik et al. 2009; Dinckaya et al. 2010), antibodies–antigens (Rogers 2000; Luppa et al. 2001;

Mello and Kubota 2002), microorganisms (Ellis and Goodacre 2001; Patel 2002; Venugopal 2002; Leonard et al. 2003; Sezginturk and Dinckaya 2003; Akyilmaz and Dinckaya 2005; Lei et al. 2006; Rasooly and Herold 2006), DNA and DNA segments (Zhang et al. 2007; Sassolas et al. 2008), aptamers (Kandimalla and Ju 2004; Papamichael et al. 2007; Tombelli et al. 2007; Cruz-Aguado and Penner 2008; Mairal et al. 2008; Wang et al. 2008; Tombelli and Mascini 2009; Kim et al. 2010), and molecular imprinted matrices (Holthoff and Bright 2007; Sadik et al. 2009). Furthermore, some other studies are carried out to limit the diffusion of macro-molecules into the recognition layer (Maestre et al. 2001; Mikeladze et al. 2002; Davis and Higson 2005). Some polymeric structures were developed or modified for repelling the potential interferents that may affect the signal positively (constructive interference) or negatively (destructive interference). It is quite normal that all types of interferences are not desired in amperometric transduction systems, and this sub-ject is discussed in Section 1.4.

Another approach in construction of amperometric biosensors is to focus mainly on shorter response time. For that purpose, in order to avoid mass transfer limita-tions created by the outer membrane and immobilized biomolecule matrix, only the inner membrane with an ion-selective property against interferents is modified for biomolecule immobilization (Biederman 2001; Boyaci et al. 2002; Akbayirli and Akyilmaz 2007; Carelli et al. 2007; Mita et al. 2007; Chen et al. 2009; Arecchi et al. 2010; Dinckaya et al. 2010). This group of electrochemical sensors is also a good example of immobilized whole DNA or DNA segments (Zhang et al. 2007; Palchetti et al. 2008; Sassolas et al. 2008; Galandova et al. 2009).

Figure 1.3 shows a comparative illustration of the mechanism of sandwich-type and single-layer enzyme electrodes (Mutlu et al. 1998). The linearity of the sand-wich-type enzyme electrode with an unmodified outer membrane is calculated to be 60 mmol glucose (Mutlu et al. 1994). After treating the surface of the outer mem-brane by plasma polymerization of a precursor, hexametyldisiloxane (HMDS), due to mass transfer limitation of the substrate through the outer membrane, extends the

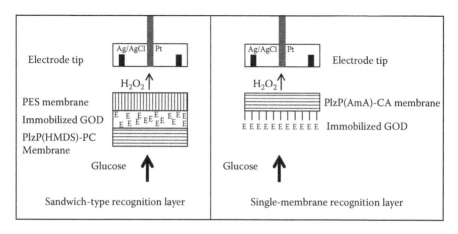

FIGURE 1.3 Comparative illustration of the mechanism of sandwich-type and single-layer enzyme electrodes.

linear range of the electrode up to 600 mmol glucose. Certainly, such an operation limits the mass transfer rate, and the response time increases up to 120 seconds. Controversially, the linear range of the single-layer glucose electrode is extended to 1000 mmol glucose concentration. Furthermore, the single-layer enzyme electrode also facilitates the transport of electroactive compound through the membrane layer 10-fold faster than the sandwich-type recognition layer. In this case, there is only one mass transfer barrier for hydrogen peroxide to reach the electrode surface.

In the literature, a group of researchers are exploiting the semipermeable membranes, which are permitting only the product of electroactive enzymatic reaction but no other electroactive substances in order to improve fast responsive electrodes (Alp et al. 2000; Muguruma et al. 2007; Ahuja et al. 2008; Gunaydin et al. 2010).

Numerous biosensor techniques have been reported that allow researchers to better study the kinetics, structure, and (solid/liquid) interface phenomena associated with protein-ligand binding interactions. The design and structural features of these devices—composed of a biological affinity element interfaced with a signal transducer—primarily determine their operational characteristics (Rogers 2000; Cosnier 2005).

Recently, conducting polymers have attracted much interest in the development of biosensors. The electrically conducting polymers are known to possess numerous features, which allow them to act as excellent materials for immobilization of biomolecules and rapid electron transfer for the fabrication of efficient biosensors (Gerard et al. 2002; Cosnier 2003; Vidal et al. 2003; Ahuja et al. 2007; Cosnier 2007; Singh 2009).

Biosensors based on screen-printed electrodes are miniature systems enabling *in vivo* and *in vitro* analysis of real samples by use of biodevices. Thus, the screen-printing (thick film) technique is widely used for large-scale fabrication of disposable biosensors with several advantages including low cost, versatility, and miniaturization. Screen-printed electrodes have enabled the production of modern sensors that can be incorporated in portable systems, an important requirement of analytical methods for direct analysis of a sample in its environment without alteration of the "natural" environmental conditions. The recent possibility of design and fabrication of screen-printed electrodes, including microelectrodes, chemically modified electrodes, and so forth, and incorporating these in a variety of highly sensitive biosensors has increased industrial, clinical, and environmental interest in this field (Vidal et al. 2003; Tudorache and Bala 2007).

Aptamers are single-stranded DNA or RNA ligands that can be selected from large libraries containing randomly created nucleic acids and used to target various molecules, ranging from proteins to small organic dyes. Because of the unique characteristics (e.g., versatility, specificity, and synthetic nature) of aptamers, which are mainly derived from their nucleic acid composition, these molecules represent ideal ligands for bioanalytical applications (Mairal et al. 2008; Tombelli and Mascini 2009).

Nanotechnology is also playing an increasingly important role in the development of biosensors. Various approaches to exploit the advances in nanoscience and nanotechnology for bioanalyses have recently been reported (Haruyama 2003; Jain 2003; Wang 2005; Chen et al. 2004; Zhang 2009). Such bioanalytical nanosensors can be used for detection of pathogens, toxins, nutrients, environmental characteristics, heavy metals, particulates, allergens, and so forth. Sensitivity, specificity, rapidity of testing, and other necessary attributes of biosensors can be improved by using nanomaterials in their construction. Guna and coworkers have developed a colorimetric nanobiosensor based on the chromogenic effect of latex microsphere hybridization with gold nanoparticles (Ko et al. 2010). A group of studies that prepared biosensors with nanoparticles was reported but none of them have yet utilized amperometric transducers (Baeumner 2004; Bhattacharya 2007; Adrian 2009).

1.4   THE INTERFERENCE-FREE BIOSENSORS

Despite the specificity of the biocomponents used in biosensors, these devices often suffer from some limitations such as long response time, short stability, poor reproducibility, and inefficient elimination of interferences.

As a well-known fact, one of the main problems to be tackled in the development of amperometric biosensors for food analysis is electroactive interferents that are commonly present in these matrices. A list of electroactive chemicals naturally present in the food or produced during the processing, fermentation, maturing, ripening, controlled or uncontrolled oxidizing, and so forth is summarized in Table 1.4.

In the general context of the amperometric enzyme electrode studies, in order to eliminate this problem, many attempts are made by employing natural barriers, such as gastric mucus or native porcine mucus (Vadgama and Alberti 1983; Desai et al. 1991, 1992; Ahmed et al. 2005).

Additionally, intensive studies were also carried out to stop the transfer of interfering components to the electrode surface by means of newly developed polymeric structures such as polyethersulphone (PES), polyethylethersulphone (PEES), and PVC, which are used as an inner membrane in sandwich-type enzyme electrodes (Mutlu et al. 1991, Christie et al. 1992, Higson et al. 1993).

In conventional enzyme electrodes, at least two membranes are placed on the electrode surface: One is an enzyme membrane and the other is a semipermeable membrane that eliminates electrochemical interferences. However, the required semipermeable membrane complicates the structure of enzyme electrode and also increases its cost. Therefore, the immobilization of the enzyme in a semipermeable membrane seems to be of particular interest. Since the molecular weights of the interferents are greater than 150 g/mole, semipermeable membranes with molecular weight cutoff of <150 g/mole are useful for preparation of enzyme electrodes to the substrates with relatively low molecular weights, for example, L-lactate- and ethanol-sensing electrodes (molecular weights of the analytes, 90 g/mole and 46 g/mole, respectively). Plasma modification of the surfaces of ion-selective membranes such as polyethylethersulphone or cellulose acetate membranes (to immobilize directly the enzyme for preparing the recognition layer of the biosensor) is one of the methods used commonly in literature (Mizutani et al. 1998; Biederman et al. 2001; Boyaci et al. 2002; Gunaydin et al. 2010). This approach not only solves the interference problem but also helps to overcome long response time, low sensitivity, and limited linearity problems. A successful correlation is obtained between glucose concentration and response time by controlling the mass transfer properties of glucose and

TABLE 1.4

Potential Electroactive Chemicals Present in the Food

Interferant for Amperometric Biosensing System/ Molecular Formula	Source	Determination Technique	References
Acetic Acid/ CH_3COOH	Acetic acid is one of the major components of fermentation of products such as *wines, vinegar, soy sauces, fruit juices*. Also, it is an aroma compound in *blueberries*.	1. Flow Injection Analysis (FIA) with Amperometric Trienzyme Sensor Detection 2. Gas Permeation Flow Injection Analysis 3. High-Performance Liquid Chromatography (HPLC) 4. Redox Reaction 5. Acid–Base Titration 6. Capillary Electrophoresis (CE)	Araujo et al. 2005 Hettiarachchi and Ridge 1998 Lenghor et al. 2002 Mizutani et al. 2003 Su and Chien 2010
Ascorbic Acid/ $C_6H_8O_6$	Ascorbic acid is found in blood plasma as well as many foods such as *vegetables, fruit, processed foods, milk products, whole diets, supplemented special diets, red and white wines, beer and various fruit juices, multivitamin tablets*, etc.	1. Dual Channel Flow Injection Electrochemical Detection System 2. Ion-Exclusion Chromatography and Amperometric Detection 3. Reversed-Phase Liquid Chromatography–Electrochemical (Lc–Ec) System Using a Glassy Carbon Electrode 4. Ion Exchange Chromatography Using a Platinum Electrode 5. Titrimetry 6. Fluorimetry 7. Spectrophotometry 8. High-Performance Liquid Chromatography (HPLC) 9. Carbon Ionic Liquid Electrode 10. Spectrofluorimetry 11. Polarography 12. Chemiluminescence 13. Polystyrene–Divinylbenzene Polymer Column (PLRP-S)	Cardwell and Christophersen 2000 Thangamuthu et al. 2007 Safavi et al. 2006 Barrales et al. 1998 Tabata and Morita 1997 Kall and Andersen 1999

(continued)

| Citric Acid/$C_6H_8O_7$ | Citric acid is widely used in the pharmaceutical and food industry as an additive. In the food industry, it is used as a pH stabilizer, as an acidifier to prevent the growth of microorganisms, as a sequestering agent for metal ions, and as a flavor and aroma agent. Its content meets the quality criteria for many kinds of food, mainly *fruit juices, soft drinks, and other citrus products.* | 1. Pulse Injection Analysis with Chemiluminescence Detection
2. Photospectrometric Analysis
3. Electroanalysis
4. Chromatography
5. High-Performance Liquid Chromatography (HPLC)
6. Reagent-Injection Spectrophotometric Determination
7. Fluorimetry
8. Spectrophotometry
9. Pyrolysis Mass Spectrometry (Py/Ms)
10. Gravimetric Analysis
11. Gas Chromatography–Mass Spectrometry (Gc–Ms)
12. Potentiometry and Biosensors
13. Enzymatic Flow Injection
14. Chemiluminescence Methods
15. Capillary Electrophoresis
16. Flow-Injection Spectrophotometry
17. Thermometric methods | Moreno-Cid et al. 2004
Zhike et al. 1998
Ghassempour et al. 2003
Themelis and Tzanavaras 2001
Volotovsky and Kim 1998
Luque-Perez 1998 |
| Ethyl Alcohol/C_2H_5OH | Ethanol in foods can be present as a product of fermentation, a residue after *baking*, an additive in packed products, and an ingredient in *candies, confectionery products, sweets*, etc. This is the case for various foods, including *bakery products* and specialty alcoholic beverages such as *Madeira wine* where ethanol can be produced by fermentation or added to improve | 1. Enzymatic–Spectrophotometric
2. Gas Chromatographic
3. High-Performance Liquid Chromatography (HPLC)
4. Refractive Index Method
5. Enzyme Biosensors
6. Microbial Biosensors | Nakamura et al. 2009
Pellegrino et al. 1999
Rotariu et al. 2004
Lerici et al. 1999 |

TABLE 1.4 (CONTINUED)
Potential Electroactive Chemicals Present in the Food

Interferant for Amperometric Biosensing System/ Molecular Formula	Source	Determination Technique	References
	the microbial stability and the sensorial properties. Alcoholic beverages such as *beer, wine, liquor,* and *spirits* contain ethyl alcohol.		
Lactic Acid/$C_3H_6O_3$	Lactic acid in foodstuffs is produced by bacterial fermentation and is an essential component related to manufacturing of *cheese, yogurt, fermented meat products, buttermilk,* etc.	1. Conventional Chemical Conversion of Lactic Acid 2. Enzymatic Reactions 3. Flow Injection Analysis (FIA) 4. Sequential Injection Analysis (SIA) 5. High-Performance Liquid Chromotography (HPLC) 6. Chemiluminescence 7. Fluorimetry 8. Spectrophotometry 9. Amperometry 10. Potentiometry 11. Gas Chromatography (GC) 12. Immobilized Enzyme Fluorescence Capillary Analysis (IE-FCA)	Yong-Sheng et al. 2008 Gomez-Alvarez et al. 1999 Martelli et al. 2001 Tan et al. 2005 Wu et al. 2005 Shu and Wu 2001
Malic Acid/$C_4H_6O_5$	Malic acid is a component of *soft drinks, fruit juices, wine,* and *honey.* Together with tartaric acid, L-malic acid is a fundamental component of	1. Chromatography 2. Enzymatic Reactions (Photometric and Electrochemical) 3. Microcalorimetric Technique 4. Multibeam Circular Dichroism (CD) Detector	Palma and Barroso 2002 Campo et al. 2006 Mazzei et al. 2007 Mato et al. 1997

	wine, representing about 90% of the total acidity.	5. High-Performance Liquid Chromatography (HPLC) 6. Biosensors 7. Proton Nuclear Magnetic Resonance Spectroscopy 8. Ultrasound-Assisted Extraction (UAE)	Yamamoto 2001 Antonelli 2008
Methyl Alcohol/CH_3OH	Since methanol resembles ethanol in odor and taste and is tax-free, it has been used as an adulterant in alcoholic beverages, which causes accidental and intentional intoxication. Methanol is cheap and easily accessible; it has been used in the production of imitation *spirits* and *wine*.	1. Gas Chromatography 2. Flow Injection 3. Enzymatic 4. Colorimetric 5. High-Performance Liquid Chromatography (HPLC) 6. Chromotropic Acid Colorimetric Method 7. Titrimetric Method 8. Enzymatic Method 9. Biosensor Detection 10. Near-Infrared Spectroscopy	Sekine et al. 1993 Mezcua et al. 2003 Buckley et al. 2001 Wang et al. 2004 Chen et al. 1998
Orthophosphoric Acid/H_3PO_4	Phosphoric acid is widely used in a great number of food and beverage products as a flavoring agent. Also, *healthier beverages and colas* contain phosphoric acid.	1. Ion Chromatography 2. Acid–Base Titration	Raposo Jr. et al. 2008 Kijkowska et al. 2002 Shirakawa et al. 2004 Tucker et al. 2006 http://www.titrations.info/acid-base-titration-phosphoric-acid
Oxalic Acid/$C_2H_2O_4$	*Rhubarb, carrots, peas, cocoa powder, fennel,* and *tea* contain oxalate. Oxalic acid naturally exists in *spinach, ginger, chocolate, beet root, mushrooms,* and *tea leaves.*	1. HPLC-Enzyme-Reactor Method 2. Catalytic Kinetic Spectrophotometric Mehtod 3. Permanganimetric Titration Method 4. Chromatography 5. Extractive Photometry 6. High-Performance Liquid Chromatography (HPLC)	Zhai et al. 2006 Zhai 2008 Ensafi et al. 2001 Zheng et al. 2009 Hönow et al. 1997

(continued)

TABLE 1.4 (CONTINUED)

Potential Electroactive Chemicals Present in the Food

Interferant for Amperometric Biosensing System/Molecular Formula	Source	Determination Technique	References
Sodium Benzoate/ $NaC_6H_5CO_2$	Sodium benzoate and/or potassium sorbate are the preservatives that are used in *juices* to inhibit mold growth, prevent spoilage, and preserve freshness. It is also in *fruit juices, soft drinks, synthetic syrups.*	7. Spectrophotometry 8. Gas Chromatography 9. Liquid Chromatography 10. Flow-Injection Catalytic Spectrophotometry 11. Ion Exclusion Chromatography 12. Multiwall Carbon Nanotubes (MWNTs) Modified Electrodes 13. Enzymatic Methods	Pylypiw, Jr. and Grether 2000 Pezza et al. 2001
Tartaric Acid/ $C_4H_6O_6$	Tartaric acid and malic acid are the most abundant organic acids in *grapes.* Tartaric acid normally presents in *wine* and affects its taste.	1. Titrimetry 2. UV-Spectrophotometry 3. Gas–Liquid Chromatography (GLC) 4. Thin-Layer Chromatography (TLC) 5. High-Performance Liquid Chromatography (HPLC) 6. Potentiometric Sensor 1. Titrimetry 2. UV-vis Spectrophotometry 3. Chromatography 4. Capillary Electrophoresis 5. Infrared Spectrophotometry 6. Potentiometry 7. Chemiluminescence 8. Flow Injection Analysis (FIA) Process	Mallet et al. 1999 Fernandes and Reis 2006 Palma and Barroso 2002 Mori et al. 1999

oxygen (Koochaki et al. 1993; Mutlu et al. 1995) or extended linearity and shortened response time (Mutlu et al. 1997, 1998; Alp et al. 2000).

A valid approach to overcome the interference problem is presented by the development of nonconducting polymeric films with built-in permselectivity (Malitesta et al. 1990; Centonze et al. 1992; Palmisano et al. 1995a,b, 2000; Guerrieri et al. 1998; Ciriello et al. 2000), which have been used successfully for an efficient elimination of common interfering species (i.e., ascorbate, urate, paracetamol, and cysteine) during glucose and lactate determination in food or biological matrices (Palmisano et al. 1996; Centonze et al. 1997). A further improvement of the sensor selectivity can be obtained by means of electrosynthesized multilayered structures (Palmisano et al. 1995a; Guerrieri et al. 2006). Selection of monomers and deposition procedures are critical steps in the production of efficient systems. For example, as evident in a recently published paper (Badea et al. 2003), the anti-interferent properties toward ascorbic acid and acetaminophen of two new monomers (2-(4-aminophenyl)-ethylamine and 2,6-dihydroxynaphtalene) copolymerized onto Pt surface are poor with respect to other bilayer films (Palmisano et al. 2000).

Another source of interference is represented by the analyte itself (Carelli et al. 2006), which can be directly oxidized at the transducer before being converted by the enzyme. This phenomenon, which gives interference currents and a time-dependence deterioration of the biosensor response due to the adsorption of the relevant electrode oxidation by-products, depends also on the electrode material used as the transducer. A successful study was performed to minimize the fouling and the interference caused by both the direct electrochemical oxidation of the biogenic amines and common interferents usually present in food products (Carelli et al. 2007).

1.5   APPLICATIONS OF AMPEROMETRIC BIOSENSORS

Food quality control is an important stage in food processing. Quality of raw material, contamination of food during processing, quality of final product, and the changes in the food during shelf life are important factors for the food producer and consumer. For that reason, it is necessary to determine the levels of some markers related to properties of the food. Enzyme-based amperometric biosensors are one of the good choices for monitoring these markers. The requirements of clinical monitoring are the major driving forces behind the enzyme electrode. On the other hand, the application potential of enzyme electrodes in food processing increases rapidly. Some of the target molecules, measured by enzyme electrodes, are glucose, dextrose, sucrose, lactose, alcohol, some organic acids, and others. It is also possible to determine some of the contaminants such as microorganisms, pesticides, and toxins using an amperometric enzyme electrode. Besides determination of food components and contamination, enzyme electrodes are used for the measurement of the food additives. Before starting the application of an enzyme electrode, the requirements for the sensor used in food processing should be known.

A rapid, selective, and highly sensitive analytical device is needed for screening food samples. Actually, these requirements are common in the other application areas. However, the target molecule is generally searched in the complex, especially in food samples. Matrix structures of food samples vary so much, and this causes difficulties in the development of enzyme electrodes. To overcome this difficulty, several designs have been developed to measure target molecules in different sample matrices.

1.5.1 Determination of Food Components

The main components of foods are carbohydrates, proteins, fats, minerals, and vitamins. The studies on carbohydrate analysis are designed most commonly using amperometric enzyme electrodes. There are also some studies about detection of proteins and fats, but studies about determination of minerals and vitamins are rare.

1.5.1.1 Carbohydrates

Carbohydrate analysis is the main driving force for development of an amperometric enzyme electrode. The most important one is glucose, which is the energy source of most living organisms. Glucose is a monosaccharide and selectively reactive with some enzymes, such as glucose oxidase and glucose dehydrogenase. These enzymes are selective to only one of the optic isomers of glucose and do not react with the others. This selectivity of the enzymes gives an opportunity to develop a selective biosensor for monitoring glucose level in health care and also in food processing.

Although the enzymatic recognition can be performed easily, the interferences, such as ascorbic acid, uric acid, and others in the sample, cause difficulties in the construction of the electrode. Different strategies and electrode designs were developed to minimize the signal of the interferences. Some examples of the amperometric glucose electrodes are given in Table 1.5.

Glucose is also the main component of the other carbohydrates such as maltose, lactose, sucrose, and starch. There are also some other enzymes (glucoamylase, beta galactosidase, invertase, and amylase) that hydrolyze these carbohydrates into monosaccharides, one of which is glucose. The combinations of the glucose-selective enzymes with these hydrolysis enzymes give an opportunity to develop selective amperometric enzyme electrodes toward maltose, lactose, sucrose, starch, and others. In this section, some brief information is given about carbohydrate sensors.

Fructose, an optic isomer of glucose, is the other important monosaccharide found in many foods. Fructose is the major component of honey, tree fruits, berries, melons, and some root vegetables. Due to its high relative sweetness value, fructose is commercially used in the food and beverage industry. Generally, fructose is found with glucose molecules in food samples. For that reason, it becomes difficult to determine the fructose and glucose molecules in the sample. Enzyme specificities toward the substrates give an opportunity to analyze fructose using amperometric enzyme electrodes. Two common procedures were followed for development of the fructose biosensor. One of them is the usage of fructose dehydrogenase (FDH) to produce an electrochemical product. The other is the conversion of the fructose into glucose and then measuring the glucose content using the glucose electrode. The first procedure has some advantages, which are minimum enzymatic stages and no interference of glucose present into the fructose measurement. On the other

TABLE 1.5

Detection of Glucose in Food Matrices Using Amperometric Enzyme Electrodes

Enzyme	Design	Food Matrix	Linearity Range	Detection Limit	References
Glucose oxidase	Graphite rod electrode modified by gold nanoparticles	Beverages	0.1–10 mM	0.08 mM	German et al. 2010
Glucose oxidase	Nylon nanofibrous membrane	Food and beverages	1–10 mM	6×10^{-6} M	Scampicchio et al. 2010
Glucose oxidase	Colloidal gold nanoparticles on a glassy carbon electrode by Nafion film	—	1–20 mM	0.37 mM	Thibault et al. 2008
Glucose oxidase	Prussian blue mediator film	Wine and yogurt	10–800 μM	1 μM	Luca et al. 2005
Glucose oxidase	Prussian blue mediator film	Soluble coffee	0.15– 2.5 mM	0.03 mM	Mattos et al. 2005
Glucose oxidase	Carbon paste electrodes with ferrocene as redox mediator	Wine Processing	0.02–50 g/L	—	Serban et al. 2004
Glucose oxidase	Copper-deposited films	Starch-containing samples	1.8–180 mg/L	0.64 mM	Bourais et al. 2004
Glucose oxidase	Polyion complex layer coated electrode	Beverages	0.01–3 mM	—	Mizutani et al. 1998
Glucose dehydrogenase	Electrochemical mediator	Honey	5×10^{-6} – 2×10^{-4} M	1×10^{-7} mM	Antiochia et al. 1997

hand, in the second procedure glucose becomes one of the main interferences in the fructose measurement. Therefore, the glucose content of the sample has to be measured first, and then the glucose produced from enzymatic conversion of the fructose is calculated to determine fructose concentraion. Bassi and coworkers have developed an amperometric fructose electrode using FDH. The enzyme was immobilized using nonconducting electropolymer film of 1,3 phenylene diamine-resorcinol (Bassi et al. 1998). Two different types of electrochemical mediators, soluble hexacyanoferrate (III) (FeCN) and the water-insoluble tetracyanoquinodi-methane (TCNQ) were used as redox mediators for the amperometric measurement of fructose. It was reported that fructose electrode has a detection limit of 10 μM and it is linear up to 1 mM fructose concentration. The electrode was also successfully applied for determination of fructose in honey. A disposable fructose sensor based upon the enzyme FDH has been developed by Trivedi and coworkers (Trivedi et al. 2009). FDH enzyme has been immobilized in a polymer matrix of polyethylenemine (PEI) and poly(carbamoylsulphonate) (PCS) hydrogel on a platinum (Pt) tip of a screen-printed graphite electrode. A ferricyanide mediator has been used as the electron acceptor. It was reported that the sensitivity of the fructose electrode was 0.62 ± 0.10 nA/M and it was usable in real samples of fruit juice, soft drinks, and honey.

A biosensor for fructose was developed using carbon paste modified with silica gel coated with Meldola's blue and fructose 5-dehydrogenase (Garcia et al. 1996). Linearity range of the fructose sensor was given as 0.1 to 0.8 mM.

Kinnear and Monbouquette (1997) developed a prototype amperometric fructose biosensor based on membrane-bound FDH and the coenzyme ubiquinone-6 immobilized in a membrane mimetic layer on a gold electrode. The detection limit of the developed system was reported as 10 μM. Interference of the ascorbic acid was also investigated, and as a result 4% positive error was reported in the presence of ascorbic acid at 5% of the fructose concentration (2 mM). These examples and others indicated that the amperometric fructose electrode could be successfully applied for fructose measurements in food samples in laboratories and also in the field using portable biosensors.

One of the important carbohydrates in food processing is lactose. Lactose is the characteristic carbohydrate of milk and dairy products, and determination of lactose is routinely carried out in the dairy industry to ensure effective process and product control. Lactose, a disaccharide sugar, is hydrolyzed by beta-galactosidase into glucose and galactose, which are further metabolized. Selectivity of beta-galactosidase toward the lactose hydrolysis is commonly used in detection of lactose concentration. The lactose level in food products is also important for the patient who has congenital lactose intolerance. Lactose intolerance is an inability to digest and absorb lactose that results in gastrointestinal symptoms when milk or products containing milk are consumed. To minimize the gastrointestinal symptoms, the lactose level has to be reduced or totally removed from the patient's diet. At this point, a lactose electrode becomes a critical instrument for measuring the lactose level in food samples. Different amperometric lactose electrodes were developed by following different strategies and enzyme configurations. It is possible to categorize the lactose biosensor into five groups based on the enzymes used. They are (1) three-enzyme configuration with beta galactosidases, glucose oxidase, and peroxidase (Eshkenazi et al. 2000); (2) two-enzyme configuration with beta galactosidases, glucose oxidase (Boyacı and Mutlu 2002; Göktuğ et al. 2005; Loğoğlu et al. 2006); (3) galactose oxidase (Tkac et al. 2000); (4) oligosaccharide dehydrogenase (Ikeda et al. 1990); and (5) cellobiose dehydrogenases (Stoica et al. 2006). Different strategies such as immobilization of the enzyme on gelatin film, immobilization of enzymes onto a glassy carbon electrode coated with a thin mercury film, microdialysis-coupled flow injection amperometric sensor, packed-bed column reactors integrated with amperometric enzyme electrode, and others were done to perform lactose electrodes that are usable in food samples (Boyacı and Mutlu 2002; Gulce et al. 2002; Rajendran and Irudayaraj 2002; Göktuğ et al. 2005).

Sucrose is the other carbohydrate commonly used in the food industry as a sweetener. Sucrose is a disaccharide formed from glucose and fructose. Invertase enzyme hydrolyzes the sucrose into its monosaccharides selectively; and three enzymes—invertase, mutaratase, and glucoseoxidase—are commonly used for amperometric sucrose electrode development (Boyacı and Mutlu 2002). The other three enzyme configurations—sucrose phosphorylase, phosphoglucomutase and glucose-6-phosphate 1-dehydrogenase—are also used for this purpose (Maestre et al. 2001). Protein layer (Majer-Baranyi et al. 2008), graphite paste (Kennedy et al. 2007), clay matrix (Mohammadi et al. 2005), Prussian blue films (Haghighi et al. 2004), and packed-bed column were used for enzyme immobilization and construction of the sucrose electrode.

Starch, the most important carbohydrate in the human diet, consists of a large number of glucose units joined together by glycosidic bonds. Starch and its hydrolysis products are widely used in the food industry. Alpha-amylase and glucoamylase are the two common enzymes used for hydrolyzing the gelatinized starch into glucose. Produced glucose is measured by glucose electrode. Several biosensors were developed for determination of starch in food samples by immobilizing enzymes on the electrochemical sensor surface (Vrbova et al. 1993; Hu et al. 1999; Marconi et al. 2004).

Maltose electrode with glucose oxidase, alpha-glucosidase, and mutarotase enzymes; inulin electrode with FDH and inulinase enzymes; and trehalose with trehalase and glucose oxidase enzymes were developed using amperometric electrodes for measuring maltose, inulin and trehalose in food samples (Zajoncova et al. 2004; Manso et al. 2008; Antonelli et al. 2009).

1.5.1.2 Proteins

Protein, the main building block in the body, is the primary component of most cells and has an important function virtually in every process performed in the cell. Besides its importance in living cells, it is also an essential nutrient in the human diet. Proteins such as casein from milk, gluten from wheat, and albumin from egg are the main components of these foods. Proteins are macromolecules made of amino acids arranged in a linear chain and folded into a globular form. Twenty-two amino acids are the building blocks of the proteins, and eight of them are essential amino acids. Essential amino acids cannot be synthesized from other compounds in the human body. For that reason, these amino acids must be supplied from the diet. In addition to these, some of the food proteins (casein, gluten) and their hydrolyzing products (such as biogenic amines) have allergenic reactions in some human beings. As a result, measurements of protein and also its fragments are critical issues for quality control, shelf life check, and determination of bioavailability and of allergenic and toxic protein contents. Although detection methods of proteins and their fragments are commonly performed based on immunologic reactions, only a limited number of detections were done using enzyme–substrate interactions. In this section, some of the examples related with amperometric enzyme electrodes for detection of proteins and protein fragments are given.

Rosini and coworkers have developed an amperometric enzyme electrode for measuring D-amino acid concentration in food samples (Rosiní et al. 2008). D-amino acid oxidase enzyme was used as the biorecognizer agent to determine the concentration of D-amino acids in various biological samples. The detection limit of

TABLE 1.6

Detection of Fats and Oils in Food Matrices Using Amperometric Enzyme Electrodes

Target	Enzyme	Food Matrix	Linearity Range	Detection Limit	References
Essential fatty acids	Lipoxygenase, lipase, and esterase	Oil samples and margarine	0.01–0.2 mM	0.04 mM	Schoemaker et al. 1997
Glycerol	Glycerol oxidase	Magarach Institute of grapes and wine	0.05–25.6 mM	0.05 mM	Goriushkina et al. 2010
Glycerol and triglycerides	Glycerol dehydrogenase and lipase	—	—	—	Laurinavicius et al. 1996

the developed amperometric enzyme electrode was reported as 0.25 mM. Another amperometric enzyme electrode was developed for determination of D-amino acids in fruit juices and some milk samples (Wsiclo 2007). D-amino acid oxidase enzyme was used in the biosensing layer, and D-analine concentration was determined in the range of 5–200 µM. An L-lysine, an essential amino acid, biosensor was developed based on amperometric enzyme electrode by Kelly and coworkers (Kelly et al. 2000). L-lysine alpha-oxidase enzyme is used in the construction of enzyme electrode. A brief summary of L-lysine detection was given in this study also.

1.5.1.3 Fats and Oils

Fats and oils are the other main components of foods. These are commonly found in the food as triacylglycerol, which is an ester of glycerol and fatty acids. Besides providing some essential activities in the human body, they are the main energy sources for humans. Fats and oils are generally soluble in organic solvents and largely insoluble in water. For that reason only a limited number of studies have reported determination of fats and oils and their hydrolyzing products. Detection methods of fats and oils in a food matrix using amperometric enzyme electrodes are given in Table 1.6.

1.5.1.4 Organic Acids

Many acids are produced naturally in foods, and some of them are used as food additives and processing aids. Common organic acids are acetic, lactic, ascorbic, malic, and citric acids. Lactic, malic, and acetic acid are produced during the fermentation process. Some acids are produced as a result of uncontrolled microbial activities, and an increase in the acid concentration may cause a decrease or total loss of the economic value of the foods. Measurement of organic acids in foods is a critical issue in food quality control, and accurate, rapid, and automatic methods for determination of acids are needed. Thus, the usage of an amperometric enzyme electrode for this purpose can be advantageous because of the selectivity, possibility of fast analysis of acid, and practical usage of the electrode in laboratories and also in the field with

TABLE 1.7

Detection of Acids in Food Matrices Using Amperometric Enzyme Electrodes

Target	Enzyme	Food Matrix	Linearity Range	Detection Limit	References
L-lactate	Lactate oxidase and peroxidase	Red wine and shaken yogurt	—	1.4 and 0.9 mM	Serra et al. 1999
L-lactate	L-lactate dehydrogenase	Wine	Up to 1.3 mM	10 mM	Katrlik et al. 1999
L-lactate	Lactate oxidase	Food	—	—	Yashina et al. 2010
Acetic acid	Acetate kinase, pyruvate kinase and pyruvate oxidase	Wine	0.05–20 mM	—	Mizutani et al. 2003
L-malate	L-malate dehydrogenase	Wine	Up to 1.1 mM	10 mM	Katrlik et al. 1999
Oxalate	Oxalate oxidase	Tea leaves and strawberries	2.5–400 μM	—	Milardovic et al. 2000
Citrate	Oxaloacetate decarboxylase and pyruvate oxidase	Food samples and fermentation broths	1 μM to 1 mM	0.5 mu mol/L	Gajovic et al. 1997
Ascorbate acid	Ascorbate oxidase	Foodstuffs	1–200 μg/ml	—	Daily et al. 1991
Glutamate	Glutamate oxidase	Food seasonings	10–160 mg/L	1.7 mg/L	Beyene et al. 2003

portable models. Therefore, amperometric acid electrodes have a wide application area in food quality control. Some biosensors that have been developed to measure acids or their salts are given in Table 1.7.

1.5.1.5 Enzymes

Enzymes are a protein that react as biochemical catalyzers and accelerate one or several specific biochemical processes. Enzymes are produced by the living cell to perform reactions that are required by the organism. In enzymatic reactions, the molecules that the enzymes react with selectively are called substrates, and the outputs of the enzymatic reactions are called products. The selectivity between enzyme and substrate is used in most of the bioassay and biosensor studies. Besides the usage of the enzyme as the bioactive agent in amperometric enzyme electrode, enzyme is also used as the labeling agent in amperometric immuno and DNA sensors studies. It is possible to categorize the enzyme activity measurements into two groups. The first one is determination of enzyme activity with an amperometric enzyme electrode. A good example of this is the measurement of alpha amylase activity using glucose electrode that alpha glucosidase, mutarotase, and glucose oxidase integrated (Zajoncova et al. 2004). Alpha amy-lase hydrolyzes the starch into maltose units, and catalytic activity is related with the produced maltose concentration. Maltose is converted to alpha glucose with alpha glucosidase first and then to beta glucose with mutarotase enzyme. Finally, beta glucose concentration is determined with amperometric glucose electrode in which glucose oxidase is entrapped. Although the number of the enzymatic reac-tions are high, the response time of the maltose electrode is only 35 sec, and alpha amylase activity can be determined in a short time, depending on the interaction between alpha amylase and substrate. An amperometric glucose electrode was also used for kinetic analysis of alpha glucosidase activity successfully (Tatsumi and Katano 2004). An amperometric enzyme electrode based on colin oxidase enzyme was designed for monitoring cholinesterase enzyme activity, which is critical for diagnostic value in liver diseases and malignant tumors and also for detection of some toxins, pesticides, and heavy metals (Mizutani and Yabuki 1995). Pseudocholinesterase activity was determined in less then 10 sec with a detection limit of 2 U/L. These examples indicate that amperometric enzyme electrodes are successfully applicable for measuring enzyme activity. In the sec-ond group, an amperometric electrode is used only as a chemical sensor, and the concentration change of the substrate or product during the enzymatic reaction is monitored by the electrode. This type of usage of the amperometric electrode for determination of enzyme activity is not fitted to the amperometric enzyme elec-trode definition, which is that an enzyme has to be used as biorecognizer agent in the sensor. For that reason no information is given in this section.

1.5.1.6 Other Food Components

The determination of alcohol, especially ethanol, is important in the food industry. An alcoholic beverage contains ethanol, and the alcohol concentration needs to be measured during and at the end of the alcohol production for quality control. The alcohol oxidase enzyme provides an opportunity to prepare an amperometric alco-hol electrode and measures the alcohol concentration selectively and sensitively. Alcohol electrodes have been developed by different researchers to measure alcohol concentration in different food samples such as beer and wine (Boujtita et al. 2000; Shkotova et al. 2006).

Cholesterol is a small steroid molecule, essential to the human body. It is used for manufacturing hormones. However, a high level of blood cholesterol increases the risk of heart disease. Some of the animal foods such as egg, meat, and diary products contain a high amount of cholesterol, and overconsuming of these foods may cause an increase in the blood cholesterol level. Measuring the cholesterol level in the food sample is important for consumers to plan on their diet and cholesterol intake. Based on this requirement, the amperometric cholesterol electrode was developed by using cholesterol esterase and cholesterol oxidase enzyme as biorecognizing agents by using different immobilization techniques and electrodes (Pena et al. 2001; Ram et al. 2001; Adanyi and Varadi 2003). The usability of the amperometric cholesterol enzyme was also demonstrated in food samples.

Lecithin is an important phospholipid needed by the living organism. It is one of the most common phospholipids in nature. Lecithin can be found in many foods such as soybeans, split peas, seeds, nuts, eggs, cabbage, cauliflower, and garbanzo beans and can be easily extracted with hexane. Besides its importance in the human diet, lecithin is also important in some of the food production technologies. It is used as a food additive (an emulsifier) in the food industry especially in chocolate, biscuits, and similar products. To measure the lecithin concentration, an amperometric enzyme electrode was developed by using phospholipase D and choline oxidase enzymes; and applicaton of the sensor in several real food samples (egg yolk, soybean flour, oil, etc.) was demonstrated by Campanella and coworkers (1998).

1.5.2   Determination of Food Contaminants

1.5.2.1   Pesticides

A substance or mixture of substances intended for preventing, destroying, repelling, or mitigating any pest is called a pesticide. The term includes substances intended to be used as a plant growth regulator, defoliant, desiccant, or agent for thinning fruit or preventing the premature fall of fruit, and substances applied to crops either before or after harvest to protect the commodity from deterioration during storage and transport (FAO 2002). Some of the most commonly used pesticides inhibit acetylcholinesterase (AChE) enzyme activity, which degrades acetylcholine (ACh) into producing choline and an acetate group. ACh is a neurotransmitter found at neuromuscular junctions. Inhibition of AChE causes an increase of ACh concentration, which can cause hyperstimulation of muscle tissue leading to paralysis. AChE inhibitors at high doses can cause death resulting from respiratory failure. Because of the toxicity of the pesticide, its concentration has to be measured. One of the practical ways of measuring the pesticide concentration is to monitor the inhibition effect of pesticides on AChE enzyme. Activity of AChE enzyme can be monitored by an amperometric electrode in the presence of constant amount of ACh at standard reaction conditions. Two reaction approaches are commonly followed for monitoring AChE with amperometric electrode. The first one is the bienzyme approach in which choline oxidase is used with AChE. First, ACh is hydrolyzed into choline and actate, and then choline oxidase is used to produce hydrogen peroxide, which is an electrochemically active compound. In the second approach, only AChE enzyme is used with nonspecific substrate (such as acetylthiocholine), and the product of the reaction, thicholine, is directly monitored with the amperometric electrode.

1.5.2.2   Foodborne Pathogens

The detection and identification of pathogens in food is one of the most important requirements in food processing. The classical method for microbial identification, which includes enrichment, isolation, and identification, takes a long time. Rapid methods are needed to minimize the analysis time. On the other hand, the number of pathogens searched in food samples is generally low. Thus the method has to be sensitive to detect a low number of pathogens in the sample. The ideal detection method needs to combine qualities such as sensitivity, specificity, and low analysis time. To meet these requirements, several biosensor studies were performed. *E. coli, Salmonella typhimurium, Campylobacter jejuni, Legionella pneumophila, Staphylococcus aureus, Bacillus cereus,* and *Streptococci* are common foodborne pathogens that are searched for in food samples. Immunological reactions based on antigen–antibody interactions are commonly used in pathogen detection. However, it is difficult to measure these reactions using an amperometric electrode. Enzymes or enzyme systems are integrated into immunological reaction to monitor sensitively.

Two common approaches are used. One of them is labeling the antibody with enzymes such as alkaline phosphatase and horseradish peroxidase enzyme. An enzyme-labeled antibody is used as secondary antibody, and at the end of the immunological reaction the amount of the secondary antibody (this concentration is related to target pathogen concentration) is determined with specifically designed substrate. The second approach is designed based on the measurement of the enzyme produced by organisms. Depending on the life cycle of the organism, they produce different enzymes. Beta-galactosidase is produced by Gram-negative (e.g., strains belonging to the *Enterobacteriaceae, Vibrionaceae, Pseudomonadaceae,* and *Neisseriaceae*) and several Gram-positive bacteria, yeasts, protozoa, and fungi. Beta-glucuronidase has been found in most *E. coli* strains and also in other members of the *Enterobacteriaceae,* including some *Shigella* and *Salmonella* strains and a few *Yersinia, Citrobacter, Edwardia,* and *Hafnia* strains (Tryland and Fiksdal 1998). Although standard microbiological methods based on membrane filtration and selective growth on solid media usually take 24–72 hours to complete, it is possible to measure the activities of these enzymes in a short time by using electrochemical or chromogenic substrates. It is also possible to determine bacteria concentrations in the samples by following the activities of these enzymes. But these enzymes indicate that a minimum one of the enzymes is in the sample. Generally, it does not make sense, and the species and/or strains of the pathogens have to be known. At this stage, enzyme activity measurements are combined with the immunological recognition. First, the target microorganism is specifically isolated from sample using immunological methods (e.g., immunomagnetic separation), and then bacteria enumeration is performed based on the enzyme activity of the target. Although several methods were designed based on these approaches (Boyacı et al. 2005; Dudak et al. 2009), it is hard to define these assays as amperometric enzyme electrodes due to fact that performing the enzymatic reaction and measuring the electrochemical change are done separately. For that reason, examples about foodborne pathogen detection using amperometric electrodes are not given here.

1.5.2.3 Toxins

Aflatoxins (AFs) are naturally occurring, highly toxic, and carcinogenic secondary metabolites that are produced by the fungi *Aspergillus flavus* and *Aspergillus parasiticus.* Cereals (maize, sorghum, pearl millet, rice, wheat), spices (peppers, coriander, turmeric, ginger), oilseeds (peanut, soybean, sunflower, cotton) and tree nuts (almond, pistachio, walnut, and coconut) are subject to the growth of these molds and AFs production. The toxin can also be found in the milk of cows fed with AF-containing feeds. These foods are major sources of the human diet, and AF contamination has been known to decide whether the product is consumable or not. Commonly, immunological methods are preferred using AFs' selective antibody. It is also possible to determine AFs using amperometric enzyme electrode. Rejeb and coworkers developed an amperometric enzyme electrode for measurement of AFs based on AChE inhibition by aflatoxin B1 (AFB1) (Rejeb et al. 2009). AChE is the key enzyme in the transmission of nerve impulses, and its activity is inhibited by pesticides and heavy metals and some of the toxins. There is a high correlation between concentration of these compounds and AChE activity. AFB1 concentration was monitored in the range of 10–60 pbb in olive oil. This study is one of the key studies that determine biological toxins in food samples using amperometric enzyme electrode.

1.5.2.4   Other Contaminants

Biogenic amines are low-molecular-weight organic bases, synthesized and degraded during normal metabolism of animals, plants, and microorganisms. Histamine, putrescine, cadaverine, tyramine, tryptamine, phenylethylamine, agmatine, spermine, and spermidine are the most important biogenic amines in food products. A wide range of food products including fish, meat, wine, beer, vegetables, fruits, nuts, and chocolate contain biogenic amines. The presence of biogenic amines in food constitutes a potential public health concern because of their physiological and toxicological effects. Histamine has been implemented as the causative agent in several outbreaks of food poisoning. Tyramine and β-phenylethylamine have been proposed as the initiators of hypertensive crisis. The toxicity of histamine is increased by the presence of other amines such as cadaverine, putrescine, and tyramine. Biogenic amine concentrations increase during food fermentation, microbial growth in foodstuffs, and also aging of the food during storage. Biogenic amine concentrations are used as an indicator of the food freshness. Therefore, determination of biogenic amine contents is a critical issue for the consumers. An amine oxidase–based amperometric biogenic amine electrode was prepared by Muresan and coworkers (2008). An amine oxidase enzyme was coimmobilized with horseradish peroxidase on an electrode surface, and the developed biogenic amine electrode was successfully applied for the measurement of tyramine, putrescine, cadaverine, histamine, agmatine, and spermidine in fish samples.

1.5.3   Determination of Food Additives

Benzoic acid and its salts are used as a food preservative in various food products for their antimicrobial activity to preserve and protect food from spoilage. This can be harmful to human health if the level of the benzoic acid in the food is higher than the permitted safety levels. The maximum permitted concentrations for benzoic acid for each type of food are dictated by law (European Parliament and Council Directive No. 95/2/EC, February 1995.) and controlled by regulatory agencies. The determination of benzoic acid in food products is critical for quality control and also consumer protection. San and coworkers developed an ampero-metric enzyme electrode for monitoring benzoic acid based on its inhibiting action on tyrosinase enzyme (San et al. 2007). Glassy carbon electrodes were modified with a tyrosinase-calcium carbonate nanomaterial (nano-$CaCO_3$) complex for the measurement of benzoic acid. It was reported that the results obtained from a ben-zoic acid sensor were linear in the range of 5.6×10^{-7} to 9.2×10^{-5} M benzoic acid concentration and they can be useful to successfully measure benzoic acid in some beverage samples and yogurt.

Monosodium glutamate is a sodium salt of the naturally occurring glutamic acid which is a nonessential amino acid. Glutamate is found naturally in all protein con-taining foods such as meat, poultry, seafood and milk. Besides its natural occur-rence, it is commonly used as a flavor enhancing agent to enhance their original flavor. There is a limitation about the usage of the glutamate in food products as all the other food additives. At this point amperometric glumate biosensor was devel-oped to measure glutamate concentration in food seasonings (Beyene et al. 2003). Glutamate oxidase enzyme is used as a biorecognizer on the screen printed elec-trode and glutamate concentration was determined in the range of 10–160 mg/L, successfully.

Phytic acid is another additive used in food products. It has a strong ability to chelate multivalent metal ions, such as zinc, calcium and iron (Hurrell 2003). Phytic acid has also an antioxidant effect. Phytic acid reduces the uptake of the minerals due to the ability of chelate forming and decreasing bioavailability of the minerals, which have some beneficial health effects such as reducing the risk of colon cancer and heart disease (Vucenik and Shamsuddin 2003, Jariwalla et al. 1990). However, the low intakes of essential minerals have an undesirable effect for especially young children and the people who live in developing countries. As a result, phytic acid concentration in food has to be known. Mak and coworkers (2004) have developed an amperometric enzyme electrode for phytic acid by fol-lowing the sequentially acting enzyme phytase and pyruvate oxidase. Hydrogen peroxide is produced at the end of the enzymatic reaction and it was measured by using a hydrogen peroxide electrode. Phytic acid concentration in the range of 0.2 to 2.0 mM was measured using a bi-enzyme electrode with a detection limit of 0.002 mM.

1.6 COMMERCIAL AVAILABILITY OF AMPEROMETRIC BIOSENSORS FOR FOOD

Only few amperometric biosensors targeted to foodstuff and food processing engi-neering are commercially available. Although the third generation sensors employ-ing redox polymers are technologically and commercially available on the market, most of the biosensors are still working in the basic principles of oxygen and hydro-gen peroxide detection. The main characteristics and performance of these devices are presented in Table 1.8.

TABLE 1.8
Commercial Amperometric Biosensors Used in Foodstuff Analyses

Manufacturer, Instrument, Contact Data	Enzymic Pathway	Analyte	Applied Matrix	General Comments
Bioanalytical Systems, Inc. (Indiana)				
Peroxidase Redox Polymer Wired Enzyme Electrode Kit http://www.basinc.com	Glucose oxidase Acetylcholine esterase/ Choline esterase Lactate oxidase	Glucose Acetylcholine/ Choline Lactate	General applications	Enzymes are covalently bounded within a postcolumn reactor (IMER); shipped in cold containers H_2O_2 oxidation at Pt anode 0.1 V; mediated by Osmiun (III) Polyvinylpyridine redox polymer
BioFutura S.r.l (Gorizia, Italy)				
PerBacco2000 PerBacco2002 http://www.biofutura.com	Glucose oxidase Fructose dehydrogenase/ diaphorase/ferricyanide Lactate dehydrogenase Malate dehydrogenase/ diaphorase/ferricyanide	Glucose Fructose Lactate Malate	Wine Must	Enzymatic; Amperometric Solid binding matrices bearing the biocatalysts Dialysis membranes for preventing interferences Dilution and decolorization of the samples are recommended In red wines or in high content of polyphenols
Chemel AB (Lund, Sweden)				
SIRE P200 Series http://www.chemel.com	No details	L-Ascorbic acid D-Glucose L-Lactate	Fruit and vegetable juices Beer, wine, soft drinks Dairy products Tomato paste Baby food Beet sugar production	Enzyme; soluble and injected in the buffer flow Sample volume: 25 ml Analysis time: 2–3 min/sample Optimal temperature: 20–30°C

(continued)

TABLE 1.8 (CONTINUED)
Commercial Amperometric Biosensors Used in Foodstuff Analyses

Manufacturer, Instrument, Contact Data	Enzymic Pathway	Analyte	Applied Matrix	General Comments
Nova Biomedical Corporation (Waltham, Massachusetts)				
BioProfile Analyzers http://www.novabiomedical.com	No details	Glucose Lactate Glutamine Glutamate	Determining the consumption and production of key metabolites in cell culture and fermentation processes	Auto sampling, fully automated Immobilized enzyme-membrane Sample size: 500 µL–2 mL Eppendorf tubes 500 µL–4 mL sample cups 5 mL–7 mL test tubes Measuring range: Glutamine: 0.2–6.0 mmol/L Glutamate: 0.2–6.0 mmol/L Glucose: 0.2–15.0 g/L Lactate: 0.2–5.0 g/L Imprecision/Resolution 5%
Analox Instruments Ltd. (London, United Kingdom)				
AM2 Industrial Alcohol Analyzer	Alcohol oxidase	Ethanol	Beers and lagers, low-level beers, ciders, wines and spirits, alcoholic soft drinks, waste-line monitoring in distilleries and industrial ethanol plants, yeast production, quality control of alcohol-containing consumer product	Enzyme-based biosensor; O_2 cathode reduction using a Clark-type amperometric electrode Sample size: 3.5–25 µL depending on the assay Response time: 20 seconds Reagent stability: 6–12 months unopened at 0–5°C
AM3 Industrial Alcohol Analyzer	Alcohol oxidase	Ethanol		
AM5 Methanol Analyzer	Alcohol oxidase	Methanol		

GL6 Multiassay Analyzer for Industrial Applications	Alcohol oxidase	Ethanol	Measurement of methanol in Pichia pastoris expression systems, quality control for Dimethyl dicarbonate (DMDC) preservative dosing	Lactate: Up to 20 mM; Alcohol: 0–40% v/v; Glucose: 0–20% w/v; Methanol: 0–500 ppm; Sucrose/Lactose: Up to 50 mM total glucose; Glutamine/NH_3: 0.5–20 mM
GM8 Micro-Stat Multiassay Analyzer	Lactate oxidase; Glucose oxidase; Glycerol kinase/glycerol-3-phosphate oxidase	Methanol; Lactate; Glucose; Glycerol; Ethanol/	Food and beverage, pharmaceutical, biotechnology and other industrial production/process control facilities	Enzyme-based biosensor; O_2 cathode reduction using a Clark-type amperometric electrode; Sample size: 3.5–25 µL depending on the assay; Response time: 20 seconds; Reagent stability: 6–12 months unopened at 0–5°C
GM10 Industrial Glucose Analyzer	Alcohol oxidase; Lactate oxidase; Glucose oxidase; Glycerol kinase/glycerol-3-phosphate oxidase; Invertase/glucose oxidase	Methanol; Lactate; Glucose; Glycerol; Sucrose; Ammonia	Fermentation broths, cell culture media, and other microbial applications; Food and beverage, biopharmaceutical, and biotechnology industries; Dairy (directly on milk) and food and beverage industries	Lactate: Up to 20 mM; Alcohol: 0–40% v/v; Glucose: 0–20% w/v; Methanol: 0–500 ppm; Sucrose/Lactose: Up to 50 mM total glucose; Glutamine/NH_3: 0.5–20 mM
LM5 Lactate Analyzer http://www.analox.com	Glutamate dehydrogenase/NADH/α-ketoglutarate/peroxidase; Glutaminase/glutamate dehydrogenase/NADH/α-ketoglutarate/peroxidase	Glutamine; Lactose; Pyruvate; Glucose; Sucrose; Lactose; Lactate		

(continued)

TABLE 1.8 (CONTINUED)
Commercial Amperometric Biosensors Used in Foodstuff Analyses

Manufacturer, Instrument, Contact Data	Enzymic Pathway	Analyte	Applied Matrix	General Comments
	β-Galactosidase/glucose oxidase			
	Lactate dehydrogenase/ peroxidase			
	Glucose oxidase			
	Invertase/glucose oxidase			
	β-Galactosidase/glucose oxidase			
	Lactate oxidase			
Applied Enzyme Technology Ltd. (Pontypool, United Kingdom) *Stabilization of Biosensors* http://www.gwent.org/Aet/ index.html	Alcohol oxidase	Alcohol	Beers and wines	Alcohol: High levels of enzyme activity in the dry state at room temperature and indefinitely at 4°C
	Glutamate oxidase	Glutamate	Food and drink	Glutamate: High levels of enzyme activity
	Lactate oxidase	Lactate	Wine industry	Lactate: High levels of enzyme activity in the dry state at room temperature and up to a year at 4°C
Gwent Sensors Ltd. (Pontypool, United Kingdom)				
The Answer 8000	Horseradish Peroxidase/ glucose oxidase/ ferrocene	Glucose	Potatoes and industrial applications	Flow analysis system Graphite enzyme-cartridge Cartridge lifetime: 50 runs Chronoamperometric measurements at –0.1 V Build-in dilutor (for high concentrations)

MC2 Multisensor http://www.gwent.org/Gsl/ index.html	Gucose oxidase Invertase β-Galactosidase	Glucose Sucrose Lactose	Industrial applications	Amperometric enzymatic biosensor Immobilized enzyme-membrane Glucose: Applied potential 0.9 volts Measurement time: 20 seconds Glucose: 0.5–3.0 g/L Sucrose: 1.0–6.0 g/L Lactose: 1.0–6.0 g/L Glucose+Sucrose: 0.5–3.0 g/L Glucose+Lactose: 0.5–3.0 g/L
Bioanalytical Systems, Inc. (Indiana) Distributor for Sycopel International Ltd. Microdialysis Biosensor by Sycopel http://www.biotechproducts. com	Glucose oxidase Glutamate oxidase Glutaminase/glutamate oxidase Glycerol kinase/ glycerol-3-P oxidase Pyruvate oxidase Xanthine oxidase ACh esterase/choline oxidase Cholesterol esterase/ cholesterol oxidase Alcohol oxidase Galactose oxidase Sarcosine oxidase D/L-amino acid oxidase Lysine oxidase Lactate oxidase	Glucose Glutamate Glutamine Glycerol Pyruvate Xanthine Acetyl choline Cholesterol ester Alcohol Galactose N-Me-glycine D/L-amino acids Lysine Lactate	General applications	Electropolymerized poly(o-phenylenediamine) is used either as a molecular barrier to electroactive species or to immobilize the enzyme onto the Pt-working electrode Ascorbate oxidase is used for the elimination of ascorbates interference Enables real-time measurements of analytes *in vivo* Specifications for glutamate analysis: Linear range: 0.5 µM to 1 mM glutamate Response time: 20 s for 90% steady-state response

(continued)

TABLE 1.8 (CONTINUED)

Commercial Amperometric Biosensors Used in Foodstuff Analyses

Manufacturer, Instrument, Contact Data	Enzymic Pathway	Analyte	Applied Matrix	General Comments
Yellow Springs Instruments (Ohio)				
YSI 2700 SELECT	Glucose oxidase	Glucose	Dextrose and sucrose; potatoes, cereal	5–65 µL sample size
Biochemistry Analyzer	Invertase/mutarotase/ glucose oxidase	Sucrose	products, and molasses	Immobilization on polymeric membranes
	Glucose oxidase	Dextrose	Choline; infant formulations	H_2O_2 anode oxidation
	Galactose oxidase	Lactose	Lactate and ethanol; tomato processing	Response time: 60 seconds
	L-Lactate oxidase	L-Lactate	Beer and wine	Precision CV (n = 10) ~2%
	L-Glutamate oxidase	L-Glutamate	Cheese, whey	
	Choline oxidase	Choline	Lunch meat	
	Glutaminase/L- Glutamate oxidase	L-Glutamine	Chicken broth, spice mix	
	Alcohol oxidase	Ethanol	Pet foods	
	Alcohol oxidase	Methanol	Starch in foods	
	Amyloglucosidase/ glucose oxidase	Starch		
	Galactose oxidase	Galactose		
		H_2O_2		

YSI 7100MBS Analyzer http://www.ysi.com	Glucose oxidase	Glucose	Fermentation	Immobilization between polycarbonate and cellulose acetate
	Glutamate oxidase	Glutamate	Food and beverage quality assurance	H_2O_2 anode oxidation
	Glutaminase/glutamate oxidase	Glutamine	Food and beverage processing	10–50 μL sample size
	Lactate oxidase	Lactate		Typical membrane working life: 5–21 days
	Xylose oxidase	Xylose		Response time: 1.5–3.5 minutes
	Alcohol oxidase	Ethanol		Precision CV (n = 10) ~2% (Glutamine ~4%)
	Alcohol oxidase	Methanol		
	Galactose oxidase	Galactose		
	Invertase/mutarotase/glucose oxidase	Sucrose		

1.7   CONCLUSION

In this chapter a brief perspective is given on the amperometric biosensors that are used in food processes. Amperometric biosensors are categorized according to the target molecules looked for in food samples. Different approaches developed by various research groups to detect food components and contaminants are mentioned. The literature survey indicates that amperometric biosensors are the most common biosensors developed to detect target molecules either in food or other types of samples. Clark and Lyons first proposed the initial concept of biosensors in 1962. The first enzymatic glucose electrode was developed by construction of glucose oxidase immobilized dialysis membrane on oxygen electrode. This technology was commercialized by Yellow Springs Instruments in Dayton, Ohio, in 1975. Since then, different biosensors were designed using different transducers and different biorecognition agents. But there is no other commercial biosensor on the market that reaches to the success of the amperometric glucose electrode. Surface plasmon resonance biosensor is the other technology, which is commercialized in the market, but it is not as common as the glucose electrode. The success of the amperometric biosensor is also seen in the number of scientific articles. The number of publications in 1980, 1990, 2000, and 2009 are 1, 90, 832, and 2511, respectively, according to the statistics of ISI-WOS. Since May 2010, the total number of biosensor publications was 20,405. The number of amperometric biosensor articles in Web science databases totals 4000. This means that amperometric biosensors cover almost one-fifth of the total biosensor studies. There are several reasons for the commonness of the amperometric biosensors. These include:

- Amperometric biosensors can be more easily built up than the other biosensors.
- The cost of the amperometric biosensor is lower than other biosensors.
- Labor cost for the amperometric system is lower than the others.
- Amperometric biosensors can be easily adapted to measure other target molecules or organisms.
- It is possible to produce low-cost disposable products using amperometric biosensor technology.
- It is possible to prepare portable amperometric biosensors at a reasonable price.
- Amperometric biosensor technology can be integrated with the other biosensor technologies.

Today, nanotechnology is also playing an increasingly important role in the development of biosensors. Various approaches to exploit the advances in nanoscience and nanotechnology for bioanalyses have been reported. The present information and experience on amperometric systems is potentially the biggest candidate to integrate with nanotechnology to increase performance of the biosensors. This integration can be considered as a new challenge on "nanobiosensor" technology for determining single molecule/cell detection and nanopore-based DNA sequencing using electrochemistry.

REFERENCES

Adanyi, N., and M. Varadi. 2003. Development of organic phase amperometric biosensor for measuring cholesterol in food samples. *European Food Research and Technology* 218:99–104.

Adrian, J., S. Pasche, and D.G. Pinacho. 2009. Wavelength-interrogated optical biosensor for multi-analyte screening of sulfonamide, fluoroquinolone, beta-lactam and tetracycline antibiotics in milk. *Trac-Trends in Analytical Chemistry* 28(6):769–777.

Ahmed, S., C. Dack, G. Farace, G. Rigby, and P. Vadgama. 2005. Tissue implanted glucose needle electrodes: Early sensor stabilization and achievement of tissue-blood correlation during the run period. *Analytica Chimica Acta* 537:153–161.

Ahuja, T., D. Kumar, and Rajesh. 2008. Polymer based urea biosensors: A brief overview. *Sensor Letters* 6:663–674.

Ahuja, T., I.A. Mir, and D. Kumar. 2007. Biomolecular immobilization on conducting polymers for biosensing applications. *Biomaterials* 28:791–805.

Akbayirli, P., and E. Akyilmaz. 2007. Activation-based catalase enzyme electrode and its usage for glucose determination. *Analytical Letters* 40:3360–3372.

Akmal, N., and A.M. Usmani. 2002. Redox monomers and polymers for biosensors. In *Biomedical diagnostic science and technology*, ed. W.T. Law, A.M. Usmani, and N. Akmal, 63–79. New York: Marcel Dekker Inc.

Akyilmaz, E., and E. Dinckaya. 2005. An amperometric microbial biosensor development based on *Candida tropicalis* yeast cells for sensitive determination of ethanol. *Biosensors and Bioelectronics* 20:1263–1269.

Alaejos, M.S., and F.J.G. Montelongo. 2004. Application of amperometric biosensors to the determination of vitamins and α-amino acids. *Chemical Reviews* 104:3239–3265.

Albery, W.J., and D.H. Craston. 1987. Amperometric enzyme electrodes: Theory and experiment. In *Biosensors: fundamentals and applications*, ed. A.P.F. Turner, I. Karube, and G.S. Wilson, 180–210. Oxford: Oxford Science Publications.

Albery, W.J., and J.R. Knowles. 1976. Evolution of enzyme function and development of catalytic efficiency. *Biochemistry* 15:5631–5640.

Albery, W.J., P.N. Bartlett, and M. Bycroft. 1987a. Amperometric enzyme electrodes. 3. A conducting salt electrode for the oxidation of four different flavoenzymes. *Journal of Electroanalytical Chemistry* 218:119–126.

Albery, W.J., P.N. Bartlett, A.E.G. Cass et al. 1987b. Amperometric enzyme electrodes. *Philosophical Transactions of the Royal Society of London Series B-Biological Sciences* 316:107–119.

Alp, B., S. Mutlu, and M. Mutlu. 2000. Glow-discharge-treated cellulose acetate (CA) membrane for a high linearity single-layer glucose electrode in the food industry. *Food Research International* 33:107–112.

Amine, A., H. Mohammadi, and I. Bourais. 2006. Enzyme inhibition-based biosensors for food safety and environmental monitoring. *Biosensors and Bioelectronics* 21:1405–1423.

Andreescu, S., and J.L. Marty. 2006. Twenty years research in cholinesterase biosensors: From basic research to practical applications. *Biomolecular Engineering* 23:1–15.

Antiochia, R., and G. Palleschi. 1997. A tri-enzyme electrode probe for the sequential determination of fructose and glucose in the same sample. *Analytical Letters* 30:683–697.

Antonelli, M.L., F. Arduini, A. Lagana, D. Moscone, and V. Siliprandi. 2009. Construction, assembling and application of a trehalase-GOD enzyme electrode system. *Biosensors and Bioelectronics* 24:1382–1388.

Arecchi, A., M. Scampicchio, and S. Drusch. 2010. Nanofibrous membrane based tyrosinase-biosensor for the detection of phenolic compounds. *Analytica Chimica Acta* 659:133–136.

Araujo, C.S.T., J.L. de Carvalho, D.R. Mota, C.L. de Araujo, and N.M.M. Coelho. 2005. Determination of sulphite and acetic acid in foods by gas permeation flow injection analysis. *Food Chemistry* 92:765–770.

Badea, M., A. Curulli, and G. Palleschi. 2003. Oxidase enzyme immobilisation through electropolymerised films to assemble biosensors for batch and flow injection analysis. *Biosensors and Bioelectronics* 18:689–698.

Baeumner, A. 2004. Nanosensors identify pathogens in food. *Food Technology* 58:51–55.

Barrales, P.O., M.L.F. de Cordova, and A.M. Diaz. 1998. Indirect determination of ascorbic acid by solid-phase spectrophotometry. *Analytica Chimica Acta* 360:143–152.

Bassi, A.S., E. Lee, and J.X. Zhu. 1998. Carbon paste mediated, amperometric, thin film biosensors for fructose monitoring in honey. *Food Research International* 31:119–127.

Beyene, N.W., H. Moderegger, and K. Kalcher. 2003. Stable glutamate biosensor based on MnO_2 bulk-modified screen-printed carbon electrode and Nafion film-immobilized glutamate oxidase. *South African Journal of Chemistry* 56:54–59.

Bhattacharya, S., J.S. Jang, and L.J. Yang. 2007. Biomems and nanotechnology-based approaches for rapid detection of biological entities. *Journal of Rapid Methods and Automation in Microbiology* 15:1–32.

Biederman, H., I.H. Boyacı, P. Bilkova et al. 2001. Characterization of glow-discharge treated cellulose acetate membrane surfaces for single-layer enzyme electrode studies *Journal of Applied Polymer Science* 81:1341–1352.

Boujtita, M., J.P. Hart, and R. Pittson. 2000. Development of a disposable ethanol biosensor based on a chemically modified screen-printed electrode coated with alcohol oxidase for the analysis of beer. *Biosensors and Bioelectronics* 15:257–263.

Bourais, I., A. Amine, and C.M.A. Brett. 2004. Combination of gold-modified electrode and α-amyloglucosidase for simultaneous determination of starch and glucose. *Analytical Letters* 37:1529–1543.

Boyaci, I.H., and M. Mutlu. 2002. Measurement of glucose, sucrose and lactose in food samples with enzyme-immobilised packed-bed column reactors integrated to an amperometric enzyme electrode. *Nahrung/Food* 46:174–178.

Boyaci, I.H., U.Ö.S. Seker, and M. Mutlu. 2002. Determination of ß-glucan content of cereals with an amperometric glucose electrode. *European Food Research and Technology* 215:538–541.

Boyaci, İ.H., Z.P. Aguilar, M. Hossain, H.B. Halsall, C. J. Seliskar, and W.R. Heineman. 2005. Amperometric enumeration of *Escherichia coli* using antibody-coated paramagnetic beads. *Analytical and Bioanalytical Chemistry* 382:1234–1241.

Buckley, T.J., J.D. Pleil, J.R. Bowyer, and J.M. Davis. 2001. Evaluation of methyl *tert*-butyl ether (MTBE) as an interference on commercial breath-alcohol analyzers. *Forensic Science International* 123:111–118.

Campanella, L., F. Pacifici, M.P. Sammartino, and M. Tomassetti. 1998. A new organic phase bienzymatic electrode for lecithin analysis in food products. *Biosensors and Bioelectronics* 47:25–38.

Cardwell, T.J., and M.J. Christophersen. 2000. Determination of sulfur dioxide and ascorbic acid in beverages using a dual channel flow injection electrochemical detection system. *Analytica Chimica Acta* 416:105–110.

Carelli, D., D. Centonze, C. Palermo et al. 2007. An interference free amperometric biosensor for the detection of biogenic amines in food products. *Biosensors and Bioelectronics* 23:640–647.

Carelli, D., D. Centonze, A. De Giglio, M. Quinto, and P.G. Zambonin. 2006. An interference-free first generation alcohol biosensor based on a gold electrode modified by an overoxidised non-conducting polypyrrole film. *Analytica Chimica Acta* 565:27–35.

Cass, A.E.G., G. Davis, G.D. Francis et al. 1984. Ferrocene-mediated enzyme electrode for amperometric determination of glucose. *Analytical Chemistry* 56:667–671.

Centonze, D., A. Guerrieri, C. Malitesta, F. Palmisano, and P.G. Zambonin. 1992. Interference-free glucose sensor based on glucose-oxidase immobilized in an overoxidized nonconducting polypyrrole film. *Fresenius Journal of Analytical Chemistry* 342:729–733.

Centonze, D., C.G. Zambonin, and F. Palmisano. 1997. Determination of glucose in nonalcoholic beverages by a biosensor coupled with microdialysis fiber samplers. *Journal of AOAC International* 80:829–833.

Chen, X.H., H. Guo, J. Yi et al. 2009. Fabrication of AucoreCo$_3$O$_4$shell/PAA/HRP composite film for direct electrochemistry and hydrogen peroxide sensor applications. *Sensors and Materials* 21:433–444.

Chen, J.R., Y.Q. Miao, N.Y. He, X.H. Wu, and S.J. Li. 2004. Nanotechnology and biosensors. *Biotechnology Advances* 22:505–518.

Chen, S.-H., H.-L. Wu, C.-H. Yen, S.-M. Wu, S.-J. Lin, and H.-S. Kou. 1998. Trace determination of methanol in water–ethanol solution by derivatization and high-performance liquid chromatography. *Journal of Chromatography A* 799:93–99.

Christie, I.M., P.H. Treloar, and P. Vadgama. 1992. Plasticized poly(vinyl chloride) as a permselective barrier membrane for high-selectivity amperometric sensors and biosensors. *Analytica Chimica Acta* 269:65–73.

Ciriello, R., T.R.I. Cataldi, D. Centonze, and A. Guerrieri. 2000. Permselective behavior of an electrosynthesized, nonconducting thin film of poly(2-naphthol) and its application to enzyme immobilization. *Electroanalysis* 12:825–830.

Clark, L.C., and C. Lyons. 1962. Electrode systems for continuous monitoring in cardiovascular surgery. *Annals of the New York Academy of Sciences* 102:29–45.

Cokeliler, D., and M. Mutlu. 2002. Performance of amperometric alcohol electrodes prepared by plasma polymerisation technique. *Analytica Chimica Acta* 469:217–223.

Cosnier, S. 2003. Biosensors based on electropolymerized films: New trends. *Analytical and Bioanalytical Chemistry* 377:507–520.

Cosnier, S. 2005. Affinity biosensors based on electropolymerized films. *Electroanalysis* 17:1701–1715.

Cosnier, S. 2007. Recent advances in biological sensors based on electrogenerated polymers: A review. *Analytical Letters* 40:1260–1279.

Cruz-Aguado, J.A., and G. Penner. 2008. Determination of Ochratoxin A with a DNA aptamer. *Journal of Agricultural and Food Chemistry* 56:10456–10461.

Cunningham, A.J. 1998. *Introduction to bioanalytical sensors.* New York: Wiley.

Daily, S., S.J. Armfield, B.G.D Haggett, and M.E.A. Downs. 1991. Automated enzyme packed-bed system for the determination of vitamin-C in foodstuffs. *Analyst* 116:569–572.

Davis, F., and S.P.J. Higson. 2005. Structured thin films as functional components within biosensors. *Biosensors and Bioelectronics* 21:1–20.

Davis, G., I.J. Higgins, and H.A.O. Hill. 1984. Ferrocene-based electrode for glucose. *Journal of Biomedical Engineering* 6:174–175.

Del Campo, G., I. Berregi, R. Caracena, and J.I. Santos. 2006. Quantitative analysis of malic and citric acids in fruit juices using proton nuclear magnetic resonance spectroscopy. *Analytica Chimica Acta* 556:462–468.

Desai, M.A., M. Mutlu, and P. Vadgama. 1992. A study of macromolecular diffusion through native porcine mucus. *Experientia* 48:22–26.

Desai, M.A., C.V. Nicholas, and P. Vadgama. 1991. Electrochemical determination of the permeability of porcine mucus to model solute compounds. *Journal of Pharmacy and Pharmacology* 43:124–127.

Diamond, D. 1998. Ed. *Principles of chemical and biological sensors.* New York: Wiley.

Dinckaya, E., E. Akyilmaz, and M.K. Sezginturk. 2010. Sensitive nitrate determination in water and meat samples by amperometric biosensor. *Preparative Biochemistry and Biotechnology* 40:119–128.

Dudak, F.C., I.H. Boyacı, A. Jurkevica et al. 2009. Determination of *Escherichia coli* using antibody-coated paramagnetic beads with fluorescence detection. *Analytical and Bioanalytical Chemistry* 393:949–956.

Dzyadevych, S.V., V.N. Arkhypova, A.P. Soldatkin, A.V. El'skaya, C. Martelet, and N. Jaffrezic-Renault. 2008. Amperometric enzyme biosensors: Past, present and future. *IRBM* 29:171–180.

Eggins, B. 1996. *Biosensors: An introduction.* New York: Wiley-Teubner.

Ellis, D.I., and R. Goodacre. 2001. Rapid and quantitative detection of the microbial spoilage of muscle foods: Current status and future trends. *Trends in Food Science and Technology* 12:414–429.

Ensafi, A.A., S. Abbasi, and B. Rezaei. 2001. Kinetic spectrophotometric method for the determination of oxalic acid by its catalytic effect on the oxidation of safranine by dichromate. *Spectrochimica Acta Part A* 57:1833–1838.

Eshkenazi, I., E. Maltz, B. Zion, and J. Rishpon. 2000. A three-cascaded-enzymes biosensor to determine lactose concentration in raw milk. *Journal of Dairy Science* 83:1939–1945.

European Parliament and Council Directive No. 95/2/EC, February 1995, http://eur-lex.europa.eu/LexUriServ/LexUriServ.do?uri=CONSLEG:1995L0002:20060815:EN:PDF; (accessed September 20, 2010).

Fernandes, E.N., and B.F. Reis. 2006. Automatic spectrophotometric procedure for the determination of tartaric acid in wine employing multicommutation flow analysis process. *Analytica Chimica Acta* 557:380–386.

Food and Agriculture Organization of the United Nations (FAO). International Code of Conduct on the Distribution and Use of Pesticides (Revised version). Rome, 2002. http://www.fao.org/fileadmin/templates/agphomel/documents/Pests_Pesticides/code/code.pdf (accessed September 20, 2010).

Gajovic, N., A. Warsinke, and F.W. Scheller. 1997. Comparison of two enzyme sequences for a novel L-malate biosensor. *Journal of Chemical Technology and Biotechnology* 68:31–36.

Galandova, J., J. Labuda, and I. Palchetti. 2009. Polymer interfaces used in electrochemical DNA-based biosensors. *Chemical Papers* 63:1–14.

Garcia, C.A.B., G.D. Neto, L.T. Kubota, and L.A. Grandin. 1996. A new amperometric biosensor for fructose using a carbon paste electrode modified with silica gel coated with Meldola's Blue and fructose 5-dehydrogenase. *Journal of Electroanalytical Chemistry* 418:147–151.

Gerard, M., A. Chaubey, and B.D. Malhotra. 2002. Application of conducting polymers to biosensors. *Biosensors and Bioelectronics* 17:345–359.

German, N., A. Ramanaviciene, J. Voronovic, and A. Ramanavicius. 2010. Glucose biosensor based on graphite electrodes modified with glucose oxidase and colloidal gold nanoparticles. *Microchimica Acta* 168:221–229.

Ghassempour, A., N.M. Najafi, and A.A. Amiri. 2003. Determination of citric acid in fermentation media by pyrolysis mass spectrometry. *Journal of Analytical and Applied Pyrolysis* 70:251–261.

Goktug, T., M.K. Sezginturk, and E. Dinckaya. 2005. Glucose oxidase-b-galactosidase hybrid biosensor based on glassy carbon electrode modified with mercury for lactose determination. *Analytica Chimica Acta* 551:51–56.

Gomez-Alvarez, E., E. Luque-Perez, A. Rios, and M. Valcarcel. 1999. Flow injection spectrophotometric determination of lactic acid in skimmed milk based on a photochemical reaction. *Talanta* 50:121–131.

Goriushkina, T.B., L.V. Shkotova, G.Z. Gayda et al. 2010. Amperometric biosensor based on glycerol oxidase for glycerol determination. *Sensors and Actuators B* 144:361–367.

Günaydin, B., N. Şir, S. Kavlak, A. Güner, and M. Mutlu. 2010. A new approach for the electrochemical detection of phenolic compounds. Part I: Modification of graphite surface by plasma polymerization technique and characterization by Raman spectroscopy. *Journal of Food and Bioprocess Technology* 3:473–479.

Guerrieri, A., G.E. De Benedetto, F. Palmisano, and P.G. Zambonin. 1998. Electrosynthesized non-conducting polymers as permselective membranes in amperometric enzyme electrodes: A glucose biosensor based on a co-crosslinked glucose oxidase/overoxidized polypyrrole bilayer. *Biosensors and Bioelectronics* 13:103–112.

Guerrieri, A., V. Lattanzio, F. Palmisano, and P.G. Zambonin. 2006. Electrosynthesized poly(pyrrole)/poly(2-naphthol) bilayer membrane as an effective anti-interference layer for simultaneous determination of acethylcholine and choline by a dual electrode amperometric biosensor. *Biosensors and Bioelectronics* 21:1710–1718.

Gulce, H., A. Gulce, and A. Yildiz. 2002. A novel two-enzyme amperometric electrode for lactose determination. *Analytical Sciences* 18:147–150.

Haghighi, B., S. Varma, F.M. Alizadeh Sh., Y. Yigzaw, and L. Gorton. 2004. Prussian blue modified glassy carbon electrodes—Study on operational stability and its application as a sucrose biosensor. *Talanta* 64:3–12.

Hartmeier, W. 1988. *Immobilized biocatalysts: An introduction.* Berlin: Springer Verlag.

Haruyama, T. 2003. Micro- and nanobiotechnology for biosensing cellular responses. *Advanced Drug Delivery Reviews* 55:393–401.

Hettiarachchi, K., and S. Ridge. 1998. Capillary electrophoretic determination of acetic acid and trifluoroacetic acid in synthetic peptide samples. *Journal of Chromatography A* 817:153–161.

Higson, S.P.J., M.A. Desai, S. Ghosh, I. Christie, and P. Vadgama. 1993. Amperometric enzyme electrode biofouling and passivation in blood—Characterization of working electrode polarization and inner membrane effects. *Journal of the Chemical Society, Faraday Transactions* 89:2847–2851.

Holthoff, E.L., and F.V. Bright. 2007. Molecularly imprinted xerogels as platforms for sensing. *Accounts of Chemical Research* 40:756–770.

Hönow, R., D. Bongartz, and A. Hesse. 1997. An improved HPLC-enzyme-reactor method for the determination of oxalic acid in complex matrices. *Clinica Chimica Acta* 261:131–139.

Hu, T., X. Zhang, and Z. Zhang. 1999. Disposable screen-printed enzyme sensor for simultaneous determination of starch and glucose. *Biotechnology Techniques* 13:359–362.

Hurrell, R.F. 2003. Influence of vegetable protein sources on trace element and mineral bioavailability. *The Journal of Nutrition* 133:2973–2977.

Ikeda, T., T. Shibata, S. Todoriki, M. Senda, and H. Kinoshita. 1990. Amperometric response to reducing carbohydrates of an enzyme electrode based on oligosaccharide dehydrogenase—detection of lactose and alpha-amylase. *Analytica Chimica Acta* 230:75–82.

Jain, K.K. 2003. Nanodiagnostics: Application of nanotechnology in molecular diagnostics. *Expert Review of Molecular Diagnostics* 3:153–161.

Jariwalla, R.J., R. Sabin, S. Lawson, and Z.S. Herman. 1990. Lowering of serum cholesterol and triglycerides and modulation of divalent cations by dietray phytate. *Journal of Applied Nutrition* 42:18–28.

Kall, M.A., and C. Andersen. 1999. Improved method for simultaneous determination of ascorbic acid and dehydroascorbic acid, isoascorbic acid and dehydroisoascorbic acid in food and biological samples. *Journal of Chromatography B* 730:101–111.

Kandimalla, V.B., and H.X. Ju. 2004. New horizons with a multi dimensional tool for applications in analytical chemistry—Aptamer. *Analytical Letters* 37:2215–2233.

Karan, H.I., H.L. Lan, and Y. Okamoto. 1994. Viologen derivative containing polysiloxane as an electron-transfer mediator in amperometric glucose sensors. In *Diagnostic biosensor polymers*, ed. A.M. Usmani and N. Akmal, (556):169–179. Washington DC: ACS Books.

Kartlik, J., A. Pizzarielloa, V. Mastihubaa, J. Svorc, M. Stredansky, and S. Miertus. 1998. Biosensors for L-malate and L-lactate based on solid binding matrix. *Analytica Chimica Acta* 379:193–200.

Karube, I., and M.E.S.M. Chang. 1991. Microbial biosensors. In *Biosensors principles and applications*, ed. L.J. Blum and P.R. Coulet, 267–301. New York: Marcel Dekker.

Kelly, S.C., P.J. O'Connell, C.K. O'Sullivan, and G.G. Guilbault. 2000. Development of an interferent free amperometric biosensor for determination of l-lysine in food. *Analytica Chimica Acta* 412:111–119.

Kennedy, J.F., M.C.B. Pimentel, E.H.M. Melo, and J.L. Lima-Filho. 2007. Sucrose biosensor as an alternative tool for sugarcane field samples. *Journal of the Science of Food and Agriculture* 87:2266–2271.

Kijkowska, R., D. Pawlowska-Kozinska, Z. Kowalski, M. Jodko, and Z. Wzorek. 2002. Wet-process phosphoric acid obtained from Kola apatite. Purification from sulphates, fluorine, and metals. *Separation and Purification Technology* 28:197–205.

Kim, Y.J., Y.S. Kim, J.H. Niazi et al. 2010. Electrochemical aptasensor for tetracycline detection. *Bioprocess and Biosystems Engineering* 33:31–37.

Kinnear, K.T., and G.H. Monbouquette. 1997. An amperometric fructose biosensor based on fructose dehydrogenase immobilized in a membrane mimetic layer on gold. *Anaytical Chemistry* 69:1771–1775.

Ko, S., S. Gunasekaran, and J. Yu. 2010. Self-indicating nanobiosensor for detection of 2,4-dinitrophenol. *Food Control* 21:155–161.

Koochaki, Z., S.P.J. Higson, M. Mutlu, and P. Vadgama. 1993. The diffusion limited oxidase-based glucose enzyme electrode: Relation between covering membrane permeability and substrate response.*Journal of Membrane Science* 76:261–268.

Krysteva, M.A., and L.K. Yotova. 1992. Multienzyme membranes for biosensors. *Journal of Chemical Technology and Biotechnology* 54:13–18.

Laurinavicius, V., B. Kurtinaitiene, V. Gureiviciene, L. Boguslavsky, L. Geng, and T. Skotheim. 1996. Amperometric glyceride biosensor. *Analytica Chimica Acta* 330:159–166.

Law, W.T., N. Akmal, and A.M. Usmani. 2002. Eds. *Biomedical diagnostic science and technology*. New York: Marcel Dekker.

Lei, Y., W. Chen, and A. Mulchandani. 2006. Microbial biosensors. *Analitica Chimica Acta* 568:200–210.

Lenghor, N., J. Jakmunee, M. Vilen, R. Sara, G.D. Christian, and K. Grudpan. 2002. Sequential injection redox or acid-base titration for determination of ascorbic acid or acetic acid. *Talanta* 58:1139–1144.

Leonard, P., S. Hearty, J. Brennan et al. 2003. Advances in biosensors for detection of pathogens in food and water. *Enzyme and Microbial Technology* 32:3–13.

Lerici, C.R., L. Manzocco, and M. Anese. 1999. Ethanol in food: Liquid–vapour partition in model systems containing Maillard reaction products. *Food Research International* 32:429–432.

Li, H.Q., Z.H. Guo, C.X. Liu et al. 2006. An L-glutamate amperometric biosensor based on redox polymer wired horseradish peroxidase and glutamate oxidase. *Chinese Journal of Analytical Chemistry* 34:S1–S4.

Loechel, C., G.C. Chemnitius, and M. Borchardt. 1998. Amperometric bi-enzyme based biosensor for the determination of lactose with an extended linear range. *Zeitschrift Fur Lebensmittel-Untersuchung Und-Forschung A, Food Research snd Technology* 207:381–385.

Loğoğlu, E., S. Sungur, and Y. Yildiz. 2006. Development of lactose biosensor based on b-galactosidase and glucose oxidase immobilized into gelatin. *Journal of Macromolecular Science* 43:525–553.

Luca, S.D., M. Florescu, M.E. Ghica et al. 2005. Carbon film electrodes for oxidase-based enzyme sensors in food analysis. *Talanta* 68:171–178.

Luppa, P.B., L.J. Sokoll, and D.W. Chan. 2001. Immunosensors—Principles and applications to clinical chemistry. *Clinica Chimica Acta* 314:1–26.

Luque-Perez, E., A. Rios, and M. Valcarcel. 1998. Flow-injection spectrophotometric determination of citric acid in beverages based on a photochemical reaction. *Analytica Chimica Acta* 366:231–240.

Maestre, E., I. Katakis, and E. Dominguez. 2001. Amperometric flow-injection determination of sucrose with a mediated tri-enzyme electrode based on sucrose phosphorylase and electrocatalytic oxidation of NADH. *Biosensors and Bioelectronics* 16:61–68.

Maines, A., D. Ashworth, and P. Vadgama. 1996. Enzyme electrodes for food analysis. *Food Technology and Biotechnology* 34:31–42.

Mairal, T., V.C. Ozalp, P.L. Sanchez et al. 2008. Aptamers: Molecular tools for analytical applications. *Analytical and Bioanalytical Chemistry* 390:989–1007.

Majer-Baranyi, K., N. Adanyi, and M. Varadi. 2008. Investigation of a multienzyme based amperometric biosensor for determination of sucrose in fruit juices. *European Food Research and Technology* 228:139–144.

Mak, K.K.W., H. Yanase, and R. Renneberg. 2005. Cyanide fishing and cyanide detection in coral reef fish using chemical tests and biosensors. *Biosensors and Bioelectronics* 20:2581–2593.

Mak, W.C., Y.M. Ng, C.Y. Chan, W.K. Kwong, and R. Renneberg. 2004. Novel biosensors for quantitative phytic acid and phytase measurement. *Biosensors and Bioelectronics* 19:1029–1035.

Mallet, S., M. Arellano, J.C. Boulet, and F. Couderc. 1999. Determination of tartaric acid in solid wine residues by capillary electrophoresis and indirect UV detection. *Journal of Chromatography A* 853:181–184.

Malitesta, C., F. Palmisano, L. Torsi, and P.G. Zambonin. 1990. Glucose fast-response amperometric sensor based on glucose oxidase immobilized in an electropolymerized poly(o-phenylenediamine) film. *Analytical Chemistry* 62:2735–2740.

Manso, J., M.L. Mena, P. Yanez-Sedeno, and J.M. Pingarron. 2008. Bienzyme amperometric biosensor using gold nanoparticle-modified electrodes for the determination of inulin in foods. *Analytical Biochemistry* 375:345–53.

Marconi, E., M.C. Messia, G. Palleschi, and R. Cubadda. 2004. A maltose biosensor for determining gelatinized starch in processed cereal foods. *Cereal Chemistry* 81:6–9.

Martelli, P.B., B.F. Reis, A.N. Araujo, M. Conceicao, and B.S.M. Montenegro. 2001. A flow system with a conventional spectrophotometer for the chemiluminescent determination of lactic acid in yogurt. *Talanta* 54:879–885.

Mato, I., J.F. Huidobro, M.P. Sanchez, S. Muniategui, M.A. Fernandez-Muino, and M.T. Sancho. 1998. Enzymatic determination in honey of L-malic acid. *Food Chemistry* 62(4):503–508.

Mattos, I.L., and M.C. Cunha Areias. 2005. Automated determination of glucose in soluble coffee using Prussian blue–glucose oxidase–Nafion® modified electrode. *Talanta* 66:1281–1286.

Mazzei, F., F. Botre, and G. Favero. 2007. Peroxidase based biosensors for the selective determination of D,L-lactic acid and L-malic acid in wines. *Microchemical Journal* 87:81–86.

Mehrvar, M., and M. Abdi. 2004. Recent developments, characteristics, and potential applications of electrochemical biosensors. *Analytical Sciences* 20:1113–1126.

Mello, L.D., and L.T. Kubota. 2002. Review of the use of biosensors as analytical tools in the food and drink industries. *Food Chemistry* 77:237–256.

Mello, L.D., M.D.P.T. Sotomayor, and L.T. Kubota. 2003. LT HRP-based amperometric biosensor for the polyphenols determination in vegetables extract. *Sensors and Actuators B-Chemical* 96:636–645.

Mezcua, M., A. Aguera, M.D. Hernando, L. Piedra, and A.R. Fernandez-Alba. 2003. Determination of methyl *tert*-butyl ether and *tert*-butyl alcohol in seawater samples using purge-and-trap enrichment coupled to gas chromatography with atomic emission and mass spectrometric detection. *Journal of Chromatography A* 999:81–90.

Mikeladze, E., A. Schulte, M. Mosbach et al. 2002. Redox hydrogel-based bienzyme microelectrodes for amperometric monitoring of L-glutamate. *Electroanalysis* 14:393–99.

Milardovic, S, Z. Grabaric, V. Rumenjak, and M. Jukic. 2000. Rapid determination of oxalate by an amperometric oxalate oxidase-based electrode. *Electroanalysis* 12:1051–1058.

Mita, D.G., A. Attanasio, F. Arduini et al. 2007. Enzymatic determination of BPA by means of tyrosinase immobilized on different carbon carriers. *Biosensors and Bioelectronics* 23:60–65.

Mizutani, F., and S. Yabuki. 1995. Rapid measurement of cholinesterase activity using an amperometric enzyme electrode based on lipid-modified choline oxidase. *Analytical Sciences* 11:127–129.

Mizutani, F., S. Yabuki, Y. Sato, and Y. Hirata. 1998. Amperometric biosensors using an enzyme-containing polyion complex. In *Polymers in sensors: Theory and practice*, ed. N. Akmal and A.M. Usmani, *American Chemical Society (ACS) Symposium Series* 690:47–56.

Mizutani, F., Y. Hirata, S. Yabuki, and S. Iijima. 2003. Flow injection analysis of acetic acid in food samples by using trienzyme/poly(dimethylsiloxane)-bilayer membrane-based electrode as the detector. *Sensors and Actuators B* 91:195–198.

Mizutani, F., Y. Sato, Y. Hirata, and S. Yabuki.1998. High-throughput flow-injection analysis of glucose and glutamate in food and biological samples by using enzyme/polyion complexbilayer membrane-based electrodes as the detectors. *Biosensors and Bioelectronics* 13:809–815.

Mohammadi, H., A. Amine, S. Cosnier, and C. Mousty. 2005. Mercury-enzyme inhibition assays with an amperometric sucrose biosensor based on a trienzymatic-clay matrix. *Analytica Chimica Acta* 543:143–49.

Moody, G.J. 1992. Role of polymeric materials in the fabrication of ion-selective electrodes and biosensors. *ACS Symposium Series* 487:99–110.

Moody, G.J., and J.D.R. Thomas. 1991. Amperometric biosensors—A brief appraisal of principles and applications. *Selective Electrode Reviews* 13:113–124.

Morales, M.D., B. Serra, A.G.V. de Prada et al. 2007. An electrochemical method for simultaneous detection and identification of Escherichia coli, Staphylococcus aureus and Salmonella choleraesuis using a glucose oxidase-peroxidase composite biosensor. *Analyst* 132:572–578.

Moreno-Cid, A., M.C. Yebra, and X. Santos. 2004. Flow injection determinations of citric acid: A review. *Talanta* 63:509–514.

Mori, I., T. Kawakatsu, Y. Fujita, and T. Matsuo. 1999. Selective spectrophotometric determination of palladium(II) with 2(5-nitro-2-pyridylazo)-5-(*N*-propyl-*N*-3-sulfopropylamino) phenol(5-NO$_2$.PAPS) and tartaric acid with 5-NO$_2$.PAPS-niobium(V) complex. *Talanta* 48(5):1039–1044.

Muguruma, H. 2007. Plasma-polymerized films for biosensors II. *Trac-Trends in Analytical Chemistry* 26:433–443.

Muresan, L., R.R. Valera, I. Frebort, I.C. Popescu, E. Csoregi, and M. Nistor. 2008. Amine oxidase amperometric biosensor coupled to liquid chromatography for biogenic amines determination. *Microchimica Acta* 163:219–225.

Mutlu, M., and S. Mutlu. 1995. The effect of crosslink density on permeability in biosensors: An unsteady-state approach. *Journal of Biotechnology Techniques* 9:277–282.

Mutlu, M., S. Mutlu, B. Alp, I.H. Boyacı, and E. Piskin. 1997. Preparation of a single layer enzyme electrode by plasma polymerization technique. In *Plasma processing of polymers*, ed. R. D'Agustino, Dordrecht, 477–485. NATO ASI Series, Kluwer Academic Publishers.

Mutlu, M., S. Mutlu, I.H. Boyacı, B. Alp, and E. Piskin. 1998. High linearity glucose enzyme electrodes for food industries by plasma polymerization technique. In *Polymers in sensors: Theory and practice*, eds. N. Akmal and A.M. Usmani. *ACS Symposium Series* 690:57–65.

Mutlu, M., S. Mutlu, M.R. Rosenberg, J. Kane, J., M.N. Jones, and P. Vadgama. 1991. Matrix surface modification by plasma polymerization for enzyme immobilization. *Journal of Materials Chemistry* 1:447–50.

Mutlu, S., M. Mutlu, and E. Piskin. 1998. A kinetic approach to oxidase based enzyme electrodes: The effect of enzyme layer formation on the response time. *The Biochemical Engineering Journal* 1:39–43.

Mutlu, S., B. Alp, R.S. Özmelles, and M. Mutlu. 1997. Amperometric determination of enzymatic activity by multienzyme biosensors. *Journal of Food Engineering* 33:81–86.

Mutlu S., M. Mutlu, P. Vadgama, and E. Piskin. 1994. Sandwich type amperometric enzyme electrodes for determination of glucose. In *Diagnostic polymeric materials*, eds. A.M. Usmani and N. Akmal. *ACS Symposium Series* 556:71–83.

Nakamura, H., R. Tanaka, K. Suzuki, M. Yataka, and Y. Mogi. 2009. A direct determination method for ethanol concentrations in alcoholic beverages employing a eukaryote double-mediator system. *Food Chemistry* 117:509–513.

Nikolaus, N., and B. Strehlitz. 2008. Amperometric lactate biosensors and their application in (sports) medicine, for life quality and wellbeing. *Microchimica Acta* 160:15–55.

Okubo, M., and M. Takahashi. 1994. Production of submicron size monodisperse polymer particles having aldehyde groups by the seeded aldol condensation polymerisation of glutaraldehyde (II). *Colloid Polymer Science* 272:422–426.

Pal, P.S., and P. Sarkar. 2002. Polymers in biosensors—A review. *J. The Indian Chemical Society* 79:211–18.

Palchetti, I., and M. Mascini. 2008. Electroanalytical biosensors and their potential for food pathogen and toxin detection. *Analytical and Bioanalytical Chemistry* 391:455–471.

Palma, M., and C.G. Barroso. 2002. Ultrasound-assisted extraction and determination of tartaric and malic acids from grapes and wine-making by-products. *Analytica Chimica Acta* 458:119–130.

Palmisano, F., A. Guerrieri, M. Quinto, and P.G. Zambonin. 1995a. Electrosynthesized bilayer polymeric membrane for effective elimination of electroactive interferents in amperometric biosensors. *Analytical Chemistry* 67:1005–1009.

Palmisano, F., C. Malitesta, D. Centonze, and P.G. Zambonin. 1995b. Correlation between permselectivity and chemical structure of overoxidized polypyrrole membranes used in electroproduced enzyme biosensors. *Analytical Chemistry* 67:2207–2211.

Palmisano, F., D. Centonze, and P.G. Zambonin. 2000. Amperometric biosensors based on electrosynthesised polymeric films. *Fresenius Journal of Analytical Chemistry* 366:586–601.

Palmisano, F., D. Centonze, M. Quinto, and P.G. Zambonin. 1996. A microdialysis fibre based sampler for flow injection analysis: Determination of L-lactate in biofluids by an electrochemically synthesised bilayer membrane based biosensor. *Biosensors and Bioelectronics* 11:419–26.

Papamichael, K.I., M.P. Kreuzer, and G.G. Guilbault. 2007. Viability of allergy (IgE) detection using an alternative aptamer receptor and electrochemical means. *Sensors and Actuators B-Chemical* 121:178–186.

Patel, P.D. 2002. (Bio)sensors for measurement of analytes implicated in food safety: A review. *Trac-Trends in Analytical Chemistry* 21:96–115.

Pellegrino, S., F.S. Bruno, and M. Petrarulo. 1999. Liquid chromatographic determination of ethyl alcohol in body fluids. *Journal of Chromatography B* 729:103–110.

Pena, N., G. Ruiz, A.J. Reviejo, and J.M. Pingarron. 2001. Graphite-teflon composite bienzyme electrodes for the determination of cholesterol in reversed micelles. Application to food samples. *Analytical Chemistry* 73:1190–1195.

Pezza, L., A.O. Santini, H.R. Pezza, C.B. Melios, V.J.F. Ferreira, and A.L.M. Nasser. 2001. Benzoate ion determination in beverages by using a potentiometric sensor immobilized in a graphite matrix. *Analytica Chimica Acta* 433:281–288.

Prodromidis, M.I., and M.I. Karayannis. 2002. Enzyme based amperometric biosensors for food analysis. *Electroanalysis* 14:241–261.

Pylypiw, Jr., H.M., and M.T. Grether. 2000. Rapid high-performance liquid chromatography method for the analysis of sodium benzoate and potassium sorbate in foods. *Journal of Chromatography A*, 883:299–304.

Rajendran, V., and J. Irudayaraj. 2002. Detection of glucose, galactose, and lactose in milk with a microdialysis-coupled flow injection amperometric sensor. *Journal of Dairy Science* 85:1357–61.

Ram, M.K., P. Bertoncello, H. Ding, S. Paddeu, and C. Nicolini. 2001. Cholesterol biosensor prepared by layer by layer technique. *Biosensors and Bioelectronics* 16:849–856.

Ramsay, G. 1998. Ed. *Commercial biosensors.* New York: Wiley.

Raposo, Jr., J.L., S.R. Oliveira, J.A. Nobrega, and J.A.G. Neto. 2008. Internal standardization and least-squares background correction in high-resolution continuum source flame atomic absorption spectrometry to eliminate interferences on determination of Pb in phosphoric acid. *Spectrochimica Acta Part B* 63:992–995.

Rasooly, A., and K.E. Herold. 2006. Biosensors for the analysis of food- and waterborne pathogens and their toxins. *Journal of AOAC International* 89:873–883.

Rejeb, I.B., F. Arduinib, A. Arvinted, et al. 2009. Development of a bio-electrochemical assay for AFB1 detection in olive oil. *Biosensors and Bioelectronics* 24:1962–1968.

Ricci, F., and G. Palleschi. 2005. Sensor and biosensor preparation, optimisation and applications of Prussian Blue modified electrodes. *Biosensors and Bioelectronics* 21:389–407.

Rogers, K.R. 2000. Principles of affinity-based biosensors. *Molecular Biotechnology* 14(2):109–129.

Rogers, K.R., and M. Mascini. 1998. Biosensors for field analytical monitoring. *Field Analytical Chemistry and Technology* 2:317–331.

Rosini, E., G. Molla, C. Rossetti, M. S. Pilone, L. Pollegioni, and S. Sacchi. 2008. A biosensor for all d-amino acids using evolved d-amino acid oxidase. *Journal of Biotechnology* 135:377–384.

Rotariu, L., C. Bala, and V. Magearu. 2004. New potentiometric microbial biosensor for ethanol determination in alcoholic beverages. *Analytica Chimica Acta* 513:119–123.

Rotariu, L., and C. Bala. 2003. New type of ethanol microbial biosensor based on a highly sensitive amperometric oxygen electrode and yeast cells. *Analytical Letters* 36:2459–2471.

Sadik, O.A., A.O. Aluoch, and A.L. Zhou. 2009. Status of biomolecular recognition using electrochemical techniques. *Biosensors and Bioelectronics* 24:2749–2765.

Safavi, A., N. Maleki, O. Moradlou, and F. Tajabadi. 2006. Simultaneous determination of dopamine, ascorbic acid, and uric acid using carbon ionic liquid electrode. *Analytical Biochemistry* 359:224–229.

San, N., M. Kilic, Z. Tuiebakhova, and Z. Cinar. 2007. Enhancement and modeling of the photocatalytic degradation of benzoic acid. *Journal of Advanced Oxidation Technologies* 10:43–50.

Sassolas, A., B.D. Leca-Bouvier, and L.J. Blum. 2008. DNA biosensors and microarrays. *Chemical Reviews* 108:109–139.

Scampicchio, M., A. Arecchi, N.S. Lawrence, and S. Mannino. 2010. Nylon nanofibrous membrane for mediated glucose biosensing. *Sensors and Actuators B* 145:394–397.

Schoemaker, M., R. Feldbrugge, B. Grundig, and F. Spener, 1997. The lipoxygenase sensor, a new approach in essential fatty acid determination in foods. *Biosensors and Bioelectronics* 12:1089–1099.

Sekine, Y., M. Suzuki, T. Takeuchi, E. Tamiya, and I. Karube. 1993. Selective flow-injection determination of methanol in the presence of ethanol based on a multi-enzyme system with chemiluminescence detection. *Analytica Chimica Acta* 280(2):179–184.

Serban, S., A.F. Danet, and N. El Murr. 2004. Rapid and sensitive automated method for glucose monitoring in wine processing. *Journal of Agricultural and Food Chemistry* 52:5588–592.

Serra, B., A.J. Reviejo, C. Parrado, and J.M. Pingarron. 1999. Graphite-Teflon composite bienzyme electrodes for the determination of L-lactate: Application to food samples. *Biosensors and Bioelectronics* 14:505–513.

Sezginturk, M.K., and E. Dinckaya. 2003. A novel amperometric biosensor based on spinach (Spinacia oleracea) tissue homogenate for urinary oxalate determination. *Talanta* 59:545–551.

Shirakawa, N., C. Saini, B. De Borba, and R. Kiser. 2004. Determination of phosphoric acid and citric acid in soft drink products by reagent-free ion chromatography. *The Applications Book* 19–20.

Shkotova, L.V., A.P. Soldatkin, M.V. Gonchar, W. Schuhmann, and S.V. Dzyadevych. 2006. Amperometric biosensor for ethanol detection based on alcohol oxidase immobilised within electrochemically deposited Resydrol film. *Materials Science and Engineering C* 26:411–414.

Shu, H.-C., and N.-P. Wu. 2001. A chemically modified carbon paste electrode with D-lactate dehydrogenase and alanine aminotranferase enzyme sequences for D-lactic acid analysis. *Talanta* 54:361–368.

Singh, M., N.Verma, A.K. Garg, et al. 2008. Urea biosensors. *Sensors and Actuators B-Chemical* 134:345–351.

Singh, M., P.K. Kathuroju, and N. Jampana. 2009. Polypyrrole based amperometric glucose biosensors. *Sensors and Actuators B-Chemical* 143:430–443.

Stoica, L., R. Ruzgas, R. Ludwig, D. Haltrich, and L. Gorton. 2006. Direct electron transfer—A favorite electron route for cellobiose dehydrogenase (CDH) from Trametes villosa. Comparison with CDH from Phanerochaete chrysosporium. *Langmuir* 22:10801–10806.

Su, M.-S., and P.-J. Chien. 2010. Aroma impact components of rabbit-eye blueberry (*Vaccinium ashei*) vinegars. *Food Chemistry* 119:923–928.

Tatsumi, H., and H. Katano. 2004. Kinetic analysis of enzymatic hydrolysis of raw starch by glucoamylase using an amperometric glucose sensor. *Chemistry Letters* 33:692–693.

Tabata, M., and H. Morita. 1997. Spectrophotometric determination of a nanomolar amount of ascorbic acid using its catalytic effect on copper(II) porphyrin formation. *Talanta* 44:151–157.

Tan, L., Y. Wang, X. Liu, H. Ju, and J. Li. 2005. Simultaneous determination of L- and D-lactic acid in plasma by capillary electrophoresis. *Journal of Chromatography B* 814:393–398.

Thangamuthu, R., S.M.S. Kumar, and K.C. Pillai. 2007. Direct amperometric determination of l-ascorbic acid (vitamin C) at octacyanomolybdate-doped-poly(4-vinylpyridine) modified electrode in fruit juice and pharmaceuticals. *Sensors and Actuators B* 120:745–753.

Thevenot, D.R., K. Toth, R.A. Durst, et al. 2001. Electrochemical biosensors: Recommended definitions and classification. *Analytical Letters* 34:635–659.

Themelis, D.G., and P.D. Tzanavaras. 2001. Reagent-injection spectrophotometric determination of citric acid in beverages and pharmaceutical formulations based on its inhibitory effect on the iron(III) catalytic oxidation of 2,4-diaminophenol by hydrogen peroxide. *Analytica Chimica Acta* 428:23–30.

Thibault, S., H. Aubriet, C. Arnoult, and D. Ruch. 2008. Gold nanoparticles and a glucose oxidase based biosensor: An attempt to follow-up aging by XPS. *Microchimica Acta* 163:211–117.

Tkac, J., E. Sturdik, and P. Gemeiner. 2000. Novel glucose non-interference biosensor for lactose detection based on galactose oxidase–peroxidase with and without co-immobilised b-galactosidase. *Analyst* 125:1285–1289.

Tombelli, S., and M. Mascini. 2009. Aptamers as molecular tools for bioanalytical methods. *Current Opinion in Molecular Therapeutics* 11:179–188.

Tombelli, S., M. Minunni, and M. Mascini. 2007. Aptamers-based assays for diagnostics, environmental and food analysis. *Biomolecular Engineering* 24:191–200.

Trivedi, U.B., D. Lakshminarayana, I.L. Kothari, P.B. Patel, and C.J. Panchal. 2009. Amperometric fructose biosensor based on fructose dehydrogenase enzyme. *Sensors and Actuators B* 136:45–51.

Tryland, I., and L. Fiksdal. 1998. Enzyme characteristics of beta-D-galactosidase- and beta-D-glucuronidase-positive bacteria and their interference in rapid methods for detection of waterborne coliforms and Escherichia coli. *Applied and Environmental Microbiology* 64:1018–1023.

Tucker, K.L., K. Morita, N. Qiao, M.T. Hannan, L.A. Cupples, and D.P. Kiel. 2006. Colas, but not other carbonated beverages, are associated with low bone mineral density in older women: The Framingham osteoporosis study. *The American Journal of Clinical Nutrition* 84:936–942.

Tudorache, M., and C. Bala. 2007. Biosensors based on screen-printing technology, and their applications in environmental and food analysis. *Analytical and Bioanalytical Chemistry* 388:565–578.

Turner, A.P.F., I. Karube, and G.S. Wilson. 1987. Eds. *Biosensors: Fundamentals and applications.* Oxford: Oxford Science Publications.

Vadgama, P., and P.W. Crump. 1992. Biosensors-recent trends-a review. *Analyst* 117:1657–1670.

Vadgama, P., and K.G.M.M. Alberti. 1983. The effect of a gastric mucus barrier on the dynamic-response of a pH electrode. *Experientia* 39:573–576.

Venugopal, V. 2002. Biosensors in fish production and quality control. *Biosensors and Bioelectronics* 17:147–57.

Vidal, J.C., G.R. Esperanza, and J.R. Castillo. 2003. Recent advances in electropolymerized conducting polymers in amperometric biosensors. *Microchimica Acta* 143:93–111.

Volotovsky, V., and N. Kim. 1998. Determination of glucose, ascorbic and citric acids by two-ISFET multienzyme sensors. *Sensors and Actuators B* 49:253–257.

Vrbová, E., J. Peckova, and M. Marek. 1993. Biosensor for determination of starch. *Starch-Starke* 45:341–44.

Vucenik, I., and A.M. Shamsuddin. 2003. Cancer inhibition by inositol hexaphosphate (IP6) and inositol: From laboratory to clinic. *The Journal of Nutrition* 133:3778–3784.

Wang, J. 2005. Carbon-nanotube based electrochemical biosensors: A review. *Electroanalysis* 17:7–14.

Wang, M.-L., J.-T. Wang, and Y.-M. Choong. 2004. A rapid and accurate method for determination of methanol in alcoholic beverage by direct injection capillary gas chromatography. *Journal of Food Composition and Analysis* 17:187–196.

Wang, Y.Y., Y.S. Wang, and B. Liu. 2008. Fluorescent detection of ATP based on signaling DNA aptamer attached silica nanoparticles. *Nanotechnology* 19:415605.

Wcisło, M., D. Compagnone, and M. Trojanowicz. 2007. Enantioselective screen-printed amperometric biosensor for the determination of D-amino acids. *Bioelectrochemistry* 71:91–98.

Wijayawardhana, C.A., and W.R. Heineman. 2002. Electrochemical biosensors. In *Biomedical diagnostic science and technology*, eds. W.T. Law, N. Akmal, and A.M. Usmani, 1–27. New York: Marcel Dekker.

Wilson, G.S. 1987. Fundamentals of amperometric sensors. In *Biosensors: Fundamentals and applications*, ed. A.P.F. Turner, I. Karube, and G.S. Wilson, 165–179. Oxford: Oxford Science Publications.

Wu, F., Y. Huang, and C. Huang. 2005. Chemiluminescence biosensor system for lactic acid using natural animal tissue as recognition element. *Biosensors and Bioelectronics* 21:518–522.

Yamamoto, A., N. Akiba, S. Kodama, A. Matsunaga, K. Kato, and H. Nakazawa. 2001. Enantiomeric purity determination of malic acid in apple juices by multi-beam circular dichroism detection. *Journal of Chromatography A* 928:139–144.

Yashina, E.I., A.V. Borisova, E.E. Karyakina, et al. 2010. Sol–Gel immobilization of lactate oxidase from organic solvent: Toward the advanced lactate biosensor. *Analytical Chemistry* 82:1601–1604.

Yong-Sheng, L., J. Xiang, G. Xiu-Feng, Z. Yuan-Yuan, and W. Yan-Fei. 2008. Immobilization enzyme fluorescence capillary analysis for determination of lactic acid. *Analytica Chimica Acta* 610:249–256.

Zajoncova, L., M. Jilek, V. Beranova, and P. Pec. 2004. A biosensor for the determination of amylase activity. *Biosensors and Bioelectronics* 20:240–245.

Zeravik, J., A. Hlavacek, and K. Lacina. 2009. State of the art in the field of electronic and bioelectronic tongues—towards the analysis of wines. *Electroanalysis* 21:2509–2520.

Zhai, Q.-Z. 2008. Determination of trace amount of oxalic acid with zirconium(IV)–(DBS-arsenazo) by spectrophotometry. *Spectrochimica Acta Part A* 71:332–335.

Zhang, J., Y. Wan, L.H. Wang, et al. 2007. The electrochemical DNA biosensor. *Progress in Chemistry* 19:1576–1584.

Zhang, X.Q., Q. Guo, and D.X. Cui. 2009. Recent advances in nanotechnology applied to biosensors. *Sensors* 9:1033–1053.

Zhai, Q.-Z., X.-X. Zhang, and Q.-Z. Liu. 2006. Catalytic kinetic spectrophotometry for the determination of trace amount of oxalic acid in biological samples with oxalic acid–rhodamine B–potassium dichromate system. *Spectrochimica Acta Part A* 65:1–4.

Zheng, Y., C. Yang, W. Pu, and J. Zhang. 2009. Determination of oxalic acid in spinach with carbon nanotubes-modified electrode. *Food Chemistry* 114:1523–1528.

Zhike, H., G. Hua, Y. Liangjie et al. 1998. Pulse injection analysis with chemiluminescence detection: Determination of citric acid using tris-(2,2%-bipyridine) ruthenium(II). *Talanta* 47:301–304.

2 Basic Principles of Optical Biosensors in Food Engineering

Ebru Akdoğan and Mehmet Mutlu

CONTENTS

2.1 INTRODUCTION

In the food industry, quality, freshness, as well as safety of the products, are major concerns to both consumers and producers. During production of a food material, these parameters are ensured by routine conventional analytic techniques such as chromatography, spectrophotometry, electrophoresis, titration, and microbial analysis, of which some are a part of the Hazard Analysis and Critical Control Points (HACCP) routine. The periodic evaluation of these parameters via chemical and microbial tests is not always suitable for continuous *in situ* process control and monitoring, for they need sample pretreatment and/or purification and are expensive and time-consuming. In addition, most of these methods require experienced staff. The need for fast, accurate, and cheap analysis methods has revived the possible use of biosensors in the food industry with the advantage of working with real samples.

Biosensors are devices that use specific biochemical reactions mediated by isolated enzymes, immunosystems, tissues, organelles, or whole cells to detect chemical compounds, usually by electrical, thermal, or optical signals. A biosensor consists of a biological recognition element that is coupled to a transducer. The biological recognition elements may be enzymes, immunosystems, tissues, organelles, DNA, RNA, PNA, and whole cells; and transducers may be electrochemical, mass, optical, and thermal-based transducers.

Using biosensors in the food industry offers several advantages such as high specificity, selectivity, short response times, lower costs, diversity, potential miniaturization, simplicity of operation, and potential adaptation to automation. In addition, biosensors may be used as portable devices and give the opportunity to work with real samples, thus eliminating the sample pretreatment steps. Biosensors may be used in the food industry for the determination of the composition and contamination of raw materials and/or final products; for *in situ* and real-time monitorization of the process, such as fermentation; for the detection of pathogens, pesticides, and toxins; and in the HACCP and process control applications (Luong et al. 1991; Mello and Kubota 2002).

According to the type of the transducer used, biosensors may be classified as optical, electrochemical, mass sensitive, acoustic wave, or calorimetric biosensors. This chapter reviews the basic optical biosensor systems and their operating principles with the examples of their use in the food industry.

2.2 OPTICAL BIOSENSING

Optical transducers have become an important and potent tool in biosensor applications due to the rapid progresses in optical fibers, laser technology, and nanotechnology. In addition, optical biosensors have a better possibility for miniaturization than other types of biosensors. Optical biosensors are widely used in biomedical research, health care, pharmaceuticals, environmental monitoring, homeland security, and military applications (Narayanaswamy and Wolfbeis 2004). Maybe the most important advantage of optical biosensors is that they are not prone to electromagnetic interferences, which is a major parameter in practice.

Optical biosensors offer many advantages compared with both the conventional analysis techniques and other types of biosensors. They are more versatile; they offer the possibility of remote sensing, which is especially important for the detection of hazardous materials; they have relatively low costs, especially the fiber optic sensors; they are quite suitable for miniaturization and integration; they are immune to electromagnetic interferences in contrast to the other sensor systems; they can be operated in multiple operating modes such as absorbance, fluorescence/phosphorescence, reflectance, bio/chemiluminescence; and they make possible multiple detection with a single device. However, there are also some restrictions for the usage of optical biosensors. For example, if not properly isolated, the signal can be interfered with by the ambient light; optical systems mostly use high-energy light sources (however, by LED, or light-emitting diode, technology this restriction may be overcome); and miniaturization can affect the magnitude of the signal (but optical biosensors are more suitable for miniaturization than other biosensor systems, e.g., mass sensitive biosensors) (Mello and Kubota 2002; Fan et al. 2008).

Basically, optical biosensors operate in response to the optical changes caused by the ligand–analyte interaction such as UV adsorption, fluorescence/phosphorescence, reflectance, bio/chemiluminescence, scattering, or refractive index changes. Most of the optical biosensors work on the evanescent field of detection principle, which allows the direct detection on the change of the optical properties and biomolecule interactions without sample pretreatment. However, indirect or labeled methods (fluorescence, radiolabeling, or enzyme amplification) are also used, for they offer a higher sensitivity (Lechuga 2005).

2.3 PRINCIPLES OF OPTICAL DETECTION

In the application of optical biosensors, the changes in the refractive index (RI) due to the ligand–analyte interaction are monitored. For this purpose, two approaches are used: the estimation in the change of the real part of the RI, and the estimation in the change of the imaginary part of the RI (Lechuga 2005). Most optical biosensor systems deal with the change in the real part of the RI of the optical medium in the sample solution. The change in the real part of the RI causes a change in the propagation constant of a given wave-guide mode, thus producing a measurable change in the guided light characteristics. The optical sensor measurements based on the changes in the propagation of guided light can reach high sensitivity; while the optical sensor measurements based on the changes in the imaginary part of the RI (or the extinction coefficient), that is the strength of absorption loss at a particular wavelength, are less sensitive (Lechuga 2005). The sensitivity of an optical biosensor depends on the affinity of the biorecognition element to the analyte and the detection limit of the optical sensor.

2.4 TYPES OF OPTICAL BIOSENSORS

Optical biosensors can be divided into two main groups: *direct optical detection* and *labeled systems detection*. Direct optical detection systems are reflectometry (reflectometric interference spectroscopy (RIfS), ellipsometry) and evanescent field techniques (Mach–Zehnder interferometer, Young interferometer, resonant mirror sensor, surface plasmon resonance, grating couplers). Labeled systems detection is based on fluorescence intensity/lifetime, fluorescence anisotropy, total internal reflection fluorescence (TIRF), and fluorescence resonance energy transfer (FRET). Since fiber-optic (FO) biosensor systems can be classified in both direct and labeled detection systems, they will be reviewed separately in this chapter.

2.4.1 DIRECT OPTICAL DETECTION

2.4.1.1 Reflectometric Detection

Biosensors based on reflectometric detection work on the principle of light reflection. When a light beam is directed to an interface of two different media with different RIs, a change in direction of the light occurs so that the light returns into the medium from which it originated. Biosensors based on reflectometric detection are RIfS and ellipsometry. Since these systems are not frequently used for biosensor applications, they will be briefly reviewed within the scope of this chapter.

2.4.1.2 Reflectometric Interference Spectroscopy

The principle of RIfS systems is the wavelength-associated modulations occurring at transparent films. When a light beam is guided between two different media with different RI, it is partially reflected; if the reflectance of the interfaces is small (<0.05), an array of reflected beams will be formed in the transparent thin film (Lechuga 2005). This array is a consequence of the superposition of the reflected beams; thus it is called reflectometric interference spectroscopy. The changes in thickness of the layer will alter the phase difference of the reflected beams; thus the interference between the beams and the change in the intensity of the reflected light. For biosensor applications, the detection of the analyte is performed based on the change of the physical thickness of the layer at the transducer surface due to the biochemical interactions between the target and biorecognition elements. RIfS is a very sensitive detection method; the resolution of the layer's physical thickness can be as low as 1 pm, and the system is less dependent on temperature compared with evanescent wave systems.

2.4.1.3 Ellipsometry

In ellipsometric detection, the change in the reflectance and phase difference between the parallel (R_{pl}) and perpendicular (R_p) components of a polarized light upon reflection from the surface is measured. The change in polarization measured with ellipsometry is dependent on the surface thickness. The changes in thickness of the surface will alter the intensity of the reflected light. Ellipsometric detection monitors real-time thickness changes with high sensitivity (Mora et al. 2009). Different types of ellipsometry are available such as single-wavelength ellipsometry, spectroscopic ellipsometry, standard ellipsometry, imaging ellipsometry, and *in situ* ellipsometry.

2.4.1.4 Evanescent Field Techniques

It has been pointed out that most of the optical biosensors work on the evanescent field detection principle. In the evanescent wave sensors, an optical wave guide is used for the guided light, where the propagation can be both in transverse electric (TE) and transverse magnetic (TM) modes. The wave guide is a material that has a high RI, located between two materials with lower RI. A light beam with an incident angle exceeding the critical angle is focused through the wave guide so that a total internal reflection (TIR) occurs at the interface, making the light travel inside the wave guide in a guided mode, confined within the structure (Kooyman and Lechuga 1997). However, a part of the light travels about 100 nanometers outward from the wave guide to the surrounding medium (Figure 2.1). This part comprises the evanescent field that decays exponentially from the surface. This property is the basis of the sensing mechanism; a biorecognition element is immobilized on the wave guide, and the target in the surrounding medium interacts with the biorecognition element, changing the optical properties of the wave guide. This change is

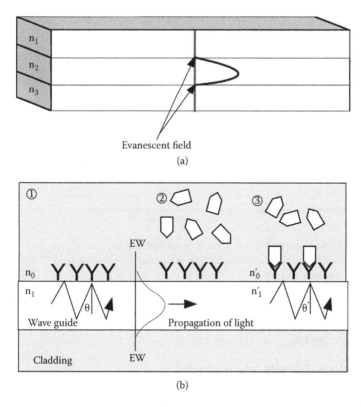

FIGURE 2.1 (a) Evanescent field in an optical wave guide. (b) Evanescent wave in a biosensor surface. (Reprinted from *Optical Biosensors*, Vol. 44 [Comprehensive Analytical Chemistry], L.M. Lechuga, Biosensors and Modern Biospecific Analytical Techniques, 209–250. Copyright 2005, with permission from Elsevier.)

detected via the evanescent wave and is dependent on the concentration of the target molecules and the affinity constant of the interaction between the biorecognition element and the analyte. This way a quantitative biosensor may be constructed. Since the evanescent wave decays exponentially through the medium, only reactions taking place on the wave guide will be detected (Kooyman and Lechuga 1997). The expansion of the evanescent field is altered by changing the physical properties of the wave guide such as RI and thickness. Different materials (glass, silicon, polymers, lithium niobate, and III–V group compounds) fabricated with different techniques (ion exchange, spin or dip coating, chemical vapor deposition, sol-gel and plasma polymerization, etc.) will have different physical properties, thus different evanescent wave fields (Kooyman and Lechuga 1997; Dominguez et al. 2003; Lechuga 2005). For the optical biosensors based on the evanescent field principle, the basic working principles are the same. The biorecognition element is immobilized to the sensor surface, the surface is interacted with the solution containing the analyte, the analyte binds to the biorecognition element, and this bioreaction alters the RI depending on the amount of the analyte. The RI changes are monitored in

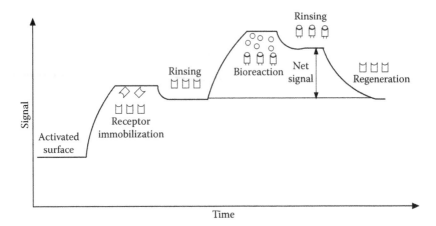

FIGURE 2.2 Representation of a basic response for an evanescent wave biosensor. (Reprinted from *Optical Biosensors*, Vol. 44 [Comprehensive Analytical Chemistry], L.M. Lechuga, Biosensors and Modern Biospecific Analytical Techniques, 209–250. Copyright 2005, with permission from Elsevier.)

real time. The analyte concentration on the solution as well as the binding constants can be determined this way (Figure 2.2).

2.4.1.5 Mach–Zehnder Interferometer

Mach–Zehnder biosensors are interferometric sensors that have higher sensitivity compared with other integrated schemes with a resolution of 1 nm, making the direct detection of biomolecules at low concentrations possible (Lechuga et al. 2003). The detection limit of interferometers is limited by noise, temperature changes, and chemical noise; and a detection limit of 10^{-7} in RI can be achieved (Lechuga 2005). Mach–Zehnder interferometer (MZI) is the most common interferometer used as a biosensor. In MZIs, a light from a laser beam is split into two identical beams by a Y junction that travels through two arms of the MZI and recombines again into a monomode channel wave guide. The beams travel through two areas: the sensor and reference areas. The detected signal is dependent on the phase difference of these two beams. If there is a difference between the sensor area in which the recognition element is immobilized and the reference area, a phase difference between the beams occurs, which changes the intensity of the recombined light (Heideman and Lambeck 1999; Lechuga 2005). Even though MZIs offer high sensitivity and *in situ* applications, their fabrication and design is complex, difficult, and expensive, so they are not preferred biosensors.

2.4.1.6 Young Interferometer

Young interferometer is a variation of MZI. It also contains a Y junction as the beam splitter, but in contrast to MZI, the light coming from the two areas is interacted at the output, not inside the structure (Brynda et al. 1999; Busse et al. 2002, 2001). The two beams form cones at the output with an interference pattern on a screen or a charged-coupled device (CCD) detector in a cosine distribution function.

The basic principle is the same as the MZI; the change in the RI of one arm with respect to the other changes in the RI of the output, depending upon the interaction between the biorecognition element immobilized in the sensor area and the analyte. The advantages of the Young interferometer are that it is relatively simple in design, the intensity distribution along the arms can be observed, and the wavelength drift and temperature effects are minimized (Lechuga 2005).

2.4.1.7   Resonant Mirror Sensor

Resonant mirror sensor is a surface plasmon resonance (SPR)-like sensor system working on the evanescent field principle. The sensor is a planar wave-guide sensor that uses TIR phenomenon. In contrast to SPR, the wave guide is not a metal; it is a dielectric resonant layer with a high RI such as hafnia or titania. The dielectric resonant layer is about 100 nm thick and is separated from glass prism by a dielectric layer with low RI such as SiO_2 of 1 μm thick so that a sandwich configuration of high-low-high RI is achieved (Buckle et al. 1993; Cush et al. 1993; Edwards et al. 1995; Lechuga 2005). From the phase change in the evanescent field occurring from the biorecognition interaction, biosensing is achieved and detected from the intensity change in real time.

2.4.1.8   Surface Plasmon Resonance

SPR has been a popular tool for unlabeled detection since it was first used as a biosensor in 1983 by Liedberg et al. (1983). SPR is an optical TIR-based biosensor where internal reflection takes place in a light guide. SPR is a quantum electro-optical phenomenon that occurs as a result of TIR of light between two media with dielectric constants of opposite sign, such as a dielectric and a metal interface. Energy carried by photons of light is coupled or transferred to electrons in a metal that results in the creation of a plasmon, a group of excited electrons along the surface of the metal. The absorption of the energy from the incident light to the metal surface causes a decrease in reflected-light intensity. The changes in the incident light can be detected from the change in the incident angle (resonance angle) or shift in the wavelength of the absorbed light, resonant intensity change, or polarization and can be expressed as a change in the SPR signal (resonance units, RUs) (Manuel et al. 1993; Kruchinin and Vlasov 1996; Mouvet et al. 1997; Dostalek et al. 2001; Wu and Pao 2004; Ho et al. 2005; Naraoka and Kajikawa 2005). The intensity of the plasmon is changed by the type of metal used and the environment around the metal surface. Changes in chemical properties of the environment within the range of the plasmon field (or evanescent field) binding cause changes in plasmon resonance.

Different methods for excitation of SPR can be employed. The four basic excitation methods are prism coupling, wave-guide coupling, fiber-optic coupling, and grating coupling (Figure 2.3) (Matsubara et al. 1988; Liedberg et al. 1993; Yu et al. 2004; Sharma et al. 2007). The prism coupling method is the most common method used for SPR excitation. In this method, the light is completely reflected at the prism–metal interface so that the evanescent field penetrates the metal layer. In the wave-guide coupling method, the light propagates in a wave guide with TIR so that an evanescent field is generated at the wave-guide–metal interface. In fiber-optic excitation, a thin layer of the fiber is removed and coated with a thin metal

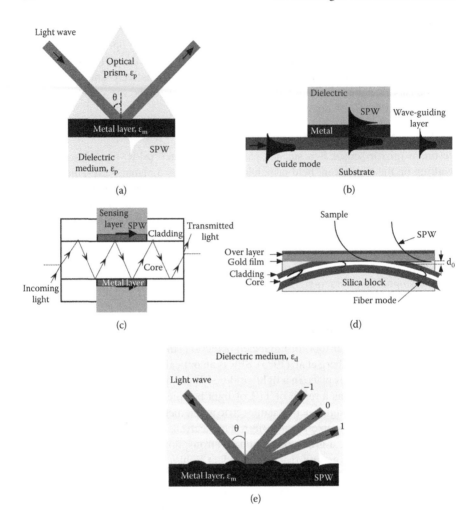

FIGURE 2.3 The four basic excitation methods for an SPR biosensor: (a) prism coupling, (b) wave-guide coupling, (c) fiber-optic coupling, and (d) grating coupling. (Reprinted from *Analytica Chimica Acta*, Vol. 620, Fan Xudong et al., Senstive Optical Biosensors for Unlabeled Targets: A Review, 8–26. Copyright 2008, with permission from Elsevier.)

layer; here, the fiber acts as the wave guide. In grating coupling excitation, the light wave is incident to a metallic grating, directly illuminating its surface (Fan et al. 2008).

For biosensing applications, an SPR device is combined with a biorecognition molecule immobilized on the metal surface. When the analyte interacts with the sensor surface, a change in the plasmon field occurs due to the increase in surface concentration, and the resonance angle shifts to greater values. This makes it possible to directly relate the concentration of the analyte in real time (Jonsson et al 1991).

SPR biosensors have advantages such as ease of use, detection of unlabeled molecules, possibility of direct analysis of samples without purification, and good detection limits. However, since in basic SPR systems the plasmon field can only penetrate about 100 nm of the sample medium, large molecules such as cells are difficult to detect, and the RI change cannot be differentiated from the RI change of the sample media or RI change of the surface, making the system unfavorable for complex samples (Fan et al. 2008). To eliminate these disadvantages, long-range surface plasmon (LRSP) and short-range surface plasmon (SRSP) modes can be used. This dual mode of the LRSP and SRSP allows the sensor to differentiate the background RI change and surface RI change, and provides longer penetration into the sample, making cell detection possible (Nenninger et al. 2001; Slavik and Homola 2007; Guo et al. 2008; Hastings 2008). Also with multichannel SPR sensor configurations, simultaneous multianalyte detections can be performed (Homola et al. 2001; Dostalek et al. 2005). The SPR imaging (SPRI) technique is also used for high-throughput detection of multianalytes where the reflected light is imaged on a CCD camera similar to the microarray technique (Smith et al. 2003; Fu et al. 2004; Shumaker-Parry and Campbell 2004; Piliarik et al. 2005; Wark et al. 2005; Zybin et al. 2005).

2.4.1.9   Grating Couplers

Grating coupler biosensors use an optical grating on a wave guide. The wave guide is a high-RI material, usually SiO_2 or TiO_2 of about 200 nm thick, deposited over a low RI material, usually glass. In grating coupler sensors, the coupling angle variation on the grating surface is used for detection for the coupling angle changes as the RI of the grating surface changes due to any interaction on the surface with an accuracy of 0.005°. Grating coupler biosensors can be used for real-time monitoring of RI changes and thickness of the surface and mass adsorbed on the surface (Schlatter et al. 1993).

Bidiffractive grating couplers have been introduced by Fattinger et al. (1995). A bidiffractive grating coupler system serves as both an input and output port for the coupling and decoupling light beams to and from the wave guide (Lechuga 2005). A bidiffractive grating biosensor uses a differential measurement of two modes of polarization, orthogonal and linear, and the sensor's response is the difference between the decoupled beam angles of these two modes. Bidiffractive grating couplers offer higher sensitivity than the grating coupler biosensors (Fattinger et al. 1995).

2.4.2   Labeled Systems Detection

In optical biosensing systems, two approaches are considered: fluorescence-based and label-free detection. Fluorescence-based detection is considered as labeled detection where either the biorecognition molecule or the analyte is labeled with a fluorescent tag, such as a dye or a quantum dot (Fan et al. 2008). Fluorescent labeled systems offer detection limits down to a single molecule, and they can be easily integrated to FO systems. However the labeling process is intense, and the tag may change the configuration of the biomolecules. Labeled systems detection over fluorescence is the most common method used in biosensing, and it is quite versatile; fluorescence-based biodetection can be performed via fluorescence intensity/lifetime, fluorescence anisotropy, TIRF, and FRET; the parameters to be measured can be fluorescence intensity, lifetime (decay time), anisotropy, quenching efficiency, and luminescence energy transfer; and the optical layouts can be in the form of stripes, wave-guide optical systems, capillary sensors, arrays, and FOs (Borisov and Wolfbeis 2008).

The most common parameter used in fluorescence-based sensing and biosensing in general is fluorescence intensity, which is easy to measure with routine instrumentation and a single label. However, fluorescence intensity measured at a single wavelength is not reliable for obtaining reproducible data. Thus two wavelengths' measurement (ratiometric) is generally preferred. For this purpose, an additional reference fluorophore or the usage of a donor and acceptor dye (FRET system) is used. For eliminating the disadvantage of single-wavelength measurement, fluorescence lifetime (decay time) can also be measured. Fluorescence lifetime measurement is performed in µs or ms and is particularly useful for the detection of biomolecular interactions and affinity measurements. The FRET system can also be used for fluorescence lifetime measurement with a suitable donor label (Borisov and Wolfbeis 2008). Another parameter to be measured for eliminating the constraints of fluorescence intensity measurement is fluorescence anisotropy. Fluorescence anisotropy is calculated from the intensities of fluorescence observed through polarizers parallel and perpendicular to the polarization of the exciting light. For biomolecular interactions it can be concluded that a fluorophore that binds to a biomolecule or experiences a reduction in its fluorescence lifetime results in an apparent increase in its emissive rate and exhibits an increased fluorescence anisotropy. Fluorescence anisotropy is widely used in biophysical studies of macromolecules and fluorescence polarization immunoassays (Thompson et al. 1998).

TIRF-based systems are particularly useful as optical biosensors. In TIRF-based biosensor systems the molecules near the sensor surface are excited with the incident light, creating a fluorescent wave that in turn couples back to the wave guide. The fluorescent wave formation or the change in its intensity is transmitted via the wave guide to the detector. In TIRF-based biosensors the basic phenomenon is attenuated total reflection (ATR). The sensitivity is further increased by the use of fluorescent labeled molecules that reemit the absorbed evanescent photons as fluorescent. This way the sensitivity of TIRF biosensors reaches the level of fM (Lechuga 2005).

2.4.3 FIBER-OPTIC BIOSENSORS

The FO biosensor principle is the transmission of light along a fiber strand to a detector. FO biosensors offer high sensitivity, high detection speed, and the possibility to be used in a large number of assay conditions. The optical fibers can be used for both evanescent wave and labeled systems detection (Boisde and Harmer 1996; Mehrvar et al. 2000; Wolfbeis 2000; Anderson and Nerurkar 2002).

FO biosensors can be classified in two main groups as extrinsic and intrinsic FO biosensors. In extrinsic FO biosensors a single fiber with an immobilized biomolecule on its distal end is used as the guide for the incoming and outgoing light. The emitted light is directed to a detector. In this type of biosensor both absorbance and fluorescence measurements can be performed. In a different configuration of extrinsic FO biosensor, the wave guide for the reflected scattered or emitted light is a second fiber.

In intrinsic FO biosensors, the sensing is based on the evanescent wave technique. Here the biorecognition element is immobilized on a tapered fused-silica fiber, and the change in absorbance, luminescence, polarization, or refractive index is detected. However, fluorescent labels with competitive formats such as sandwich format competitive immunoassay technique are usually used in this configuration instead of direct measurement.

FO biosensors can also be classified as direct and indirect FO biosensors according to the type of biomolecular interactions. In direct FO biosensors no labels are used and the detection relies on the interaction between the biorecognition element and the target molecule. This type of FO sensor provides the advantage of reagentless detection, but the sensitivity may be compromised due to the nonspecific binding format (Paddle 1996; Squillante 1998).

In the indirect FO biosensor systems, labels are used such as dyes or fluorophores and can be used in competitive format. The label provides increased sensitivity and selectivity, reducing nonspecific binding and amplifying the sensor's response. Furthermore, an increased number of analyte can be detected with indirect FO sensors compared with the direct ones. The use of a label can be regarded as a disadvantage to these systems (Paddle 1996; Squillante 1998).

FO biosensors in general offer a great number of advantages compared with other biosensor systems. The rapid development in fiber communication technology provides standard fibers with low cost and high stability. Advances in nanotechnology and materials science made submicron FO-based biosensors to be used inside a single cell possible (Vo-Dinh and Cullen 2000). FO biosensors can be easily miniaturized and integrated to a compact and portable design, making them suitable for commercial products (Wolfbeis 2000). FO biosensors can sense analytes down to a detection limit of femtomolar. With arrays of individual fibers, detection of multianalyte is possible. Maybe one of the most important properties of FO biosensors is that fibers are not prone to electronic/magnetic waves, and a large amount of signal can be transferred in long distances and harsh environments (Aizawa 1994; Mehrvar et al. 2000).

2.5 OPTICAL BIOSENSORS FOR FOOD QUALITY AND FOOD SAFETY

The use of optical biosensors for foods has been mostly focused on the detection of foodborne pathogens and their toxins. A lesser effort has been made for monitoring other components in foods such as the fermentation process.

The main focus on the detection of foodborne pathogens and their toxins is on the detection of *Escherichia coli* O157:H7, *Salmonella*, *Campylobacter*, *Listeria monocytogenes*, Staphylococcal enterotoxin B (SEB), and *Clostridium botulinum* toxin. However when foodborne or waterborne pathogens are mentioned, bacteria, viruses, protozoa, and molds should be considered. Among bacterial pathogens, live cells, spores, and toxins are included. Some important foodborne and waterborne pathogens are given in Table 2.1.

One of the most important optical biosensors for foodborne pathogen detection is a commercial FO biosensor called RAPTOR (Research International), designed for the multiple detection of foodborne and bioterrorism agents. RAPTOR's principle is based

TABLE 2.1

Some Important Foodborne/Waterborne Pathogens

Bacterial Pathogens	Examples
Live cells	*Escherichia coli* O157:H7
	Listeria monocytogenes
	Clostridium perfingens
	Salmonella enterica
	Vibrio, Shigella, Campylobacter, and *Yersinia* species
Toxins	Botulinum toxin (*Clostridium botulinum*)
	Staphylococcal enterotoxin (*Staphylococcus aureus*)
	Epsilon toxin (*Clostridium perfringens*)
	Diarrheagenic/emetic enterotoxins (*Bacillus cereus*)
Spores	*Bacillus anthracis* spores
Viral Pathogens	Norwalk virus
	Hepatitis A virus
Protozoan Pathogens	*Giardia lamblia*
	Entamoeba histolytica
	Toxoplasma gondi
	Cryptosporidium parvum
	Cyclospora cayetanensis
	Isospora belli
Molds	Aflatoxin (*Aspergillus flavus* and *A. parasiticus*)
	Mycotoxins of *Fusarium* species
	Ochratoxin (*Aspergillus ochraceus* and *Penicillium verrucosum*)
	Patulin (*e.g., Aspergillus* and *Penicillium* species)

on a sandwich immunoassay where Cy5 dye and Alexa-flour 647-labeled antibody are used. RAPTOR is used for the detection of SEB in water with a 0.1–0.5 ng/ml limit of detection; *Bacillus anthracis* spores in water with a 5×10^4 CFU/mL limit of detection; *E. coli* O157:H7 from hamburger slurry with a 100–1000 CFU/mL limit of detection; *Yersinia pestis* F1 antigen in water with a 1 ng/mL limit of detection; *Salmonella typhimurium* cells from water with a 2×10^4 CFU/mL limit of detection; and Cholera toxin in water with a 0.1–1 ng/ml limit of detection (RAPTOR Portable, Multianalyte Bioassay Detection System, http://www.resrchintl.com/raptor-detection-system.html).

The main commercial SPR sensor systems are BIACORE from Biacore AB (Uppsala, Sweden), Spreeta™ from Texas Instruments (Dallas, Texas), SPR Spectroscope from Optrel (GbR, Germany), and Reichert SR7000 from Reichert Analytical Instruments (Depew, New York). The sensitivity of SPR-based biosensors is considered the same as enzyme-linked immunosorbent assay (ELISA)-based detection systems and can be used for the detection of both cells and toxins.

Perkins and Squirrell (2000) have detected *Bacillus cereus* spores from a liquid buffer medium by using combined SPR and light scattering with a 10^7 spores/mL sensitivity and a BIACORE 2000–SPR system with a >10^7 spores/mL sensitivity. Oh et al. (2003) have used an SPR biosensor system for the detection of *E. coli* O157:H7 at 10^2 cfu/mL. Waswa et al. (2007) used a sandwich-based SPR biosensor system for the detection of *E.coli* O157:H7 from milk, apple juice, and ground beef patties at 10^2–10^3 cfu/ml concentrations.

SPR has also been used for the detection of *Salmonella* species and *L. monocytogenes* at different detection limits varying from 10^2 to 10^7 cfu/ml (Koubova et al. 2001; Bokken et al. 2003; Oh et al. 2004; Leonard et al. 2004, 2005). The important use of SPR sensors for toxin detection are for fumonisin B1 at 50 ng/mL detection limit and with a response time of 10 min (Mullett et al. 1998), aflatoxin B1 at 3 ng/ml detection limit (Daly et al. 2000), zearalenone at 0.01 ng/g, ochratoxin A at 0.1 ng/g, deoxynivalenol, and fumonisin B1at 0.1 ng/g detection limits (Tudos et al. 2003). Nedelkov et al. (2000) and Nedelkov and Nelson (2003) used the Biacore SPR system coupled with mass spectrometry for the detection of SEB at 1 ng/ml detection limit in milk and mushroom. Homola et al. (2002) used a two-channel SPR to detect SEB with a 0.5–5 ng/ml detection limit in liquid buffer and milk as media.

The Multi-Analyte Array Biosensor (MAAB) developed by the Naval Research Laboratory (NRL) has been used for the simultaneous detection of foodborne pathogens. MAAB is based on a fluoro-immunoassay format and is used for the detection of *Salmonella typhimurium* at 4×10^3 cfu/ml and 8×10^3 cgu/g (Rowe-Taitt et al. 2000; Taitt et al. 2004). The array biosensor is also used for various toxin detection such as fumonisin at 0.5 ng/ml detection limit from various foods and beverages (Ligler et al. 2003), ochratoxin A at 4–100 ng/g detection limit for cereal samples and 7–38 ng/g for beverages (Ngundi et al. 2005), aflatoxin B1 at 0.3 ng/g in buffer and at 0.6–5.1 ng/g in nut samples (Sapsford et al. 2006), deoxynivalenol at 0.2 ng/mL in buffer and at 50 ng/mL for oats samples (Ngundi et al. 2006).

Other than foodborne pathogens, optical biosensors are also used for the detection of carbohydrates, alcohols, pesticides, and antibiotic residues. Optical luminescent biosensors are used for the control of fermentation processes and for the detection of alcohol and carbohydrates. Folic acid and biotin were detected by Indyk et al. (2000) via an SPR biosensor using antifolic acid antibody and antibiotin antibody at a range of 2–70 ng/ml in infant formulas and milk. Andres and Narayanaswamy (1997) have used acetylcholinesterase (AChE) for the detection of pesticide residues in synthetic samples with an FO biosensor; they were able to detect carbofuran at a range of 5×10^{-8} M to 5×10^{-7} M and paroxon at a range of 5×10^{-7} to 5×10^{-6} M.

Optical biosensors have been developed for the determination of veterinary drug residues on food samples. Those veterinary drugs are chloramphenicol, tylosin, nicarbazin, fenicol antibiotic residues, and β-lactam antibiotic. An SPR immunochemical screening assay has been used for chloramphenicol and chloramphenicol glucuronide residues at 1.1 μg/kg of chloramphenicol in poultry muscle, honey, prawn, and cow milk with a competitive assay format by Ferguson et al. (2005). An

SPR system was used for the detection of tylosin (Caldow et al. 2005), nicarbazin by McCarney et al. (2003), fenicol antibiotic residues (Dumont et al. 2006), and β-Lactam antibiotic (van Eenennaam et al. 1995; Cacciatore et al. 2004).

2.6 CONCLUSION

Many published biosensor systems, optical or not, hold promise for food safety and quality control applications, but few biosensor systems are commercially available. Commercial biosensor systems have many problems such as the limited shelf life of the biological components in the biosensor system, cost of mass production, requirement of trained users, stability, sensitivity, and storability. However, with the tremendous development in the area of nanotechnology, commercialization of biosensors seems to be more feasible in the near future. Rapid developments in the communication area leading to better FO systems will probably make FO biosensors more advantageous. However other aspects of biosensors such as the selection of the biorecognition molecule (DNA, antibodies, RNA, molecular beacons, aptamers, etc.), immobilization procedure, and kinetic modeling of biosensors systems should also be considered.

REFERENCES

Aizawa, M. 1994. Immunosensors for clinical analysis. *Advances in Clinical Chemistry* 31:247–275.

Anderson, G.P., and N.L. Nerurkar. 2002. Improved fluoroimmunoassays using the dye Alexa Fluor 647 with the RAPTOR, a fiber optic biosensor. *Journal of Immunological Methods* 271:17–24.

Andres, R.T., and R. Narayanaswamy. 1997. Fiber-optic pesticide biosensor based on covalently immobilized acetylcholinesterase and thymol blue. *Talanta* 44:1335–1352.

Boisde, G., and A. Harmer. 1996. *Chemical and biochemical sensing with optical fibers and waveguides.* Boston: Artech House Publishers.

Bokken, G.C.A.M., R.J. Corbee, F. van Knapen, and A.A. Bergwerff. 2003. Immunochemical detection of *Salmonella* groups B, D, and E using an optical surface plasmon resonance biosensor. *FEMS Microbiology Letters* 222:75–82.

Borisov, S.M., and O.S. Wolfbeis. 2008. Optical biosensors. *Chemical Reviews* 108:423–461.

Brynda, E., M. Houska, A. Brandenburg, A. Wikerstal, and J. Skvor. 1999. The detection of human β₂-microglobulin by grating coupler immunosensor with three dimensional antibody networks. *Biosensors and Bioelectronics* 14:363–368.

Buckle, P.E., R.J. Davies, T. Kinning, et al. 1993. The resonant mirror: A novel optical sensor for direct sensing of biomolecular interactions. II: Applications. *Biosensors and Bioelectronics* 8:355–363.

Busse, S., M. DePaoli, G. Wenz, and S. Mittler. 2001. An integrated optical Mach–Zehnder interferometer functionalized by β-cyclodextrin to monitor binding reactions. *Sensors and Actuators B* 80:116–124.

Busse, S., V. Scheumann, B. Menges, and S. Mittler. 2002. Sensitivity studies for specific binding reactions using the biotin/streptavidin system by evanescent optical methods. *Biosensors and Bioelectronics* 17:704–710.

Cacciatore, G., M. Petz, S. Rachid, R. Hakenbeck, and A.A. Bergwerff. 2004. Development of an optical biosensor assay for detection of β-lactam antibiotics in milk using the penicillin-binding protein 2x. *Analytica Chimica Acta* 520:105–115.

Caldow, M., S.L. Stead, J. Day, M. Sharman, C. Situ, and C.J. Elliott. 2005. Development and validation of an optical SPR biosensor assay for tylosin residues in honey. *Journal of Agricultural and Food Chemistry* 53:7367–7370.

Cush, R., J.M. Cronin, W. Stewart, C.H. Maule, J. Molloy, and N.J. Goddard. 1993. The resonant mirror: A novel optical biosensor for direct sensing of biomolecular interactions Part I: Principle of operation and associated instrumentation. *Biosensors and Bioelectronics* 8:347–354.

Daly, S.J., G.J. Keating, P.P. Dillon et al. 2000. Development of surface plasmon resonance-based immunoassay for aflatoxin B1. *Journal of Agricultural and Food Chemistry* 48:5097–5104.

Dominguez, C., L.M. Lechuga, and J.A. Rodriguez. 2003. Integrated optical transducers for (bio)chemical sensing. In *Integrated analytical systems*, ed. S. Alegret, 541–584, Amsterdam: Elsevier.

Dostalek, J., J. Ctyroky, J. Homola et al. 2001. Surface plasmon resonance biosensor based on integrated optical wave guide. *Sensors and Actuators B Chemical* 76:8–12.

Dostalek, J., H. Vaisocherova, and J. Homola. 2005. Multichannel surface plasmon resonance biosensor with wavelength division multiplexing. *Sensors and Actuators B Chemical* 108:758–764.

Dumont, V., A.C. Huet, I. Traynor, C. Elliott, and P. Delahaut. 2006. A surface plasmon resonance biosensor assay for the simultaneous determination of thiamphenicol, florefenicol, florefenicol amine and chloramphenicol residues in shrimps. *Analytica Chimica Acta* 567:179–183.

Edwards, P.R., A. Gill, D.V. Pollard-Knight et al. 1995. Kinetics of protein–protein interactions at the surface of an optical biosensor. *Analytical Biochemistry* 231:210–217.

Fan, X., I.M. White, S.I. Shopova, H. Zhu, J.D. Suter, and Y. Sun. 2008. Sensitive optical biosensors for unlabeled targets: A review. *Analytica Chimica Acta* 620:8–26.

Fattinger, C., C. Mangold, M.T. Gale, and H. Schuetz. 1995. Bidiffractive grating coupler: Universal transducer for optical interface analytics. *Optical Engineering* 34:2744–2753.

Ferguson, J.P., G.A. Baxter, P. Young et al. 2005. Detection of chloramphenicol and chloramphenicol glucuronide residues in poultry muscle, honey, prawn and milk using a surface plasmon resonance biosensor and Qflex® kit chloramphenicol. *Analytica Chimica Acta* 529:109–113.

Fu, E., T. Chinowsky, J. Foley, J. Weinstein, and P. Yager. 2004. Characterization of a wavelength-tunable surface plasmon resonance microscope. *Review of Scientific Instruments* 75:2300–2306.

Guo, J., P. Keathley, and J. Hastings. 2008. Dual-mode surface-plasmon-resonance sensors using angular interrogation. *Optics Letters* 33:512–514.

Hastings, J.T. 2008. Optimizing surface-plasmon resonance sensors for limit of detection based on a Cramer–Rao bound. *IEEE Sensors Journal* 8:170–175.

Heideman, R.G., and P.V. Lambeck. 1999. Performance of a highly sensitive optical waveguide Mach–Zehnder interferometer immunosensor. *Sensors and Actuators B* 61:100–127.

Ho, H.P., W.C. Law, S.Y. Wu, C. Lin, and S.K. Kong. 2005. Real-time optical biosensor based on differential phase measurement of surface plasmon resonance. *Biosensors and Bioelectronics* 20:2177–2180.

Homola, J., H.B. Lu, G.G. Nenninger, J. Dostalek, and S.S. Yee. 2001. A novel multichannel surface plasmon resonance biosensor. *Sensors and Actuators B Chemical* 76:403–410.

Homola, J., J. Dostalek, S. Chen, A. Rasooly, S. Jiang, and S.S. Yee. 2002. Spectral surface plasmon resonance biosensor for detection of staphylococcal enterotoxin B in milk. *International Journal of Food Microbiology* 75:61–69.

Indyk, H.E., E.A. Evans, M.C.B. Caselunghe et al. 2000. Determination of biotin and folate in infant formula and milk by optical biosensor-based immunoassay. *Journal of AOAC International* 83:1141–1148.

Jonsson, U., L. Fagerstam, and B. Ivarsson. 1991. Real-time biospecific interaction analysis using surface plasmon resonance and a sensor chip technology. *Biotechniques* 11:620–627.

Kooyman, R.P.H., and L.M. Lechuga. 1997. Immunosensors based on total internal reflection. In *Handbook of biosensors: Medicine, food and the environment*, ed. E. Kress-Rogers, 169–196, Boca Raton, FL: CRC Press.

Koubova, V., E. Brynda, L. Karasova et al. 2001. Detection of foodborne pathogens using surface plasmon resonance biosensors. *Sensors and Actuators B Chemical* 74:100–105.

Kruchinin, A.A., and Y.G. Vlasov. 1996. Surface plasmon resonance monitoring by means of polarization state measurement in reflected light as the basis of a DNA-probe biosensor. *Sensors and Actuators B Chemical* 30:77–80.

Lechuga, L.M. 2005. Optical biosensors. In *Biosensors and modern biospecific analytical techniques*, ed. L. Gorton, 209–250. London: Elsevier.

Lechuga, L.M., F. Prieto, and B. Sepfilveda. 2003. Interferometric biosensors for environmental pollution detection. In *Optical sensors: Industrial, environmental and diagnostic applications*, eds. R. Narayanaswamy, O.S. Wolfbeis, 227–248. Heidelberg: Springer.

Leonard, P., S. Hearty, J. Quinn, and R. O'Kennedy. 2004. A generic approach for the detection of whole Listeria monocytogenes cells in contaminated samples using surface plasmon resonance. *Biosensors and Bioelectronics* 19:1331–1335.

Leonard, P., S. Hearty, G. Wyatt, J. Quinn, and R. O'Kennedy. 2005. Development of a surface plasmon resonance-based immunoassay for *Listeria monocytogenes*. *Journal of Food Protection* 68:728–735.

Liedberg, B., C. Nylander, and I. Lunstrom. 1983. Surface plasmon resonance for gas detection and biosensing. *Sensors and Actuators* 4:299–304.

Liedberg, B., I. Lundstrom, and E. Stenberg. 1993. Principles of biosensing with an extended coupling matrix and surface plasmon resonance. *Sensors and Actuators B Chemical* 11:63–72.

Ligler, F., C.R. Taitt, L.C. Shriver-Lake, K.E. Sapsford, Y. Shubin, and J.P. Golden. 2003. Array biosensor for detection of toxins. *Analytical and Bioanalytical Chemistry* 377:469–477.

Luong, J.H.T., C.A. Groom, and K.B. Male. 1991. Potential role of biosensors in the food and drink industries. *Biosensors and Bioelectronics* 6:547–554.

Manuel, M., B. Vidal, R. Lopez, et al. 1993. Determination of probable alcohol yield in musts by means of an SPR optical sensor. *Sensors and Actuators B Chemical* 11:455–459.

Matsubara, K., S. Kawata, and S. Minami. 1988. Optical chemical sensor based on surface plasmon measurement. *Applied Optic* 27:1160–1163.

McCarney, B., I.M. Traynor, T.L. Fodey, S.R.H. Crooks, and C.T. Elliott. 2003. Surface plasmon resonance biosensor screening of poultry liver and eggs for nicarbazin residues. *Analytica Chimica Acta* 483:165–169.

Mehrvar, M., C. Bis, J.M. Scharer, M. Moo-Young, and J.H. Luong. 2000. Fiber-optic biosensors-trends and advances. *Analytical Sciences* 6:677–692.

Mello, L.D., and L.T. Kubota. 2002. Review of the use of biosensors as analytical tools in the food and drink industries. *Food Chemistry* 77:237–256.

Mora, M.F., J.L. Wehmeyer, R. Synowicki, and C.D. Garcia. 2009. Investigating protein adsorption via spectroscopic ellipsometry. In *Biological interactions on materials surfaces, understanding and controlling protein, cell, and tissue responses*, eds. D.A Puleo, R. Bizios, 20–35. London: Springer.

Mouvet, C., R.D. Harris, C. Maciag et al. 1997. Determination of simazine in water samples by wave guide surface plasmon resonance. *Analytica Chimica Acta* 338:109–117.

Mullett, W., E.P. Lai, and J.M. Yeung. 1998. Immunoassay of funonisins by a surface plasmon resonance biosensor. *Analytical Biochemistry* 258:161–167.

Naraoka, R., and K. Kajikawa. 2005. Phase detection of surface plasmon resonance using rotating analyzer method. *Sensors and Actuators B Chemical* 107:952–956.

Narayanaswamy, R., and O.S. Wolfbeis. 2004. *Optical sensors*. New York: Springer.

Nedelkov, D., and R.W. Nelson. 2003. Detection of staphylococcal enterotoxin B via biomolecular interaction analysis mass spectrometry. *Applied and Environmental Microbiology* 69:5212–5215.

Nedelkov, D., A. Rasooly, and R.W. Nelson. 2000. Multitoxin biosensor-mass spectrometry analysis: A new approach for rapid, real-time, sensitive analysis of staphylococcal toxins in food. *International Journal of Food Microbiology* 60:1–13.

Nenninger, G.G., P. Tobiska, J. Homola, and S.S. Yee. 2001. Long-range surface plasmons for high-resolution surface plasmon resonance sensors. *Sensors and Actuators B Chemical* 74:145–151.

Ngundi, M.M., L.C. Shriver-Lake, M.H. Moore, F.S. Ligler, and C.R. Taitt. 2006. Multiplexed detection of mycotoxins in foods with a regenerable array. *Journal of Food Protection* 69:3047–3051.

Ngundi, M.M., L.C. Shriver-Lake, M.H. Moore, M.E. Lassman, F.S. Ligler, and C.R. Taitt. 2005. Array biosensor for detection of ochratoxin A in cereals and beverages. *Analytical Chemistry* 77:148–154.

Oh, B.-K., W. Lee, W.H. Lee, and J.-W. Choi. 2003. Nano-scale probe fabrication using self-assembly technique and application to detection of *Escherichia coli* O157: H7. *Biotechnology and Bioprocess Engineering* 8:227–232.

Oh, B.-K., Y.-K. Kim, K.W. Park, W.H. Lee, and J.-W. Choi. 2004. Surface plasmon resonance immunosensor for the detection of *Salmonella typhimurium*. *Biosensors and Bioelectronics* 19:1497–1504.

Paddle, B.M. 1996. Biosensors for chemical and biological agents of defence interest. *Biosensors and Bioelectronics* 11:1079–1113.

Perkins, E.A., and D.J. Squirrell. 2000. Development of instrumentation to allow the detection of microorganisms using light scattering in combination with surface plasmon resonance. *Biosensors and Bioelectronics* 14:853–859.

Piliarik, M., H. Vaisocherova, and J. Homola. 2005. A new surface plasmon resonance sensor for high-throughput screening applications. *Biosensors and Bioelectronics* 20:2104–2110.

Rowe-Taitt, C.A., J.W. Hazzard, K.E. Hoffman, J.J. Cras, J.P. Golden, and F.S. Ligler. 2000. Simultaneous detection of six biohazardous agents using a planar waveguide array biosensor. *Biosensors and Bioelectronics* 15:579–589.

Sapsford, K.E., C.R. Taitt, S. Fertig, et al. 2006. Indirect competitive immunoassay for detection of aflatoxin B1 in corn and nut products using the array biosensor. *Biosensors and Bioelectronics* 21:2298–2305.

Schlatter, D., R. Barner, C. Fattinger, et al. 1993. The difference interferometer: Application as a direct affinity sensor. *Biosensors and Bioelectronics* 8:109–116.

Sharma, A.K., R. Jha, and B.D. Gupta. 2007. Fiber-optic sensors based on surface plasmon resonance: A comprehensive review. *IEEE Sensors Journal* 7:1118–1129.

Shumaker-Parry, J.S., and C.T. Campbell. 2004. Quantitative methods for spatially resolved adsorption/desorption measurements in real time by surface plasmon resonance microscopy. *Analytical Chemistry* 76:907–917.

Slavik, R., and J. Homola. 2007. Ultrahigh resolution long range surface plasmon-based sensor. *Sensors and Actuators B Chemical* 123:10–12.

Smith, E.A., M.G. Erickson, A.T. Ulijasz, B. Weisblum, and R.M. Corn. 2003. Surface plasmon resonance imaging of transcription factor proteins: Interactions of bacterial response regulators with DNA arrays on gold films. *Langmuir* 19:1486–1492.

Squillante, E. 1998. Applications of fiber-optic evanescent wave spectroscopy. *Drug Development and Industrial Pharmacy* 24:1163–1175.

Taitt, C.R., J.P. Golden, Y.S. Shubin et al. 2004. A portable array biosensor for detecting multiple analytes in complex samples. *Microbial Ecology* 47:175–185.

Thompson, R.B., B.P. Maliwal, and V.L. Feliccia. 1998. Determination of picomolar concentrations of metal ions using fluorescence anisotropy: Biosensing with a "reagentless" enzyme transducer. *Analytical Chemistry* 70:4717–4723.

Tudos, A.J., E.R. Lucas van den Bos, and E.C. Stigter. 2003. Rapid surface plasmon resonance-based inhibition assay of deoxynivalenol. *Journal of Agricultural and Food Chemistry* 51:5843–5848.

vanEenennaam, A.L., I.A. Gardner, J. Holmes et al. 1995. Financial analysis of alternative treatments for clinical mastitis associated with environmental pathogens. *Journal of Dairy Science* 78:2086–2095.

Vo-Dinh, T., and B. Cullen. 2000. Biosensors and biochips: Advances in biological and medical diagnostics. *Fresenius' Journal of Analytical Chemistry* 366:540–551.

Wark, A.W., H.J. Lee, and R.M. Corn. 2005. Enzymatically amplified surface plasmon resonance imaging detection of DNA by exonuclease III digestion of DNA microarrays. *Analytical Chemistry* 77:5096–5100.

Waswa, J., J. Irudayaraj, and C. DebRoy. 2007. Direct detection of *E. Coli* O157:H7 in selected food systems by a surface plasmon resonance biosensor. *LWT—Food Science and Technology* 40:187–192.

Wolfbeis, O.S. 2000. Fiber-optic chemical sensors and biosensors. *Analytical Chemistry* 72:81–90.

Wu, C.M., and M.C. Pao. 2004. Sensitivity-tunable optical sensors based on surface plasmon resonance and phase detection. *Optics Express* 12:3509–3514.

Yu, F., S. Tian, D. Yao, and W. Knoll. 2004. Surface plasmon enhanced diffraction for label-free biosensing. *Analytical Chemistry* 76:3530–3535.

Zybin, A., C. Grunwald, V.M. Mirsky, J. Kuhlmann, O.S. Wolfbeis, and K. Niemax. 2005. Double-wavelength technique for surface plasmon resonance measurements: Basic concept and applications for single sensors and two-dimensional sensor arrays. *Analytical Chemistry* 77:2393–2399.

3 Mass Sensitive Biosensors
Principles and Applications in Food

Selma Mutlu

CONTENTS

3.1 INTRODUCTION

Applications of biosensors increased significantly in parallel with the development of the microprocessor-applied technology and rapid growth of biotechnology. These applications vary in different areas such as medical, food, environmental, military, and so forth. Considering food technology in particular, the biological and chemical contaminants should be determined, for they are of crucial importance for food safety and quality. Conventional methods applied for the determination of these components have some drawbacks when compared with biosensors. These procedures involve a series of time-consuming steps and bulky devices including separative techniques coupled to various detectors such as gas chromatography (GC), high-performance liquid chromatography (HPLC), mass spectrometer (MS), ultraviolet

(UV) and so forth. Furthermore, skilled staff is required to operate these conventional systems. By virtue of the demand for rapid, sensitive, and accurate methods to detect biological and chemical contaminants, the development of biosensors became crucial (Ricci et al. 2007).

In the food industry, biosensors exist in different transducing systems such as electrochemical, piezoelectric, thermometric, and optical biosensors. Sensing principles of these sensors based on the immunological reactions depend on the measurement of the signal, such as shift in the resonance of wavelength, frequency, potential, and conductivity. In immunosensors, it is possible to detect the antibodies (Ab) by the antigens (Ag) immobilized sensor, while the detection of Ag is carried out by reacting with the Ab immobilized sensor. In detection schemes, the labeled Ab* and Ag* are used in competition with free Ag and Ab will be measured. This type of detection scheme is called the indirect method. Transducers, such as piezoelectric sensors (quartz crystal microbalance, QCM) and modern optical sensors based on surface plasma resonance (SPR), allow the label-free detection with a direct quantification of the immunocomplex (Ab–Ag). QCM is one of the most favorable immunosensors in comparison with SPR as a label-free method for the detection of immunological reactions (Aberl et al. 1994; Laricchia-Robbio and Revoltella 2004; O'Sullivan and Guilbault 1999). QCM and SPR sensors represent some similarity in physical principles, but many fundamental differences exist (Kosslinger et al. 1995).

In QCM applications, adsorption of material on the surface of the crystals results with an oscillation frequency change. It is reported that the adsorption of mass about 1 ng causes a frequency shift of 1 Hz when an AT-cut quartz crystal of 9 MHz is used (Kurosawa et al. 1990, Muratsugu et al. 1993). To obtain a quartz crystal immunosensor with good performance, a properly designed immobilization of the biological component is essential. It has been reported that various coating methods including silanization, immobilization on the precoated crystal, and immobilization via entrapment or cross-linking are used for antibody binding onto the quartz crystal (Park et al. 2004). In addition to these techniques, plasma polymerization systems have proven to be useful for surface modification of the quartz crystal immunosensors (Nakanishi et al. 1996; Mutlu et al. 1999, 2008; Wu et al. 2000). These plasma-polymerized (PlzP) films covering the crystal surfaces are extremely thin and homogenous. The obtained layer adheres strongly to the crystal and is highly resistant to chemical and physical treatments. The sensors produced by this method are more reproducible from sample to sample and exhibit lower noise than sensors produced by the conventional immobilization methods.

In the literature, QCM immunosensors used in food analysis have been presented for pathogenic bacteria such as *Escherichia coli*, Staphylococcal enterotoxins, and *Salmonella*. A flow-type immunosensor system, which uses *anti–E. coli* antibody based on a quartz crystal microbalance, has been used (Kim and Park 2003). A study describing an immunoassay based on a piezoelectric crystal immunosensor for Staphylococcal enterotoxins (SETs), which are a significant cause of food poisoning, recently appeared in the literature (Lin and Tsai 2003). A QCM immunosensor is described for the detection of *Salmonella typhimurium* (Su and Li 2005). Another harmful component in food is histamine, which is a biogenic amine. It can be found as a consequence of microbial activity in certain foods (e.g., wine, beer, fermented meat, fish products, cheese, and fermented vegetables) and can be hazardous due to its toxic effect. Therefore, the role of QCM immunosensors is critical to indicating the level of toxic components in some foods (Rice et al. 1976; Izoquierdo-Pulido et al. 1993; Halasz et al. 1994; Ayhan et al. 1999; Shakila et al. 2001).

A number of methods have been developed for the determination of histamine, such as an enzyme-linked immunosorbent assay (ELISA) method and a radio immunoassay (RIA) method for the diagnosis of allergies as well as for research in various fields of allergy. However, these methods have some disadvantages; for example, although the ELISA method provides high analytical throughput, tedious and time-consuming steps such as washing, separation of bound and free antigens, and so forth constitute major disadvantages. Another drawback of this method is that it requires a long incubation time because the slow diffusion process of an immunoreaction on an immunoplate is the rate-determining step.

In this chapter, the principle of detection, assay format, and applications of QCM as a mass sensitive immunosensor in food analysis are summarized in concept of the theory and the literature in this field. Furthermore, our study is also presented, which is carried out by QCM for the detection of histamine levels in food.

3.2 DETECTION SCHEME OF QUARTZ CRYSTAL MICROBALANCE (QCM)

In QCM immunosensors, the immunocomplex Ab-Ag is quantified by measuring the frequency change, which corresponds to a mass change of the sensor surface. This assay concept comes forward as a method, eliminating any use of potentially hazardous-labeled materials (Xiaodi et al. 2000).

Quartz crystal placed in a QCM device as a transducer is formed with a quartz crystal wafer sandwiched between two metal electrodes such as gold, silver, aluminum, or nickel (Richard and Jerome 1996). An external oscillator circuit connecting the electrodes drives the quartz crystal at its resonant frequency. The resonant frequency of the crystal depends on the crystal properties, such as the cutting edge, thickness, and so forth. The crystals used as a biosensor are commonly 5, 9, or 10 MHz quartz in the form of AT-cut disks.

The QCM detection scheme is based on the measurement of the mass changes and physical properties of thin layers deposited on the crystal surfaces (Ballantine et al. 1997; Janshoff et al. 2000; Lu et al. 2004). Hence, the quartz crystal, which is a highly precise and stable oscillator, is effectively used as an immunosensor based on the measurement of the frequency changes.

Sauerbrey (1959) described the working principle of the sensor for the gaseous phase by the relation between the mass loading on quartz crystals and the corresponding change in resonant frequency of the crystal:

$$\Delta f = -2 f_0^2 \Delta m / A (\rho_q \mu_q)^{1/2}$$

where Δf is the frequency change of the crystal resonance, f_0 is the fundamental frequency of the crystal (in Hz), A is the surface area (in cm²), Δm is the deposited mass (in g), and ρq and μq are the properties of the density and the shear modulus of crystal, respectively.

By inserting the values of properties (density, $\rho_q = 2.648$ g cm⁻³; shear modulus, $\mu_q = 2.987 \times 10^{11}$ g cm⁻¹ s⁻²) for an AT cut crystal, the above relationship can be rearranged as:

$$\Delta f = -2.27 \times 10^{-6} f_0^2 \, \Delta m / A$$

The Sauerbrey equation gives the frequency shift measured in air corresponding the coated materials. The equation is modified for the frequency measurement in liquid phase due to bulk liquid properties (conductivity, viscosity, density, and dielectric constant). In the liquid phase, the frequency shifts are determined by the Bruckenstein and Shay (1985) and Kanazawa and Gordon's equation (1985) as follows:

$$\Delta f = -f_0^{3/2} (\rho_L \eta_L / \pi \rho_q \mu_q)^{1/2}$$

where η_L and ρ_L are the absolute viscosity and density of the liquid, respectively.

3.3 ASSAY FORMAT OF QCM

QCM is based on different assay procedures depending on its construction obtained by antigen/antibody immobilization, label/label-free, and sandwich-type configurations. These assay procedures can be classified as one-step direct assay, one-step indirect/competitive assay, two-step sandwiched assay, displacement assay, and mass amplified assay (Figure 3.1). The simplest design, one-step direct assay, involves the sample incubation on the biologically active electrode surface (Xiaodi et al. 2000). In this scheme, as a result of the binding of antibody and antigen, the mass on the quartz crystal increases. Thus, the measured frequency of the crystal decreases. The change of the frequency of the crystal directly reflects the presence of the target component that the response can be determined as the amount of the target.

For the design of the quartz crystal immunosensor, the purified biomolecules are introduced onto the surface of the crystals, and also the relatively large molecule weights of the analytes are required. Thus, the construction of the QCM surface is obtained with a highly purified receptor that is the basis of the specific recognition for the target. The detection of small molecules is not easy in that the QCM can provide only several Hz of frequency change. This limitation of low sensitivity can be overcome by using the indirect assay and sandwiched assay protocols.

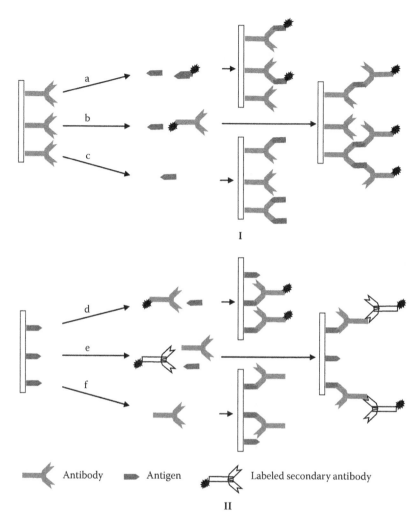

FIGURE 3.1 Steps of immunoassay pathways. (I) Antibody immobilized on the support (a) direct competitive assay, (b) sandwich assay, and (c) direct assay. (II) Antigen immobilized on the support (d) direct competitive assay, (e) indirect competitive assay, and (f) direct assay.

3.4 APPLICATIONS OF QCM IN FOOD ANALYSIS

QCM immunosensors in food analysis are mostly used for bacterial detection and toxin detection. The detection formats are dip and dry for one-step direct assay mode and flow injection assay for a mode of liquid phase detection. In Table 3.1, the applications of QCM in food analysis are summarized.

TABLE 3.1
QCM Immunoassay for Food Analysis

Analyte	Food	Reference Year
Salmonella	Raw eggs	Su et al. 2000
		Wong et al. 2002
Salmonella typhimurium		Prusak-Sochaczewski et al. 1990
		Park et al. 1998
Salmonella paratyphi A		Shihui et al. 1997
Enterobacteria	Drinking water	Plomer et al. 1992
Escherichia coli (E. coli)	Many types of food	Shen et al. 2007
		Mao et al. 2006
		Muramatsu et. al. 1989
Vibrio cholerae serotype O139		Carter et al. 1995
Listeria monocytogenes	Milk	Minunni et al. 1996
Pseudomonas aeruginosa	Milk	Bovenizer et al. 1998
Ochratoxin A	Agricultural crops	Tsai and Hsieh 2007
Antiaflatoxin B1		Akdoğan et al. 2006
Aflatoxin		Tombelli et al. 2009
Staphylococcus aureus	Many types of foods	Wei et al. 1998
Legionella pneumophila	Drinking water	Decker et al. 2000
Bacillus anthracis	Meat products	Hao et al. 2009
Candida albicans		Muramatsu et al. 1986
Hepatitis B	Many types of foods	Zhou et al. 2002
DNA (GMOs)	Tobacco plants	Karamollaoğlu et al. 2009
Pflp (ferrodoxin like protein)-gene		Passamano and Pighini 2006
		Mannelli et al. 2003
Nutrients	Many types of food	Phillips et al. 2006
Trimethylamine (TMA)	Fish	Zheng et al. 2008
Volatile compounds (ethanol, chloroform etc.)	Industrially packed bread	Bello et al. 2007

3.5 QCM IMMUNOSENSOR FOR HISTAMINE

3.5.1 Modification of Quartz Crystal Surfaces

The quartz crystals are first treated by a suitable method to obtain the surfaces that have the functionality for immobilization of the antibody/antigen. In order to get the functional groups on the surface, the chemical methods such as wet treatment in the solution and the plasma polymerization technique as a deposition in medium of the radical reactions can be applied on several surfaces. In our studies, we mainly focused on using the plasma polymerization with the monomers including the functional groups of amine($-NH_2$), carboxyl($-COOH$) for modification of the quartz crystal surfaces.

3.5.1.1 Chemical Treatment

The quartz crystal surfaces are chemically treated to improve the functionality of the surface for the immobilization of the biomolecules. Various chemical agents are used for the surface treatment to obtain the proper chemical linkers on the surface. These chemical linkers provide the immobilization of ligand to the surface via the reactive groups, such as amine (NH_2), thiol (SH), and aldheyde (COH) (Collingsy and Caruso 1997).

During the last few decades, ordered organic films of nanometer thickness have been formed to modify noble metal surfaces (Dubois and Nuzzo 1991; Ulman 1991). These ultrathin films are simply formed by the mere immersion of the noble metal surface in a dilute solution (mM) of the organic molecule at ambient conditions. This unimolecular organic film is popularly known as self-assembled monolayer (SAM). SAM formation provides one easy route toward the functionalization of the surface by organic molecules (both aliphatic and aromatic) containing suitable functional groups like –SH, –CN, –COOH, –NH_2, and silanes on selected metallic (Au, Cu, Ag, Pd, Pt, Hg, and C) as well as semiconducting surfaces (Si, GaAs, indium coated tin oxide, etc.) (Nuzzo and Allara 1986). Due to the highly ordered nature and tight packing, SAM on metallic surfaces is important for several practical applications such as chemical sensing, control of surface properties like wettability and friction corrosion protection, patterning, semiconductor passivation, and optical second harmonic generation.

SAM can be simply prepared in the laboratory by dipping the desired substrate in the required solution of mM for a specified time. In the following step, samples are thoroughly washed with the same solvent and dried. This simple procedure is an advantage of SAM for treatment of surfaces.

The good monolayers are also formed by gas-phase evaporation of the adsorbent. In this technique, the structural control is difficult. There are several factors affecting the formation and packing density of monolayers, like nature and roughness of substrate, solvent used, nature of the adsorbate, temperature, concentration of adsorbate, and so forth. Properties such as cleanliness and crystallinity of the substrate also play a crucial role in determining the compactness, often quantitatively estimated by the pinhole distribution (Chaki et al. 2001).

Another treatment mechanism is adsorption, which is divided into two classes of physical adsorption and chemical adsorption. Physical adsorption is usually weak and occurs by the formation of van der Waals bonds, occasionally accompanied by hydrogen bonds. Chemical adsorption is much stronger and involves the formation of covalent bonds. However, the biological activity of the biomaterial decreases as a result of exposing to changes in the pH, temperature, ionic strength, and substrate (Mehrvar et al. 2000). Because of these kinds of disadvantages, a novel approach is being investigated and the plasma polymerization technique has begun to be used widely for functionalization of sensor surfaces.

3.5.1.2 Plasma Polymerization

The plasma polymerization technique is one of the most effective methods for modification of the quartz crystal surfaces. Plasma-polymerized organic layers have many advantageous properties, such as good adhesion to most substrates, excellent uniformity and thickness control, and no pinhole formation. These surfaces have highly branched and cross-linked structure so that their elemental composition and physical and chemical properties (e.g., surface energy) can be varied to a wide range.

Moreover, due to their biocompatible characteristics, the biological components such as enzymes and antibodies can be loaded onto the surface of the plasma film. However, they cause some problems, such as instability of plasma of these organic vapors, which seems to twinkle and die out after a short time (Chen et al. 2007). To overcome this problem, the operation conditions are optimized for the precursors and surface exposed to plasma.

In our studies, the plasma polymerization techniques were used for surface modification of various materials such as membrane, particles, and quartz crystals. The modified surfaces were used for biosensor, separation, and adsorption studies (Mutlu et al. 1999; Çökeliler and Mutlu 2002; Saber et al. 2002; Güleç et al. 2006; Çökeliler et al. 2010). In the biosensor applications, monomers with amine group and thiols with amine and carboxyl groups were used to modify quartz crystal surfaces (Mutlu et al. 1999; Çökeliler and Mutlu 2002; Saber et al. 2002; Mutlu et al. 2008). These studies were carried out by two different plasma systems that are called rf plasma generator and electron beam generator. The plasma generator is chosen according to operation conditions such as required discharge power, monomers used as precursors, and so forth.

For the detection of histamine, quartz crystal surfaces were modified by a plasma-based electron beam generator (Göktaş et al. 2002, 2005; Mutlu et al. 2008) using cysteamine (CS), 11–Mercaptoundeconoic acid (MUA) and ethylenediamine (EDA) to obtain the functional crystal surface binding the antibody. The advantage of the electron beam generator over the conventional plasma devices is the plasma medium, which can be obtained from solid, liquid, or gas phase materials. Thiols in solid state (CS, MUA) were directly placed into the reactor; but the ethylenediamine, in liquid state, was fed from the monomer tank to the reactor under a vacuum condition. The plasma polymerization of thiols is firstly used for the surface modification of a transducer in this study. Consequently, this study can be evaluated as a good approach for the application of the plasma polymerization of thiols on the crystal surface.

The cleaned crystals were placed into the plasma polymerization reactor to obtain the amine/thiol films on the crystal surface. An intense fast-electron beam used as the plasma polymerization technique is constructed by superposing two discharges, namely, a low-pressure DC glow discharge and a high-current pulsed one. The details of the device used to produce the electron beams and its possible applications are presented elsewhere (Modreanu et al. 2000; Göktaş et al. 2002; Ogun et al. 2005).

The experiments were performed in the intense fast-electron beam reactor at different vacuum and exposure time. The conditions were established depending upon the type of precursor because the plasma can be achieved at different conditions due to the chemical and physical characteristics of the compound. The plasma conditions were 26.7 Pa vacuum and 10 min exposure time for CS treatment. The conditions in plasma applied for the MUA are 13–30 Pa and 26 min. Those conditions were changed into 66.7 Pa and 8 min for the EDA. The CS/MUA in the dry state was directly placed into the reactor, but the ethylenediamine, which is in the liquid state at the standard conditions, was fed from the monomer tank to the reactor under the vacuum due to its high vapor pressure. At the end of the process, all crystal surfaces were allowed to stand in vacuum for 10 min for the stabilization of the reactive groups, and then the plasma medium was swept with argon stream for 30 min. The frequency shifts ($\Delta f = f1-f2$) were determined as the difference between the frequencies of the unmodified ($f1$) and the modified crystal ($f2$). The depositions are determined in different values depending on the placement of the quartz crystals treated in the plasma reactor.

The measured frequency shifts (Δf) have been found as 103±10 Hz, 6486±167 Hz, and 989±36 Hz for the treated quartz crystals with EDA, CS and MUA, respectively. These frequency shifts reflect the result of the plasma treatment for the sensors having the maximum response to histamine.

3.6 ACTIVATION OF MODIFIED QUARTZ CRYSTAL SURFACES

The functional groups on the modified surfaces obtained in the first step are activated with the proper methods according to the binding between the functional groups and activation agent. The activation step provides the sensible surface to immobilize the biological component. These activation methods are improved for the immobilization strategies applied on the antibody as a recognition layer especially used in the mass sensitive biosensors. The binding of antibody is achieved by covalent coupling, self-assembling, adsorption, and so forth. In this chapter, plasma treatment is presented for the modification of the quartz crystal surfaces as an effective method. Therefore, the surfaces modified by the plasma polymerization of the monomers such as amine and thiol groups are activated through the covalent binding with glutaraldehyde for amine groups and by 1–Ethyl–3–(3–dimethylaminopropyl) carbodiimide hydrochloride (EDC)/N-hydroxysuccinimide (NHS) activation for carboxyl groups (Barlen et al. 2009).

3.6.1 FUNCTIONALIZATION OF THE SURFACES BY GLUTARALDEHYDE

The amine groups on the plasma-modified quartz crystals were activated with glutaraldehyde before the immobilization of antibody/antigen at the last step. For this activation process, the quartz crystals with amine groups were incubated into the glutaraldehyde solution of 2.5% (v/v) in a phosphate buffer at pH 7.2 for 20 hours. The experiments were performed at a temperature of 20–22°C.

After the glutaraldehyde activation of PlzP-(cysteamine) modified surfaces, the frequency shifts (Δf) were obtained in different values depending on the placement of the samples in the reactor. However, according to the performance of the sensor in the final step, the shift of 460±31 Hz was reported for the samples treated at the locations of gas input in the plasma reactor. Besides, for the PlzP–(ethylendiamine) modified surfaces, the depositions values differ due to location, and it is not possible to conclude that the high deposition rate surface provides a relatively higher frequency shift in the next antibody immobilization step. The amine functional group density on the quartz crystal surface is the most important factor for the binding of –CH=O groups of the glutaraldehyde. Hence, after the activation of PlzP–(EDA), the quartz crystals having the frequency shift about 2256±10 Hz were used to immobilize the antibody due to performance of the sensor at the final step.

3.6.2 TREATMENT WITH EDC/NHS

The activation with EDC/NHS was applied on the surfaces modified by functional groups of –COOH. In our study of histamine immunosensor, quartz crystal surfaces were first treated with MUA as presented previously. So, the carboxylic groups were created on the quartz crystal surfaces. The PlzP–(MUA) modified surfaces were activated with EDC/NHS to prepare the available surface for antibody immobilization. The quartz crystal surfaces with functional groups of –COOH were subjected to process with the solution of 1:1 involving EDC and NHS in two steps. In the first step, plasma-treated crystals with MUA were held in the 2 mM EDC solution at pH 7.4 for 1 h and then dipped into a 5-mM NHS solution for 1 h. After the activation, the crystals were washed with distilled water and buffer solution. Then, the frequencies of the activated crystals were measured. After the activation, the determined frequency shift (Δf) indicating to the deposition is 1530±95 Hz.

3.7 BIOMOLECULE IMMOBILIZATION

The principal immobilization methods of an antibody as a dominant biomolecule in the mass sensitive biosensors are the physical or chemical adsorption at solid surface (e.g., piezoelectric crystal), covalent binding to a surface, entrapment within a membrane, surfactant matrix, polymer or microcapsules, and cross-linking between molecules (Collings and Caruso 1997). In addition to these conventional methods, more recently the methods of molecularly imprinted sol-gel film entrapment, Langmuir-Blodgett deposition, and electropolymerization have all been extensively used to immobilize biological components (Bidan et al. 2000; Zhang et al. 2005; Schmidt et al. 2008).

The immobilization method employed depends on a number of factors, but in general the method needs to be compatible with the biomolecule being immobilized, the sensor surface or matrix on which immobilization is to proceed, and ultimately, the end use of the sensor.

The quartz crystals (10 MHz) treated with EDA and CS and activated with glutaraldehyde were incubated into the antihistamine solution of 20 µg/ml at 20°C for 90 min. Then, the quartz crystals with antihistamine were washed and dried in the ambient condition. After the antihistamine immobilization, the frequency shifts were determined as 578±10 Hz and 426±100 Hz for the PlzP–(EDA) and PlzP–(CS) surfaces, respectively. The measurements were done in air for the quartz crystals of 10 MHz used in the sensor based on EDA and CS treatment.

Plasma treated with thiols and activated crystal surfaces were subjected to antibody solution. The plasma treated with MUA, PlzP–(MUA) and activated crystals of 5 MHz were subjected to a process with the antihistamine solution of 20 µg/ml at 20°C for 90 min. After the process, the crystals were washed with distilled water and buffer solution to remove the unbound antibody from the surface. The frequency values of the dried crystals were measured in the buffer solution at pH 7.4, and the frequency shifts after the antibody immobilization were determined. Furthermore, the antihistamine in two different concentrations of 30 and 40 µg/ml was used for immobilization to determine the effect of the antibody loading on the performance of the sensor such as sensitivity, linearity, and response time.

The obtained frequency shifts (Δf) are 250±47 Hz for the loading of antihistamine by the process carried out in the antibody solution of 20 µg/ml. For the other immobilization processes in the antibody solutions of 30 and 40 µg/ml, the frequency shifts were determined as 30±4 Hz and 742±12 Hz, respectively. The calibration curves were plotted for these sensors and the performance parameters were determined.

3.8 ANALYSIS OF TOXINS IN FOOD

The analysis of many chemicals is based on the experimental procedures carried out in the laboratory such as ELISA. These methods are indirect measurements, which are time-consuming and consist of many experimental steps so that they need a specialist and sometimes bulky equipment. For the direct measurement, the sensors are desired to overcome such needs. In this chapter, the immunosensors have been presented for detecting the histamine in foods such as wine, beer, fermented meat, fish products, cheese, and fermented vegetables, causing a toxic effect to the human body.

3.8.1 DETERMINATION OF HISTAMINE

Plz–(MUA)-based immunosensor was tested for the histamine circulating through a flow cell integrated to the QCM system. The frequency change due to the interaction between the sensor surface and the histamine solution was monitored and the response recorded as a frequency value. The response of PlzP–(MUA)-based sensors was determined for 10 ppm histamine solution circulating at the flow rate of 0.45 ml/min, pH 7.4.

The frequency shift of the sensor responding to histamine was measured as 22±3 Hz. The response was monitored by the RQCM (Research Quartz Crystal Microbalance, Maxtek Ltd., Santa Fe Springs, California) system through the flow cell for 10 ppm histamine. The typical response of the sensor is given in Figure 3.2.

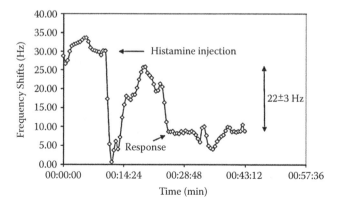

FIGURE 3.2 Response of the PlzP–(MUA)-based immunosensor to 10-ppm of histamine.

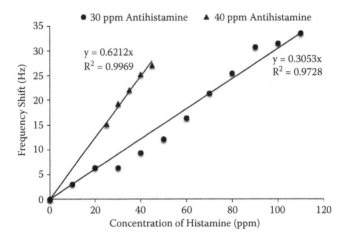

FIGURE 3.3 The calibration curves of the PlzP–(MUA)-based immunosensors; (a) 30 ppm antihistamine, (b) 40 ppm antihistamine.

3.8.1.1 Calibration of the Histamine Immunosensor

The load of antibody on the sensor surface affects the performance of the sensor. This effect is not directly related to the amount of biomolecule, but the activity of the surface should be checked for the response of the sensor. For this purpose, the PlzP–(MUA)-based sensors were prepared in antihistamine solution at two different concentrations of 30 µg/ml and 40 µg/ml. The frequency changes of the sensors exposed to histamine were measured, and the calibration curves were plotted to represent the response of the sensor versus the concentration of histamine. The calibrations of the sensors were presented in the range of 10–100 ppm histamine. The experiments were carried out through the frequency measurement system connected to the flow cell. The responses of the sensors to the histamine are given in Figure 3.3.

The saturation value of the sensors is approximately 25 Hz noted as the frequency shift. The calibration ranges of the sensors with 30 and 40 µg/ml antihistamine are 110 ppm and 45 ppm, respectively. The sensitivity of the sensor with 40 µg/ml antihistamine is higher than sensor with 30 µg/ml, but the linearity decreases.

3.8.1.2 Performance of the Histamine Immunosensor

The performance parameters of the sensors were determined from the calibration data, and the results are shown in Figure 3.4 as sensitivity, linearity, and response time. For the PlzP–(MUA)-based immunosensor prepared in 30 µg/ml antihistamine solution, the sensitivity was found as 0.30 ppm/Hz, while the linearity is 110 ppm and response time is 8 min. The performance parameters of the sensor incubated in 40 µg/ml antihistamine solution are determined as sensitivity of 0.61 ppm/Hz, linearity of 45 ppm, and response time of 5 min. When the concentration of the antihistamine solution at the binding step of antibody increases, the sensitivity of the sensor increases. This result reflects the higher linearity for the sensor having the lower sensitivity. The response of the sensor with higher antihistamine is faster than the other.

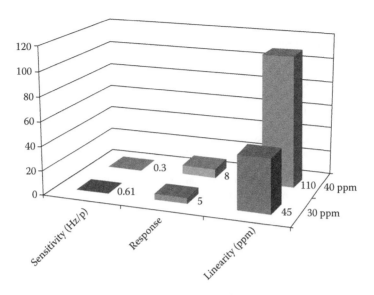

FIGURE 3.4   The performance parameters of the PlzP–(MUA)-based immunosensors.

The results represent that more antihistamine immobilization on the crystal surface causes more antigen–antibody interaction measured as the high-frequency shift of the sensor. The linear range of the sensor having the high sensitivity, 0.61 ppm/ Hz, was found as 45 ppm; this range is acceptable for the detection of histamine in food samples.

3.9   FUTURE TRENDS

In this chapter, the food analysis by QCM is briefly summarized. The analysis generally focused on the food pathogens such as *Salmonella, E. coli,* and so forth. The piezoelectric biosensors appear to be a suitable and convenient tool for detection of contaminants in food. Furthermore, this advantage can be utilized for detection of genetically modified organisms (GMOs). By virtue of this approach, it seems that QCM will be favorable tool as a DNA sensor. Future investigations will focus on improvements of the DNA sensor abilities to reveal GMO content in different food samples. Assay of QCM will be carried out to determine the features of the sensor such as sensitivity, specificity, and reproducibility.

REFERENCES

Aberl, F., H. Wolf, C. Kosslinger, S. Drost, P. Woias, and S. Koch. 1994. HIV serology using piezoelectric immunosensors. *Sensors and Actuators B* 18:271–175.
Akdoğan, E., D. Çökeliler, L. Marcinauskas, P. Valatkevicius, V. Valincius, and M. Mutlu. 2006. A new method for immunosensor preparation: Atmospheric plasma torch. *Surface and Coatings Technology* 201(6):2540–2546.
Ayhan, K., N. Kolsarıcı, and G. Alsancak-Ozkan. 1999. The effects of a starter culture on the formation of biogenic amines in Turkish soudjouks. *Meat Science* 53:183–188.

Ballantine, D.S., R.M. White, S.J. Martin et al. 1997. *Acoustic wave sensors: Theory, design and physico-chemical application.* San Diego, CA: Academic Press.

Barlen, B., S.D. Mazumdar, and M. Keusgen. 2009. Immobilization of biomolecules for biosensors. *Physica Status Solidi A* 206(3):409–416.

Bello, A., F. Bianchi, M. Careri, M. Giannetto, V. Mastria, G. Mori, and M. Musci. 2007. Potentialities of a modified QCM sensor for the detection of analytes interacting via H-bonding and application to the determination of ethanol in bread. *Sensors and Actuators B* 125:321–325.

Bidan, G., M. Billon, K. Galasso et al. 2000. Electropolymerization as a versatile route for immobilizing biological species onto surfaces. *Applied Biochemistry and Biotechnology* 89(2–3):183–193.

Bovenizer J.S., M.B. Jacobs, C.K. O'Sullivan, and G.G. Guilbault. 1998. The detection of *Pseudomonas aeruginosa* using the quartz crystal microbalance. *Analytical Letters* 31(8):1287–1295.

Bruckenstein, R., and M. Shay. 1985. Experimental aspects of use of the quartz crystal microbalance in solution. *Electrochimica Acta* 30:1295–1300.

Carter, R.M., J.J. Mekalanos, M.B. Jacobs, G.J. Lubrano, and G.G. Guilbault. 1995. Quartz crystal microbalance detection of *Vibrio cholerae* O139 serotype. *Journal of Immunological Methods* 187(1):121–125.

Chaki, N.K., M. Aslam, J. Sharma, and K. Vijayamohanan. 2001. Applications of self-assembled monolayers in materials chemistry. *Journal of Chemical Sciences* 113(5–6):659–670.

Chen, K., C. Chen, H. Lin, T. Yan, and C. Tseng. 2007. A novel technique to immobilize DNA on surface of a quartz crystal microbalance by plasma treatment and graft polymerization. *Materials Science and Engineering* 27(4):716–724.

Çökeliler, D., and M. Mutlu. 2002. Performance of amperometric alcohol electrodes prepared by plasma polymerisation technique. *Analytica Chimica Acta* 469:217–223.

Çökeliler, D., H. Göktaş H, P.D. Tosun, and S. Mutlu. 2010. Infection free titanium alloys by stabile thiol based nanocoating. *Journal of Nanoscience and Nanotechnology* 10(4):2583–2589.

Collings, A.F., and F. Caruso. 1997. Biosensors: Recent advances. *Reports on Progress in Physics* 60:1397–1445.

Decker, J., K. Weinberger, E. Prohaska et al. 2000. Characterization of a human pancreatic secretory trypsin inhibitor mutant binding to *Legionella pneumophila* as determined by a quartz crystal microbalance. *Journal of Immunological Methods* 233(1):159–165.

Dubois, L.H., and R.G. Nuzzo. 1991. Synthesis, structure, and properties of model organic surfaces. *Annual Review of Physical Chemistry* 43:437–63.

Göktaş, H, H. Kirkici, G. Oke, and A.V. Udrea. 2002. Microprocessing by intense pulsed electron beam. *IEEE Transactions on Plasma Science* 30(5):1837–1845.

Göktaş, H., M. Udrea, G. Öke, A. Alacakır, A. Demir, and J. Loureiro. 2005. Self-confinement of a fast pulsed electron beam generated in a double discharge. *Journal of Physics D: Applied Physics* 38:2793.

Güleç, H.A., K. Sarıoğlu, and M. Mutlu. 2006. Modification of food contacting surfaces by plasma polymerisation technique: Part I: Determination of hydrophilicity, hydrophobicity and surface free energy by contact angle method. *Journal of Food Engineering* 75:187–195.

Halasz, A., A. Barath, L.S. Sarkadi, and W. Holzapfel. 1994. Biogenic amines and their production by microorganisms in food. *Trends in Food Science and Technology* 51:42–49.

Hao, R., D. Wang, X. Zhang et al. 2009. Rapid detection of *Bacillus anthracis* using monoclonal antibody functionalized QCM sensor. *Biosensors and Bioelectronics* 24(5):1330–1335.

Izoquierdo-Pulido, M., M.C. Vidal-Carou, and A. Marine-Font. 1993. Determination of biogenic amines in beers and their raw material by ion-pair liquid chromatography with postcolumn derivatization. *Journal of AOAC International* 76(5):1027–1032.

Janshoff, A., H.J. Galla, and C. Steinem. 2000. Piezoelectric mass-sensing devices as biosensors—An alternative to optical biosensors? *Angewandte Chemie International Edition* 39:4004–32.

Kanazawa, K.K., and J.G. Gordon. 1985. Frequency of a quartz microbalance in contact with liquid. *Analytical Chemistry* 57:1770–1771.

Karamollaoğlu, I., H.A. Öktem, and M. Mutlu. 2009. QCM-based DNA biosensor for detection of genetically modified organisms (GMOs). *Biochemical Engineering Journal* 44(2–3):142–150.

Kim, N., and I.-S. Park. 2003. Application of a flow-type antibody sensor to the detection of *Escherichia coli* in various foods. *Biosensors and Bioelectronics* 18:1101–1107.

Kosslinger, C., E. Uttenthaler, S. Drost et al. 1995. Comparison of the QCM and the SPR method for surface studies and immunological applications. *Sensors and Actuators B* 24:107–112.

Kurosawa, S., E. Tawara, N. Kamo, and Y. Kobatake. 1990. Oscillating frequency of piezoelectric quartz crystal in solutions. *Analytica Chimica Acta* 230:41–49.

Laricchia-Robbio, L., and R.P. Revoltella. 2004. Comparison between the surface plasmon resonance (SPR) and the quartz crystal microbalance (QCM) method in a structural analysis of human endothelin–1. *Biosensors and Bioelectronics* 19:1753–1758.

Lin, H.-C., and W.-C. Tsai. 2003. Piezoelectric crystal immunosensor for the detection of staphylococcal enterotoxin B. *Biosensors and Bioelectronics* 18:1479–1483.

Lu, F., H.P. Lee, and S.P. Lima. 2004. Quartz crystal microbalance with rigid mass partially attached on electrode surfaces. *Sensors and Actuators A* 112:203–210.

Mannelli, I., M. Minunni, S. Tombelli, and M. Mascini. 2003. Quartz crystal microbalance (QCM) affinity biosensor for genetically modified organisms (GMOs) detection. *Biosensors and Bioelectronics* 18(2–3):129–140.

Mao, X., L. Yang, X. Su, and Y. Li. 2006. A nanoparticle amplification based quartz crystal microbalance DNA sensor for detection of *Escherichia coli* O157:H7. *Biosensors and Bioelectronics* 21:1178–1185.

Mehrvar, M., C. Bis, J.M. Scharer, M. Moo-Young, and J.H. Luongmehrvar. 2000. Fiber-optic biosensors-trends and advances. *Analytical Sciences* 16:677–691.

Minunni, M., M. Mascini, R.M. Carter, M.B. Jacobs, G.J. Lubrano, and G.G. Guilbault. 1996. A quartz crystal microbalance displacement assay for Listeria monocytogenes. *Analytica Chimica Acta* 325(3):169–74.

Modreanu, G., N.B. Mandache, A.M. Pointu, M. Ganciu, and I.I. Popescu. 2000. Time-resolved measurement of the energy distribution function of an electron beam created by a transient hollow cathode discharge. *Journal of Physics D: Applied Physics* 33:819–825.

Muramatsu, H., K. Kajiwara, E. Tamiya, and I. Karube. 1986. Piezoelectric immuno sensor for the detection of *Candida albicans* microbes. *Analytica Chimica Acta* 188:257–261.

Muramatsu, H., Y. Watanabe, M. Hikuma, T. Ataka, I. Kubo, E. Tamiya, and I. Karube. 1989. Piezoelectric crystal biosensor system for detection of *Escherichia coli*. *Analytical Letters* 22(9):2155–2166.

Muratsugu, M., F. Ohta, Y. Miya et al. 1993. Quartz crystal microbalance for the detection of microgram quantities of human serum albumin: Relationship between the frequency change and the mass of protein adsorbed. *Analytical Chemistry* 65(20):2933–2937.

Mutlu, S., R. Saber, C. Koçum, and E. Pişkin. 1999. An immunosensor: Immobilisation of anti-HBs antibody on glow-discharge treated piezoelectric quartz crystal for HBs-Ag detection. *Analytical Letters* 32(2):317–322.

Mutlu, S., D. Çökeliler, A. Shard, H. Goktas, B. Ozansoy, and M. Mutlu. 2008. Preparation and characterization of ethylenediamine and cysteamine plasma polymerized films on piezoelectric quartz crystal surfaces for a biosensor. *Thin Solid Films* 516:1249–1255.

Nakanishi K., H. Muguruma, and I. Karube. 1996. A novel method of immobilizing antibodies on a quartz crystal microbalance using plasma-polymerized films for immunosensors. *Analytical Chemistry* 68:1696–1700.

Nuzzo, R.G., and D.L. Allara. 1986. Adsorption of bifunctional organic disulfides on gold surfaces. *Journal of American Chemical Society* 105:4481–4483.

Ogun, S.E., H. Göktaş, H. Özkan, and N.M. Gasanly. 2005. Effect of low-energy electron irradiation on (Bi, Pb)-2212 superconductors. *Surface and Coatings Technology* 196:118–122.

O'Sullivan, C.K., and G.G. Guilbault. 1999. Commercial quartz crystal microbalances—theory and applications. *Biosensors and Bioelectronics* 14:663–670.

Park, I.-S., and N. Kim. 1998. Thiolated *Salmonella* antibody immobilization onto the gold surface of piezoelectric quartz crystal. *Biosensors and Bioelectronics* 13(10):1091–1997.

Park, I.-S., D.-K. Kim, N. Adanyi, M. Varadi, and N. Kim. 2004. Development of a direct-binding chloramphenicol sensor based on thiol or sulfide mediated self-assembled antibody monolayers. *Biosensors and Bioelectronics* 19(7):667–674.

Passamano, M., and M. Pighini. 2006. QCM DNA-sensor for GMOs detection. *Sensors and Actuators B* 118:177–181.

Phillips, K.M., K.Y. Patterson, A.S. Rasor, J. Exler, D.B. Haytowitz, J.M. Holden, and P.R. Pehrsson. 2006. Quality-control materials in the USDA National Food and Nutrient Analysis Program (NFNAP). *Analytical and Bioanalytical Chemistry* 384:1341–1355.

Plomer, M., G.G. Guilbault, and H. Bertold. 1992. Development of a piezoelectric immunosensor for the detection of enterobacteria. *Enzyme and Microbial Technology* 14(3):230–335.

Prusak-Sochaczewski, E., J.H.T. Luong, and G.G. Guilbault. 1990. Development of a piezoelectric immunosensor for the detection of *Salmonella typhimurium*. *Enzyme and Microbial Technology* 12(3):173–177.

Ricci, F., G. Volpe, L. Micheli, and G. Palleschi. 2007. A review on novel developments and applications of immunosensors in food analysis. *Analytica Chimica Acta* 605:111–129.

Rice, S.L., R.R. Eitenmiller, and P.E. Koehler. 1976. Biologically active amines in food: A review. *Journal of Milk and Food Technology* 39:353–358.

Richard, F.T., and S.S. Jerome. 1996. *Handbook of chemical and biochemical sensors*. Washington, DC: Institute of Physics Publishing.

Saber, R., S. Mutlu, and E. Pişkin. 2002. Glow-discharge treated piezoelectric quartz crystals as immunosensors for HSA detection. *Biosensors and Bioelectronics* 17(9):727–734.

Sauerbrey, G. 1959. The use of quartz oscillators for weighing thin films and for microweighing. *Zeitschrift für Physik* 206–222.

Schmidt, T.F., L. Caseli, T. Viitala, N. Osvaldo, and J. Oliveir. 2008. Enhanced activity of horseradish peroxidase in Langmuir–Blodgett films of phospholipids. *Biochimica et Biophysica Acta (BBA)—Biomembranes* 1778(10):2291–2297.

Shakila, R.J., T.S. Vasundhara, and K.V. Kumudavally. 2001. A comparison of the TLC-densitometry and HPLC method for the determination of bigenic amines in fish and fishery products. *Food Chemistry* 75:255–259.

Shen, Z., M. Huang, C. Xiao, Y. Zhang, X. Zeng, and P.G. Wang. 2007. Non-labeled QCM biosensor for bacterial detection using carbohydrate and lectin recognitions. *Analytical Chemistry* 79(6):2312–2319.

Shihui, S., F. Ren, W. Cheng, and S. Yao. 1997. Preparation of a piezoelectric immunosensor for the detection of *Salmonella paratyphi A* by immobilization of antibodies on electropolymerized films. *Fresenius' Journal of Analytical Chemistry* 357(8):1101–1105.

Su, X.-L., and Y. Li. 2005. A QCM immunosensor for *Salmonella* detection with simultaneous measurements of resonant frequency and motional resistance. *Biosensors and Bioelectronics* 21:840–848.

Su, X., F.T. Chew, and S.F.Y. Li. 2000. Design and application of piezoelectric quartz crystal-based immunoassay. *Analytical Sciences* 16(2):107–114.

Tombelli, S., M. Mascini, B. Scherm, G. Battacone, and Q. Migheli. 2009. DNA biosensors for the detection of aflatoxin producing *Aspergillus flavus* and *A-parasiticus*. *Monatshefte Fur Chemie* 140(8):901–907.

Tsai, W.-C., and C.-K. Hsieh. 2007. QCM-based immunosensor for the determination of *Ochratoxin A*. *Analytical Letters* 40:1979–1991.

Ulman, A. 1991. *An introduction to ultrathin organic films: From Langmuir-Blodgett to self-assembly*. New York: Academic Press.

Wei, W., X. Su, Q. Duan, H. Tan, and L. Bao. 1998. A rapid method for determination of *Staphylococcus aureus* based on milk coagulation by using a series piezoelectric quartz crystal sensor. *Analytica Chimica Acta* 369(1):139–145.

Wong, Y.Y., S.P. Ng, M.H. Ng, S.H. Si, S.Z. Yao, and Y.S. Fung. 2002. Immunosensor for the differentiation and detection of *Salmonella* species based on a quartz crystal microbalance. *Biosensors and Bioelectronics* 17(8):676–684.

Wu, Z., Y. Yan, G. Shen, and R. Yu. 2000. A novel approach of antibody immobilization based on n-butylamine plasma-polymerized films for sensors. *Analytica Chimica Acta* 412:29–35.

Xiaodi, S., F.T. Chew, and S.F.Y. Li. 2000. Design and application of piezoelectric quartz crystal-based immunaassay. *Analytical Sciences* 16:107–114.

Zhang, Z., H. Liao, H. Li, L. Nie, and S. Yao. 2005. Stereoselective histidine sensor based on molecularly imprinted sol-gel films. *Analytical Biochemistry* 336(1):108–116.

Zheng, J., G. Li, X. Ma, Y. Wang, G. Wu, and Y. Cheng. 2008. Polyaniline–TiO2 nano-composite-based trimethylamine QCM sensor and its thermal behavior studies. *Sensors and Actuators B* 133:374–380.

Zhou X., L. Liu, M. Hu, L. Wang, and J. Hu. 2002. Detection of hepatitis B virus by piezoelectric biosensor. *Journal of Pharmaceutical and Biomedical Analysis* 27(1):341–345.

4 Biosensing for Food Safety

María Isabel Pividori and Salvador Alegret

CONTENTS

4.1 INTRODUCTION

4.1.1 FOOD RESIDUES AND PATHOGENS IN FOOD SAFETY

Contaminated food is one of the most widespread public health problems of the contemporary world and causes considerable mortality (Nawaz 2003). In recent years, a number of high-profile food-safety emergencies have shaken consumer confidence in the production of food and have focused attention on the way food is produced, processed, and marketed. Contaminants in foods can be grouped accordingly to their origin and nature. Essentially, these are microbiological (bacteria, viruses, parasites), exogenous matter (biological, chemical, physical), natural toxins (seafood toxins, mycotoxins), other chemical compounds (such as pesticides, toxic metals, veterinary drug residues, undesirable fermentation products), and packaging materials. Most of the agents found in food are natural contaminants from environmental sources, but some are deliberate additives (Roberts 2003). While the term *contaminant* covers harmful substances or microorganisms that are not intentionally added to food, chemicals are also added during food processing in the form of *additives*. Contaminants may enter the food accidentally during growth, cultivation, or preparation; accumulate in food during storage; form in the food through the interaction of chemical components; or be concentrated from the natural components of the food. Additives were at one time a major concern, but today, microbiological issues are the greatest, followed by pesticide and animal drug residues and antimicrobial drug resistance (Todd 2003).

Although the majority of microorganisms carry out essential activities in nature and many are closely associated with plants or animals in beneficial relations, certain potentially harmful microorganisms result in numerous foodborne diseases with profound effects on animals and humans. Bacteria that cause foodborne diseases occur worldwide; *Campylobacter jejuni*, *Clostridium perfringens*, *Salmonella* spp., *Escherichia coli*, *Bacillus cereus*, *Listeria monocytogenes*, *Shigella* spp., and *Staphylococcus aureus* being the most common, among others (Todd 2003). These bacteria can multiply rapidly in moist, warm, protein-rich foods, such as meat, poultry, fish, shellfish, milk, eggs, and most food after it has been processed. Infectious organisms such as *Salmonella* and *C. perfingens* can multiply in the digestive tract and cause illness by invasion of the cell lining, production of toxins, or both. Other microorganisms are able to produce enterotoxins, such as *S. aureus* and *B. cereus*, or neurotoxins, such as *C. botulinum*, in the food during their growth and metabolism.

Many factors have contributed to recent food emergencies. The food production chain is becoming increasingly complex because of mass production. Modern farming practices are intensive and can result in microorganisms contaminating a large number of crops or infecting a large number of animals. Healthy animals may carry pathogens that cause disease in humans. Examples include *Salmonella* spp., *Listeria* spp., and some strains of *E. coli* and *Campylobacter*. Animals may be infected from feed, from other animals, or from the environment.

Although microbiological issues are the greatest concern, many consumers are worried about the long-term impact of mixtures of chemical additives (such as pesticides, toxic metals, veterinary drug residues, flavorings, and colors) and chronic as well as acute effects on vulnerable groups (Rooney and Wall 2003).

Pesticides are designed to prevent, destroy, repel, or reduce pests (animal, plant, and microbial). Pesticides are categorized according to their mode of action and include insecticides, herbicides, fungicides, acaricides, nematicides, and rodenticides. Pesticides are also used as plant growth regulators and for public-health purposes. Some are selective, impacting only target organisms, whereas others have a broad-range toxicity. Some residues may remain in both fresh produce and processed foods. Atrazine, a triazine herbicide used to control broadleaf and grassy weeds in corn, sorghum, soybeans, sugarcane, pineapple, and other crops, has a low toxicity, but animal studies indicate its potential for endocrine disruption and carcinogenicity (Todd 2003).

Concerning the veterinary drug residues, antibiotics are added to reduce disease and improve the growth of farm animals and aquaculture fish; these are of less concern for their chemical effect and more for their ability to increase antimicrobial resistance in strains that might subsequently infect humans. There is evidence that antimicrobial resistance is increasing worldwide but particularly in developing countries. The human effect of consuming foods with these drugs is still being debated, but many countries refuse to accept products derived from animals given these drugs. The withdrawal times of such drugs are critical to keep the residues in food as low as possible.

Food regulatory agencies have thus established control programs to avoid allowing food contaminants to enter the food supply. Food quality and safety can only be ensured through the application of quality-control systems throughout the entire food chain. They should be implemented at the farm level with the application of good agricultural practices and good veterinary practices at production, good manufacturing practices at processing, and good hygiene practices at the retail and catering levels. One of the most effective ways for the food sector to protect public health is to base their food management programs on Hazard Analysis and Critical Control Points (HACCP). HACCP consists of a systematic approach to the control of potential hazards in a food operation and aims to identify problems before they occur (Rooney and Wall 2003).

Application of the HACCP program is recommended for all food production processes, from small catering procedures to large-scale manufacturing. Once the sources of contamination (the hazards) and the important points at which and means by which they can be controlled (critical control points) are identified, and controls introduced, monitored, and verified, then the food manufacturers can have greater assurance that they are producing a safe product (Roberts 2003).

4.1.2 FOOD PATHOGEN DETECTION BY CULTURE AND RAPID METHODS

Conventional bacterial identification methods usually include a morphological evaluation of the microorganism as well as tests for the organism's ability to grow in various media under a variety of conditions, which involves the following basic steps: (1) preenrichment, (2) selective enrichment, (3) biochemical screening, and (4) serological confirmation (Tietjen and Fung 1995). Although standard microbiological techniques allow the detection of single bacteria, amplification of the signal is required through growth of a single cell into a colony in order to detect pathogens that typically occur in low numbers in food or water. In the case of *Salmonella*, as well as for other pathogens, is essential that initial incubation in a nonselective medium is first performed (Humphrey and Stephens 2003). This step allows the organism not only to overcome the effects of sublethal injury but also to multiply. The period of recovery or repair is often referred to as *preenrichment*. Incubation during the preenrichment step is usually performed for 18 to 24 h. After preenrichment, selective isolation is usually performed in two stages: growth in an enrichment broth followed by plating on a selective agar. Selective media make use of either dyes or chemicals. Malachite green, in combination with high concentrations of magnesium chloride and a medium pH of 5.0–5.2 in soya peptone broth (Rappaport-Vassiliadis broth, RV) has been shown to be effective for selective culture following preenrichment of *Salmonella*. The further plating in selective agars is usually performed in bismuth sulfite, xylose lysine deoxycholate (XLD), and brilliant green agar (BGA). After a further growing for 24 h, the confirmation of *Salmonella*-like colonies is mandatory. A biochemical confirmation test relies on assessment of urease and lysine decarboxylase activity, fermentation of dulcitol, indole production, growth in the presence of potassium cyanide, utilization of sodium malonate, and more recently, pyrrolindonylarylamidase (PYR) activity. These reactions, in combination with serological tests, are usually sufficient for identification, but additional tests may sometimes be necessary. As a laboratory routine, the use of commercial kits may be cost effective. Antibodies against somatic (O) and flagella (H) antigens are used to confirm and to identify *Salmonella*-like isolates.

These growing and enrichment steps are relatively time-consuming, having a total assay time of up to a week in certain food pathogens (Ivnitski et al. 1999). Although classical cultural methods can be sensitive, they are greatly restricted by the assay time. The time necessary to obtain a negative result can take up to 96 h when the whole cultural method has to be applied, while presumptive positive results may take up to 48–96 h.

Many alternative methods have been introduced in recent years to reduce analytical time and also save staff time and media requirements. These rapid methods are designed to avoid selective culturing and serological/biochemical identification.

New instrumental methods have been developed using various principles of detection, and these include chromatography, infrared or fluorescence spectroscopy, bioluminescence, flow cytometry, impedimetry, and many others. Among these methods, techniques based on electrical conductance and impedance are being used in many laboratories. When microbial growth and metabolism take place in a culture medium, changes in conductance of the medium occur but the reliability of the method largely depends on the performance of the selective medium used for the assay. Moreover, these methods are centralized in large stationary laboratories because complex instrumentations are used and require highly qualified technical staff. Furthermore, the capital cost involved in these instrumental analyses is high, which restricts its use (Ivnitski et al. 1999).

During the past decade, immunological detection of bacteria, cells, spores, viruses, and toxins has become more sensitive, specific, reproducible, and reliable with many commercial immunoassays available for the detection of a wide variety of microbes and their products in food. Advances in antibody production have stimulated this technology, since polyclonal antibodies can be now quickly and cheaply obtained and do not require the time or expertise associated with the production of monoclonal antibodies. However, polyclonal antibodies are limited both in terms of their specificity and abundance. Many test kits are available, including immunodiffusion, enzyme-linked immunosorbent assay (ELISA), and the use of specific antibodies to capture and concentrate the organism (Luk et al. 1997). Moreover, immunoassays (IAs) have shown the ability to detect not only contaminating organisms but also their biotoxins. However, IAs give total bacterial load rather than just the number of viable cells.

Nucleic acid–based detection may be more specific and sensitive than immunological-based detection. Furthermore, the polymerase chain reaction (PCR) can be easily coupled to enhance the sensitivity of nucleic acid–based assays. Target nucleic segments of defined length and sequence are amplified by PCR by repetitive cycles of strand denaturation, annealing, and extension of oligonucleotide primers by the thermostable DNA polymerase, Thermus aquaticus (Taq) DNA polymerase. Nucleic acid–based detection coupled with PCR has distinct advantages over culture and other standard methods for the detection of microbial pathogens such as specificity, sensitivity, rapidity, accuracy, and capacity to detect small amounts of target nucleic acid in a sample (Wan et al. 2000). Moreover, multiple primers can be used to detect different pathogens in one multiplex reaction (Leonard et al. 2003). This kind of method requires an enrichment period that delays results, but sensitivity can be of the order of 3 cfu per 25 g of food (Humphrey and Stephens 2003). This approach cannot be applied directly to samples because DNA is heat stable. Thus, intact DNA will be present in processed foods. DNA both in a free form and within dead cells of *Salmonella* can also survive for considerable periods in seawater. One possible way to overcome this problem is to include a culture step as a means of detecting viable cells, because this step is not sufficient to allow the growth of more severely damaged cells.

The use of nucleic acids recognition layers represents a new and exciting area in analytical chemistry that requires extensive study. Besides classical methodologies to detect DNA, novel approaches have been designed, such as the DNA chips (Bowtell 1999) and lab-on-a-chip based on microfluidic techniques (Sanders and Manz 2000). However, these technologies are still out of the scope of the food industry, since the industry requires simple, cheap, and user-friendly analytical devices.

4.1.3 Pesticide and Drug Residue Detection Methods

As pesticides and drugs produce undesirable residues, various national and international authorities regulate their use and set maximum residue levels (MRLs) in food (Nawaz 2003). An MRL is the maximum concentration of a food residue and/ or its toxic metabolites legally permitted in food commodities and animal feeds. As an example, if pesticides are properly applied at the recommended rates, and crops are only harvested after the appropriate time intervals have elapsed, residue levels are not expected to exceed MRLs. In the Eurpoean Union (EU), the regulation of the agrochemical industry and the setting of MRLs are currently being harmonized across all member states by the European Commission. Authorities have also introduced a definition for residues of veterinary medicinal drugs, which, according to EU Council Regulation 2377/90, is as follows: *All pharmacologically active substances whether active principles, excipients or degradation products, and their metabolites which remain in food stuffs obtained from animals to which the veterinary medicinal product in question has been administered.*

The monitoring of the residues in foods is often at the microgram per kilogram level or lower and has to be supported by strict analytical quality-control standards, so that the analysis produces unequivocal, precise, and accurate residue data. An analytical method to be used in food residue determination should accomplish an adequate specificity, sensitivity, linearity, accuracy, and precision at the relevant residue concentration and in appropriate food matrices.

Multiresidues analysis has been carried out using conventional chromatographic methods, such as high-performance liquid chromatography (HPLC), gas chromatography (GC), and capillary electrophoresis (CE). In the special case of veterinary drug residues, conventional microbiological methods can also be performed based on the inhibition of growing of bacteria promoted by the antibiotic.

A classical food residue analysis based on chromatography includes the following steps: (1) sampling, (2) sample preparation/subsampling, (3) extraction, (4) cleanup, and (5) chromatographic separation and instrumental determination.

Regarding the sampling in food residues, a representative sample consists of a large number of randomly collected units. Monitoring of pesticide residues for MRL compliance involves analysis of a composite sample, made up of a number of individual units (Nawaz 2003). Milk and diluted honey, but also urine serum used to trace contamination of the animal may be analyzed in some cases without extensive sample processing. However, the extraction of residues prior to the analysis of solid matrices, including fat, hair, meat, kidney, retina, and skin, is usually required. Specific solvents are mixed with the homogenized matrix to facilitate solubilization of the residues. Sample extracts not only contain the target analyte(s) but may also contain coextractives, such as plant pigments, proteins, and lipids. These coextractives may have to be removed prior to instrumental analysis to avoid possible contamination of instruments and to eliminate compounds that interfere during the determination step. Cleanup of samples is therefore necessary in most cases to overcome any interference of the analysis method. Adsorption chromatography is used in many residue laboratories for the cleanup of sample extracts.

The final stage of the residue analysis procedures involves chromatographic separation and instrumental determination. If the chromatographic properties of some food residues are affected by the sample matrix, calibration solutions should be prepared in the sample matrix. The choice of instrument depends on the physiochemical properties of the analyte(s) and the sensitivity required. GC has proved to be an excellent technique for volatile pesticides and drug residues. Thermal conductivity, flame ionization, and in certain applications electron capture and nitrogen phosphorus detectors were popular in GC analysis. In current residue GC methods, the universality, selectivity, and specificity of the mass spectrometer (MS) in combination with electron-impact ionization are by far preferred.

HPLC is increasingly being used for the determination of pesticide and drug residues, as it is especially suited to the analysis of nonvolatile, polar, and thermally labile residues that are difficult to analyze using GC. The resolution achieved on HPLC can be comparatively low, and therefore the use of selective detection systems may be necessary for reliable residue analysis. Ultraviolet (UV) spectroscopy is the most common choice for detection of residues. Although UV detection is not a very selective technique, it is commonly used for screening purposes due to its low cost, simplicity, and wide application range. Elimination of interferences and optimized chromatography are essential prior to detection in order to enhance the selectivity of UV-based methods. The use of diode array detectors can further enhance the selectivity of UV detection procedures. Fluorescence detection offers a greater selectivity and sensitivity than UV. Electrochemical detectors are used for a number of residues in relatively clean samples. The online combination of HPLC and mass spectroscopy (HPLC–MS) offers a high sensitivity and specificity. There are a number of ionization techniques used to interface HPLC with MS analyzers, of which the most widely used are electrospray and atmospheric pressure chemical ionization. Some residues require derivatization to enhance the extractability, cleanup, or subsequent chromatographic resolution and determination steps. Instead of chromatography, CE with a high resolving power may be considered as well.

For regulatory purposes, it is essential that pesticide residues are unequivocally confirmed using MS. However, if an MS method is not available, the sample extract is reanalyzed using a different chromatographic column and/or a different detection system to confirm the initial results (Nawaz 2003).

Besides physicochemical methods, the use of microbiological growth-inhibition assays to test meat and milk for the presence of antibiotics residues is popular over a long period of time. These tests use antibiotic-sensitive bacterial reporter strains, such as *Bacillus subtilis* and *Bacillus stearothermophilus* var. *calidolactis*. These bacteria are inoculated under optimal conditions with and without sample. After culturing, results are read from visible inhibition zones or from the color change of the bacterial suspension in agar gels (Bergwerff and Schloesser 2003).

Immunochemical methods are used for rapid screening of an individual or a group of closely related residues. These methods require little or no sample cleanup, require no expensive instrumentation, and are suitable for field use.

The development of immunoassays for the detection of food components and contaminants has progressed rapidly in the last few years (Chu 2003). Antibodies against almost all the important food residue compounds are currently available. Classical immunochemical methods such as immunodiffusion and agglutination methods for food analyses generally involve no labeled antigen or antibody. Concentration of the antigen–antibody complex is estimated from the secondary reaction that leads to precipitation or agglutination. These methods are not sensitive, are subject to non-specific interference, and can only be used for analysis of high-molecular-weight proteins. However, the development of radioimmunoassay (RIA) has widened the scope of immunoassays. This method combines the unique properties of specific antibody–antigen interaction and the use of a radioactive labeled marker to monitor complex formation. Thus, RIA provides specificity, sensitivity, and simplicity and can be used for analysis of both antigen and haptens. RIA involves the use of a radioactive marker that competes with an analyte in the sample for binding to an antibody. For RIA of high-molecular-weight antigen, either the antigen or antibody molecules can be labeled. It is also common to use a radiolabeled second antibody, that is, antibody against the primary antibody. In contrast, labeled hapten is typically used in RIA for low-molecular-weight substances. Although RIA is simple and sensitive, it is limited by the need for a marker with high specific radioactivity, instruments for measuring radioisotopes, and licenses for using radioactive materials and disposal of radioactive materials. Using different nonradioactive labeled markers, a variety of immunoassays, including fluorescence immunoassay (FIA), time resolved FIA, FIA polarization immunoassay, enzyme immunoassay (EIA), luminescent immunoassay (LIA), metalloimmunoassay (MIA), and viroimmunoassay (VIA), have been developed. Since no radioactive substances are used, the assays avoid the problems encountered in handling radioactivity.

EIA is a general term for immunoassays involving use of an enzyme as a marker for the detection of immunocomplex formation. Enzyme labeling can be achieved by conjugation of the enzyme to an antigen or antibody via periodate oxidation with a subsequent reductive alkylation method, cross-linking using glutaraldehyde or others. Some of the methods used in the conjugation of hapten to proteins can also be used. However, to avoid nonspecific interaction, the method for coupling of protein/hapten to enzyme should be different from the one that had been used for conjugating the hapten to protein for the purpose of immunization. Although horseradish peroxidase and alkaline phosphatase are the two enzymes most commonly used, others, such as glucose-6-phosphate dehydrogenase, coupled with oxidoreductase and luciferase, glucose oxidase, beta-galactosidase, urease, and others, have also been employed. Depending on whether or not the immunocomplex is separated from the free antigen, EIAs are further divided into two types. One type is a homogeneous system, which is called enzyme multiplied immunoassay (EMIT); it is based on the modification of enzyme activity occurring when the antibody binds with the enzyme-labeled antigen/hapten in solution. Because modification of enzymatic activity is generally not significant, this system is not very sensitive and has not been widely used in food analysis. The other is a heterogeneous system involving separation of free and bound antigen–antibody. In this system, either antigen or antibody is covalently or noncovalently bound to the solid matrix. Nonreacted antibody or antigen is simply removed by washing or centrifugation. The term ELISA is used for this type of assay. Solid phases such as microtiter plates, cellulose, nylon beads/tubes, nitrocellulose membrane, polystyrene tubes/balls, and modified magnetic beads have been used. In some cases, staphylococcal protein A or protein G is coated on the solid surface, entrapping the antibody for subsequent analysis.

This method is further divided into two major types. One type is competitive ELISA, which can be used for the analysis of both hapten and macromolecule; the other is noncompetitive sandwich-type ELISA, which is only used for divalent and multivalent antigens. Two major types are used most commonly in food analysis: direct competitive ELISA (dC-ELISA) and indirect competitive ELISA (inC-ELISA).

In the dC-ELISA specific antibodies are first coated to a solid phase. The sample or standard solution of analyte is generally incubated simultaneously with enzyme-conjugate or incubated separately in two steps. The amount of enzyme bound to the plate is then determined by incubation with a chromogenic substrate solution. The resulting color/fluorescence, which is inversely proportional to the analyte concentration present in the sample, is then measured instrumentally or by visual comparison with the standards.

A number of quick screening tests based on the ELISA principle described above have been developed. For example, microtiter plate ELISA assay can be completed in less than 20 min. Other approaches involve immobilizing the antibody on a paper disk or other membrane that is mounted either in a plastic card, in a plastic cup, or on the top of a plastic tube. In the "dipstick" assay, antibody or antigen is coated on a stick, which is then dipped in various reagents for subsequent reactions. Substrates leading to the formation of a water-insoluble chromogenic product are used in these assays. Most of these screening tests are simple and easy to perform and take 10 to 15 min to complete. They are designed to provide semiquantitative information at certain cutoff concentrations for the substance in which one is interested. The immunoscreening tests have gained wide application for monitoring residues in foods, and versatile assay kits are commercially available.

Immunoassays are available for almost all the important antibiotic residues that might be present in foods. For example, β-lactone antibiotics such as ampicillin, cloxacillin, and penicillin G could easily be measured in milk. Likewise, immunoassays for many pesticides are also available, including 2,4-D, aldicarb, carbendazim, thiabendazole, chlopyrifos, diazinon, endosulfan, and metalaxyl (Nawaz 2003). The triazine immunoassay is now available commercially, and like most immunoassays, it is specific, sensitive, rapid, and cost-effective (Au 2003).

4.2 BIOSENSING: A NOVEL STRATEGY FOR FOOD SAFETY

Traditional microbial screening methods have insufficient sensitivities to meet new regulations; and classical physiochemical techniques, such as chromatographic methods and mass spectrometry, are often precluded due to the level of experience, skills, and cost involved. Moreover, these methods require laborious extraction and cleanup steps that increase analysis time and the risk of errors.

The development of biosensors is a growing area, in response to the demand for rapid real-time, simple, selective, and low-cost techniques for food residues (Leonard et al. 2003). Biosensors are compact analytical devices, incorporating a biological sensing element either closely connected to or integrated within a transducer system. The combination of the biological receptor compounds (antibody, enzyme, nucleic acid) and the physical or physicochemical transducer produces, in most cases, real-time observation of a specific biological event (e.g., antibody–antigen interaction) (Deisingh and Thompson 2004). Depending on the method of signal transduction, biosensors can also be divided into different groups: electrochemical, optical, thermometric, piezoelectric, or magnetic (Terry et al. 2005). They allow the detection of a broad spectrum of analytes in complex sample matrices and have shown great promise in areas such as clinical diagnostics, food analysis, bioprocess, and environmental monitoring (Patel 2002; Velasco-Garcia and Mottram 2003). Hazard analysis and critical control points (HACCP) systems, which are generally accepted as the most effective system to ensure food safety, can utilize biosensors to verify that the process is under control (Mello and Kubota 2002). The sensitivity of each of the sensor systems may vary depending on the transducer's properties and the biological recognizing elements. An ideal biosensing device for the rapid detection of food contaminants should be fully automated, inexpensive, and routinely used both in the field and the laboratory. Optical transducers are particularly attractive as they can allow direct label-free and real-time detection, but they lack sensitivity. The phenomenon of surface plasmon resonance (SPR) has shown good biosensing potential, and many commercial SPR systems are now available. The Pharmacia BIAcore™ (a commercial SPR system) is by far the most reported method for biosensing of food residues in food and is based on optical transducing. As examples, the SPR biosensor was compared with existing methods (microbial inhibitor assays, microbial receptor assays, ELISA, HPLC) for detection of sulfamethazine residues in milk by an immunological reaction (Mellgren et al. 1996, Sternesjo et al. 1995). BIAcore has indicated the occurrence of sulfamethazine at a concentration below the detection limit of HPLC and offered sufficient advantages (no sample preparation, high-sensitivity rapid and full analysis in real time) to be an alternative for the control of residues and contaminants in food. A similar commercial transducer system was also reported for the determination of β-lactam antibiotics (Gaudin 2001; Gustavsson et al. 2002; Cacciatore et al. 2004), multisulfonamide residues (Haasnoot et al. 2005), and chloramphenicol residues (Ferguson et al. 2005). In the case of pesticides in food, they are mainly based on both the biosensing of the enzyme inhibition by the pesticide (Vakurov et al. 2005; Andreescu and Marty 2006) as well as the immunosensing of the pesticide performed with the specific antibody with SPR transducer (Mullett 2000; Shimomura et al. 2001; Mauriz 2006). The detection of food pathogens by SPR, however, does not reach the required limit of detection (LOD) to allow food safety without performing a preenrichment step (Barlen et al. 2007).

However, electrochemically based transduction devices are more robust, easy to use, portable, and inexpensive analytical systems (Ivnitski et al. 2000; Mehervar and Abdi 2004). Furthermore, electrochemical biosensors can operate in turbid media and offer comparable instrumental sensitivity.

Regarding the molecular recognition of food residues, immunological reagents are mainly used as molecular receptors in order to obtain a useful signal (Baeumner 2003; Nakamura and Karube 2003), while in the case of food pathogens, not only immunological reagents (Gehring et al. 1996; Croci et al. 2001) but also DNA probes (Croci et al. 2004) can be used as molecular receptors in order to obtain a useful signal after the biological reaction on the transducer element (antigen–antibody and hybridization, respectively) (Ivnitski et al. 2000).

Electrochemical immunosensors and genosensors can meet the demands of food control, offering considerable promise for obtaining information in a faster, simpler, and cheaper manner compared with traditional methods. Such devices possess great potential for numerous applications, ranging from decentralized clinical testing to environmental monitoring, food safety, and forensic investigations. The following section focuses on electrochemical biosensing for food contaminants.

4.2.1 TRANSDUCING FEATURES IN ELECTROCHEMICAL BIOSENSORS

The development of new transducing materials is a key issue in the current research efforts of electrochemical biosensors. While the immobilization of the bioreceptor and the detection of the biological event are important features, the choice of a suitable electrochemical transducer is also of great importance in the overall performance of electrochemical-based biosensors. Carbonaceous materials such as carbon paste (Wang et al. 1996), glassy carbon (Oliveira Brett 1997), and pyrolitic graphite (Hashimoto et al. 1994) are the most popular choice of electrodes used in biosensing devices. However, the use of platinum (Moser et al. 1997), gold (Pang and Abruña 1998), indium-tin oxide (Armistead and Thorp 2000), copper solid amalgam (Jelen et al. 2002), mercury (Fojta and Paleček 1997), and other continuous conducting metal substrates has been reported. Conducting polymers, such as polypyrrole and polyaniline (Teles and Fonseca 2008), and conducting composites, based on the combination of nonconducting polymers with conductive fillers (Alegret 1996), have been also continuously studied during the past few decades. Finally, nanostructured materials such as carbon nanotubes (CNTs)

(Wang 2005) and metal nanoparticles (Rajesh et al. 2009) have been also reported as a base material or fillers for conducting composites or as surface-modifiers of many types of electrochemical transducers in order to improve their electrochemical properties. Other nanostructured materials including gold nanoparticles have been intensively investigated as a component of electrochemical transducers (Pingarrón et al. 2008). Nanocomposites can be fabricated not only with nanostructured materials but also with biomolecules and redox polymers to possess unique hybrid and synergistic properties. It is expected that the combination of nanoengineered "smart" polymers with novel biocompatible nanostructured fillers—like nanoparticles and CNTs—may generate composites with new and interesting properties, such as higher sensitivity and stability of the immobilized molecules, thus constituting the basis for improved electrochemical biosensors (Teles and Fonseca 2008).

Rigid conducting graphite-epoxy composites (GECs) (Pividori et al. 2001; Pividori and Alegret 2003, 2005) and graphite-epoxy biocomposites (GEBs) (Alegret 1996, Zacco et al. 2006a) have been extensively used in our laboratories for electrochemical (bio)sensing due to their unique physical and electrochemical properties. In particular, we have used GEC made by mixing nonconducting epoxy resin (Epo-Tek, Epoxy Technology, Billerica, Massachusetts) with graphite microparticles (particle size below 50 μm).

This paste can be easily prepared by mixing graphite powder with epoxy resin in a 1:4 (w/w) ratio. The soft paste is thoroughly hand mixed to ensure the uniform dispersion of the graphite powder throughout the polymer. The moldable soft paste is put on the body of the electrodes and cured at 100°C for 2 days to obtain the rigid GEC (Pividori and Alegret 2005).

Biocomposites can also be easily prepared by adding the bioreceptor—an enzyme (Alegret 1996) and antibody (Zacco et al. 2006a), or an affinity receptor such as Protein A (Zacco et al. 2004) or avidin (Williams et al. 2003; Lermo et al. 2008).

As an example, in the case of avidin graphite-epoxy biocomposite (Av-GEB), graphite powder and epoxy resin are also hand mixed in a ratio of 1:4 (w/w). In this case, for every gram of graphite/epoxy mixture, an additional 20 mg of avidin is added—resulting in a 2% (w/w) avidin-graphite-epoxy biocomposite. This mixture is thoroughly hand mixed to ensure the uniform dispersion of the avidin and carbon throughout the polymer. The moldable soft paste is put on the body of the electrodes and cured at 40°C for 1 week to obtain the rigid Av-GEB.

Gold nanocomposites are prepared by hand mixing the following ratios of gold nanoparticles, graphite powder, and epoxy resin: 0.075/0.925/4 (w/w) for nanoAu (7.5%)-GEC. The resulting soft paste is placed in the gap of electrode and cured at 80°C for 1 week to obtain the rigid gold nanoparticles graphite-epoxy composite (nanoAu-GEC) (Oliveira Marques et al. 2009).

A magneto electrode based on GEC (m-GEC) is prepared in the same way as for the GEC transducer, but in this case, a small magnet (3 mm i.d.) is placed in the center of this electrode after the addition of a thin layer of GEC paste in order to avoid direct contact between the magnet and the electrical connector. After filling the electrode body gap completely with the soft paste, the electrode is tightly packed and then cured at the same temperature. This magneto electrode can be easily coupled with biologically modified magnetic particles (Liébana et al. 2009a,b), one of the most promising materials that has been developed for bioanalysis (Haukanes and Kvam 1993). Magnetic beads (HB) offer some new attractive possibilities in biomedicine and bioanalysis since their size is comparable to those of cells, proteins, or genes. Moreover, they can be coated with biological molecules, and they can also be manipulated by an external magnetic field gradient. As such, the biomaterial, specific cells, proteins, or DNA can be selectively bound to the HB and then separated from its biological matrix by using an external magnetic field. Moreover, HB of a variety of materials and sizes and modified with a wide variety of surface functional groups are now commercially available. They have brought novel capabilities to electrochemical immunosensing (Zacco et al. 2006b) and genosensing (Lermo et al. 2007).

Instead of the direct modification of the electrode surface, the biological reactions (as immobilization, immunological, enzymatic labeling, or affinity reactions) and the washing steps can be successfully performed on HB. After the modifications, the HB can be easily captured by magnetic forces onto the surface of m-GECs.

The GECs-based transducers present numerous advantages over more traditional carbon-based materials: higher sensitivity, robustness, and rigidity. Additionally, the surface of GEC can be regenerated by a simple polishing procedure. An ideal material for electrochemical biosensing should allow an effective immobilization of the probe on its surface, a robust hybridization of the target with the probe, a negligible nonspecific adsorption of the label, and a sensitive detection of the hybridization event. GECs fulfill all these requirements.

4.2.2 Immobilization Strategies in Electrochemical Biosensors for Food Safety

The immobilization of the biorecognition element—which specifically recognizes the target—onto the transducer is also a key issue in the construction of biosensing devices. The choice of the immobilization method depends mainly on the biomolecule to be immobilized, the nature of the solid surface, and the transducing mechanism (Cassidy et al. 1998). Beside the sensitivity, the ability of the electrochemical transducer to provide a stable immobilization environment while retaining the bioactivity must be also considered. A current problem regarding the immobilized biomolecules is the lack of stability and activity in the solid transducer, which is usually overwhelmed by the use of an *in vivo*-like environment or the use of spacer arms.

The most successful immobilization methods involve (1) multisite attachment, either electrochemical—by the application of a potential to the solid support—or physical adsorption, or (2) single-point attachment—mainly covalent immobilization, affinity linkage (such strept(avidin)/biotin binding)—and chemisorption based on self-assembled monolayers (SAMs) (Pividori et al. 2000; Zacco et al. 2006a). Among the different immobilization strategies, multisite adsorption is the simplest and most easily automated procedure, avoiding the use of pretreatment procedures based on previous activation/modification of the surface transducer and subsequent immobilization. Such pretreatment steps are known to be tedious, expensive, and time-consuming. The binding forces involving physical adsorption include hydrogen bonds, electrostatic interaction, van der Walls forces, and hydrophobic interactions if water molecules are exclude by dryness (Pividori and Alegret 2005). Wet adsorption originates a weak binding that causes easy desorption of the biomolecule from the surface, eventually leached to the sample solution during measurements. However, dry adsorption also promotes hydrophobic bonds and more stable adsorbed layers on solid surfaces (Pividori and Alegret 2005). Classical strategies such as physical entrapment in membranes and cross-linking by bifunctional reagents—such as glutaraldehyde—also represent multisite attachment methods for retaining close contact of the bioreceptor with the transducer.

Single-point attachment is beneficial for the kinetics of the biological reaction, especially if a spacer arm is used. Single-point covalent immobilization can be performed on different surface-modified electrochemical transducers, such as glassy carbon (Millan and Mikkelsen 1993), carbon paste (Millan et al. 1994), gold (Sun et al. 1998) or platinum (Moser et al. 1997), or lately, carbon nanotubes (Wang and Lin 2008), through the linkage of a carboxy moiety with an amine group by the use of the carbodiimide chemistry. Single-point affinity linkage also provides an interesting strategy for the oriented and stable immobilization of biotinilated biomolecules to solid transducers, throughout biotin/strept(avidin) binding (Zacco et al. 2006a). Nowadays, knowledge about this interaction has advanced significantly and offers an extremely versatile tool. The avidin-biotin system has gained great importance over the years as a tool for general application in the biological assays of an almost unlimited number of biological molecules (Wilchek and Bayer 1988, 1990; Wilchek et al. 2006). Historically, the development and extensive application of the avidin-biotin system was a definitive breakthrough in the area of nonradioactive labeling and detection of biologically active molecules. The avidin–biotin reaction as an immobilization strategy for biomolecules presents a variety of specific advantages over other single-point immobilization techniques. In particular, chicken egg-white avidin—a glycosylated and positively charged (pI ~10.5) protein—and its bacterial analogue streptavidin share a similarly high affinity (K_a ~10^{15} M^{-1}) for the vitamin biotin (Green 1990) similar to the formation of a covalent bonding. Despite the relatively modest sequence homology (~30% identity and 40% overall similarity), the two proteins—avidin and streptavidin—share the same tertiary fold, similar tetrameric quaternary structures, and a nearly identical arrangement of amino acid residues within the respective binding pockets. The interaction with biotin is highly resistant to a wide range of chemicals (detergents, protein denaturants), pH range variations, and high temperatures (Jones and Kurzban 1995). In addition, the avidin-biotin–based immobilization method maintains the biological activity of the biomolecule being immobilized more successfully than other commonly used methods (Darain et al. 2003; Limoges et al. 2003; Da Silva et al. 2004). Much progress has been made in the modification of biomolecules with biotin. A wide range of macromolecules including proteins (Snejdarkova et al. 1993)—enzymes or antibodies—polysaccharides, and nucleic acids or short oligonucleotides can be readily linked to biotin without serious effects on their biological, chemical, or physical properties. As such, avidin should be considered a universal affinity molecule capable of attaching to different biotinylated biomolecules (Zacco et al. 2006a).

Another immobilization strategy specific for nonmodified antibodies is based on the binding through Fc fragment to protein A or G. The bond strength between protein A (or G) and an antibody is greatly affected by the antibody classes and subclasses (Sjoquist et al. 1972a,b; Akerstrom et al. 1985; Compton et al. 1989). The affinity constant can vary from strong to weak. The interaction between protein A of *Staphylococcus aureus* and the Fc region of immunoglobulin G (IgG) has also been mapped in some detail. The data suggest a binding site spanning the Cγ2–Cγ3 junction in the Fc region. The binding involves the formation of multiple noncovalent bonds between the protein A and amino acids of the binding Fc site. Considered individually, the attractive forces (hydrogen and electrostatic bonds, van der Waals and hydrophobic forces) are weak in comparison to covalent bonds. However, the large number of interactions results in a large total binding energy in the case of IgG. These immobilization strategies allow the binding sites of the antibodies to be oriented away from the solid phase. Different from avidin, protein A is able to link the Fc region of many immunoglobulins, thus it is not necessary to have the antibody modified with biotin.

Finally, chemisorption based on SAMs has also been extensively used for single-point attachment on gold-based transducers. Highly ordered SAMs are formed by adding the biomolecule to be immobilized to a solid support and washing the excess out. SAMs are basically interfacial layers between a metal surface and a solution. The best described SAMs are those based on the strong adsorption of functionalized alkane sulfides, disulfides, and thiols onto gold surfaces (Nuzzo and Allara 1983). Sulfur and selenium compounds have a strong affinity to transition metal surfaces (Ulman 1996), coordinating strongly not only onto gold but also onto silver, platinum, or copper. Nevertheless, gold is the most favored because it is reasonably inert.

Chemisorption requires the presence of a nucleophilic group in the surface of the biomolecule to be able to be covalently linked to the gold surface, such as —SH or other gold reactive groups (such as —NH$_2$). In the case of the proteins, they are able to directly attach to the gold surface if presenting one accessible thiol (—SH) group of a cysteine at the protein surface, as in the case of Fab'-SH antibody fragments. This approach has limited application, because only few proteins present accessible sulfhydryl groups, even after disulfide reduction. Instead of the use of a natural—SH group, the second strategy relies on the introduction of this kind of moiety. As an example, it is possible to transform amine groups into sulfhydryl groups, with use of the Traut's reagent (2-iminothiolane) (Traut et al. 1973; Jue et al. 1978). DNA probes, for instance, must be chemically modified. The standard way for thiolating a strand of DNA is to attach an HS(CH2)6- linker molecule to either the 3' or the 5' end phosphate group. This method has been extensively used for the immobilization of oligonucleotides in gold surfaces (Carpini et al. 2004), gold-nanoparticles modified transducer (Cai et al. 2001), and gold nanocomposites (Oliveira Marques et al. 2009).

4.2.3 Electrochemical Detection Strategies in Electrochemical Biosensors for Food Safety

Electrochemical detection of the biorecognition event should be also considered, involving the transduction of the biological reaction into a useful and easy-to-amplify electrical signal.

The direct electrochemical detection of DNA was initially proposed by Paleček (1958, 1960) who recognized the capability of both DNA and RNA to yield reduction and oxidation signals after being adsorbed (Figure 4.1a). The oxidation of DNA

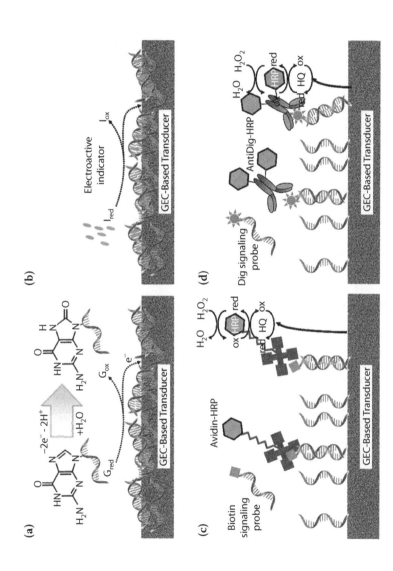

FIGURE 4.1 Schematic representation of the different detection strategies in electrochemical genosensing: (a) electrochemical genosensing based on the intrinsic DNA oxidation; (b) preconcentration on the dsDNA of electrochemical indicators and electrochemical detection of the preconcentrated indicator; (c) enzymatic labeling based on biotin-tagged signaling probe and avidin-HRP conjugate; and (d) enzymatic labeling based on digoxigenin-tagged signaling probe and antiDig-HRP conjugate. The strategies (c) and (d) require further addition of a mediator (HQ) and the enzyme substrate H_2O_2.

was shown to be strongly dependent on both the DNA adsorption conditions as well as the substrate on which DNA is being absorbed, thus requiring meticulous control of the DNA-adsorbed layer. Although simple, this strategy requires multisite attachment—such as adsorption—as an immobilization technique (Figure 4.1a).

The DNA recognition event for electrochemical transducing can be detected mostly by means of external electrochemical markers such as electroactive indicators (Carter et al. 1989; Erdem et al. 2001) (Figure 4.1b) or enzymes (Figure 4.1c,d). Enzyme labeling has been transferred from nonisotopic DNA classical methods to electrochemical genosensing. The enzyme labeling relies on the reaction between a small tag (usually biotin or digoxigenin-modified DNA probe) with the streptavidin (Pividori et al. 2001) (Figure 4.1c) or antidigoxigenin antibody (antiDig-HRP) (Pividori and Alegret 2003) (Figure 4.1d) enzyme conjugates, respectively. Although a second incubation step is usually required for labeling, higher sensitivity and specificity have been reported for the enzyme labeling method compared with the other reported methods (Alfonta et al. 2001; Paleček et al. 2002).

Enzyme labeling is also used in electrochemical immunosensing devices and is mostly based on primary and secondary antibodies labeled with an enzyme, usually horseradish peroxidase (HRP) or alkaline phosphatase (ALP) (Figure 4.3b,c,d) or antigen–enzyme conjugates in the case of direct competitive assay (Figure 4.3a).

The use of metal nanoparticles—especially gold nanoparticles (Dequaire et al. 2000; González-García et al. 2000; Wang et al. 2001a,b)—as labels for biosensing devices is also gaining importance.

4.3   ELECTROCHEMICAL IMMUNOSENSING FOR FOOD SAFETY

In order to design an immunosensor for food residues, the first issue is to immobilize the specific antibody onto an electrochemical transducer. The antibodies are a group of glycoproteins present in the serum and tissue fluids of all mammals. They differ in size, charge, amino acid composition, and carbohydrate content. Each immunoglobulin molecule is bifunctional. One region of the molecule is concerned with binding to antigen while a different region mediates so-called effector functions. The effector functions are biological activities such as complement activation and cell binding, which are related to sites that are distant from the antigen-binding sites (mostly in the Fc region). The basic structure of all immunoglobulin molecules is a unit consisting of two identical light polypeptide chains and two identical heavy polypeptide chains. These are linked together by disulphide bonds. The C-terminal half of the light chain (approximately 107 amino acid residues) is constant except for certain allotypic and isotypic variations and is called the constant light chain region, whereas the N-terminal half of the chain shows much sequence variability and is known as the variable light chain region.

The design of an immunosensor involves, as previously explained, the immobilization of an antibody on a proper electrochemical transducer. Different strategies have been reported for the immobilization of antiatrazine antibodies to GEC-based transducers, schematically outlined in Figure 4.2 (Zacco et al. 2006b, 2007a). The nonmodified antiatrazine antibodies can be easily immobilized on GEC electrodes by dry adsorption (Figure 4.2a). Although dry adsorption on GEC is the simplest and most

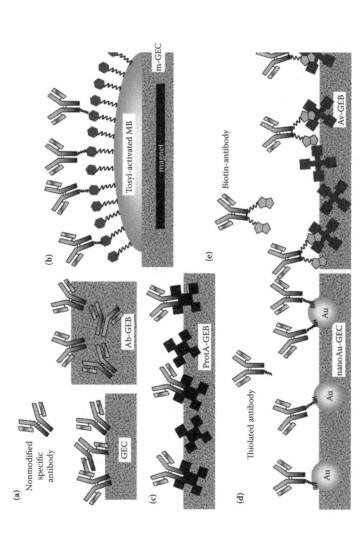

FIGURE 4.2 Schematic representation of the different strategies for the immobilization of nonmodified antibodies (a, b, and c) and chemically modified (d and e) antiatrazine antibodies: (a) physisorption of the antibodies on GEC transducer (left) and bulk-modified rigid immunocomposite (right); (b) single-point attachment based on covalent immobilization of the nonmodified antibody on tosyl-activated magnetic beads (HB) to be captured on a magnetic sensor (m-GEC); (c) oriented immobilization of the nonmodified antibody based on protein- A biocomposite (ProtA-GEB); (d) single-point immobilization of thiolated antibodies on gold nanocomposite (nanoAu-GEC); and (e) single-point immobilization of biotinylated antibodies on avidin biocomposite (Av-GEB).

easily automated procedure, it does not provide a well-oriented layer of specific anti-bodies since the antibodies are bonded through multiple sites to the GEC surface.

The antiatrazine antibodies can be directly included in the formulation of a biocomposite to obtain an Ab(antiatrazine)-GEB (Figure 4.2a). In this approach, the proper orientation of the antibody can be also an issue (Zacco et al. 2007a). The antiatrazine antibody is mostly available for the immunological reaction when included in the formulation of the immunocomposite. However, the availability of the binding site of the antibodies on the surface, after each renewal step, is not always reproducible in the immunocomposite, suggesting that the antibodies are randomly oriented and not always available in the same rate for the immunological reaction (Zacco et al. 2007a).

These multiple limitations have led to the development of alternative approaches for coupling antibodies in a controlled and oriented manner, ensuring the exposure to the aqueous environment and to the complementary sites of the target molecule.

Antiatrazine antibodies can be easily immobilized in an oriented manner on the surface of the protein A-modified transducer (protein-A graphite-epoxy biocomposite, ProtA-GEB) through the interaction between the Fc of the antibody and the protein A of the biocomposite (Figure 4.2c) (Zacco et al. 2007a).

Biotinylated antiatrazine antibodies can be easily immobilized on the surface of the avidin-modified transducer (avidin graphite-epoxy biocomposite, Av-GEB) through the avidin–biotin reaction since antibodies can be readily linked to biotin without serious effects on their biological, chemical, or physical properties (Zacco et al. 2007a) (Figure 4.2e).

The thiolated antibody (Fab'-SH fragment) obtained by mild reduction of anti-atrazine antibodies can be also immobilized on the surface of a gold nanoparticle-modified transducer (gold nanoparticle graphite-epoxy composite, nanoAu-GEC) through chemisorption, as schematically outlined in Figure 4.2d.

In all the cases explained above, a competitive immunological assay is then performed, after the modification of the transducer with the antiatrazine antibody. Briefly, the two-step experimental procedure consists of (1) competitive immunological reaction between the atrazine and the atrazine-HRP enzymatic tracer, and (2) amperometric determination based on the enzyme activity by adding H_2O_2 and using hydroquinone as a mediator.

Finally, different strategies for the attachment of specific antiatrazine antibod-ies on magnetic beads are reported (Zacco et al. 2006b): (1) covalent attachment based on carboxylated magnetic nanoparticles; (2) covalent attachment based on tosyl-activated magnetic beads (Figure 4.2b); (3) affinity interaction with protein-A modified magnetic beads. Although all the strategies for the immobilization of anti-atrazine antibodies are useful, the better immobilization performance is achieved with tosyl-activated magnetic beads.

Instead of the direct modification of the electrode surface, the biological reactions (as the immobilization and the competitive immunological step, Figure 4.3a) and the washing steps can be successfully performed on HB. After the modifications, the HB can be easily captured by magnetic forces onto the surface of m-GECs (Zacco et al. 2006b). Once immobilized on m-GEC, the immunological reaction performed on the magnetic beads can be electrochemically revealed using enzymatic labeling.

The performance of the electrochemical immunosensing strategy was successfully evaluated using spiked real orange juice samples. The competitive electrochemical

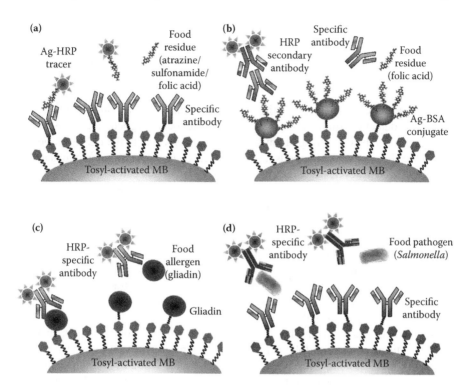

FIGURE 4.3 Schematic representation of different immunoassay formats performed on tosyl-activated magnetic beads: (a) direct competitive assay for the detection of small residues (such as atrazine, sulfonamide, folic acid); (b) indirect competitive assay with the conjugate between an inert protein (such as BSA) and the small hapten immobilized on magnetic bead for the detection of food residues (such as folic acid); (c) direct competitive assay with the antigen directly immobilized on the magnetic bead for the detection of the allergen gliadin; and (d) sandwich assay for the detection of pathogenic bacteria.

immunosensing strategy (Figure 4.3a) can easily reach the required LOD for potable water orange juice (MRL 0.1 µg L^{-1}) established by the EC directives with simple sample pretreatments. The orange juice samples spiked with atrazine are first adjusted to pH 7.5, diluted 1.5 with phosphate-buffered saline with Tween20 (PBST), and filtered through 0.22 µm filter before measurement.

A similar strategy can be performed for the immunological determination of sulfonamides in raw full cream, as well as in all varieties of ultrahigh temperature (UHT) milk, such as full cream (about 3.25% fat), semiskimmed (about 1.5–1.8% fat), and skimmed (0.1% fat), by using class-specific antisulfonamides (anti-SFM) antibodies immobilized on magnetic beads and an SFM-HRP tracer for the electrochemical detection (Zacco et al. 2007b) (Figure 4.3a). Different strategies for the attachment of specific anti-SFM antibodies on different magnetic beads have been performed: (1) covalent attachment based on carboxylated magnetic nanoparticles; (2) covalent attachment based on tosyl-activated magnetic beads (Figure 4.2b); (3) affinity interaction with protein-A modified magnetic beads. Although all the strategies for the immobilization of anti-SFM antibodies are useful, the better immo-bilization performance is achieved with tosyl-modified magnetic beads, as in the case of atrazine determination. The raw full cream samples spiked with sulfonamide are only diluted 1.5 with PBST, while all the UHT samples are processed without any treatment.

As the European legislation fix the MRL of sulfonamide in a value of 100 µg kg^{-1} (ppb), the detection limit of the novel electrochemical magneto immunosensing strategy allows the measuring of any kind of milk according to the EC legislation.

All the above reported methods rely on a direct competitive immunological assay (Figure 4.3a). However, instead of the antibody, the antigen can be immobilized in the electrochemical transducer, as schematically outlined in Figure 4.3b,c. In this case, both indirect or direct competitive assays can be performed.

An immunoassay with electrochemical detection based on magneto sensors for folic acid in fortified milk has been developed (Lermo et al. 2009). Covalent cou-pling of bovine serum albumin (BSA)–folic acid conjugate is successfully performed on tosyl-modified magnetic beads, achieving an excellent coupling efficiency. The indirect competitive electrochemical magneto immunoassay strategy (Figure 4.3b) can easily reach the folic acid values in fortified food. In the case of fortified milk the folic acid content is usually near 300 µg L^{-1}.

All the above methods are reported for small molecules (food residues, such as atrazine, sulfonamide, and folic acid). If the food residue is a protein, such as the allergen gliadin, a competitive assay can be also performed by the direct immo-bilization of the antigenic protein on tosyl-modified magnetic beads, achieving an excellent coupling efficiency. The direct competitive electrochemical magneto immunoassay strategy (Figure 4.3c) can easily reach the gliadin value of 20 mg L^{-1} to ensure food safety (free of gluten) for celiac patients (Laube et al. 2009).

If the contaminant is a bacteria (with multiple epitopes) instead of a competi-tive, a sandwich assay can be performed. A simple and rapid method for the detec-tion of *Salmonella* in milk is reported (Liébana et al. 2009b). In this approach, the bacteria is captured and preconcentrated from milk samples with magnetic beads by immunological reaction with the specific antibody against *Salmonella* (Figure 4.3d). A second polyclonal antibody labeled with HRP is used as sero-logical confirmation with electrochemical detection based on a magneto electrode. Between the different procedures, better performance is obtained with the immu-nological reactions in one-step. A limit of detection of 7.5 10^3 cfu mL^{-1} in milk is obtained in 50 min without any pretreatment. If the skimmed milk is preenriched for 6 hours, the method is able to detect as low as 1.4 cfu mL^{-1}, while if it is preen-riched for 8 hours, as low as 0.108 cfu mL^{-1} (2.7 cfu in 25 g of milk, in 5 samples of 5 mL) are detected (Liébana et al. 2009b). Moreover, the method is able to clearly distinguish between food pathogenic bacteria such as *Salmonella* and *E. coli*.

4.4 ELECTROCHEMICAL GENOSENSING FOR FOOD SAFETY

Electrochemical genosensing is mainly used in food safety for the detection of pathogenic bacteria by means of its genetic information. Furthermore, the PCR can be easily coupled to enhance the sensitivity of nucleic acid–based assays. Target nucleic segments of defined length and sequence are amplified by PCR by repetitive cycles of strand denaturation, annealing, and extension of oligonucleotide primers by the (Taq) DNA polymerase. Nucleic acid–based detection coupled with PCR has distinct advantages over culture and other standard methods for the detection of microbial pathogens such as specificity, sensitivity, rapidity, accuracy, and capacity to detect small amounts of target nucleic acid in a sample. Moreover, multiple primers can be used to detect different pathogens in one multiplex reaction.

The design of a genosensor device involves the immobilization of a DNA probe or double-stranded DNA (dsDNA) amplicon on a proper electrochemical transducer. Different strategies have been reported for the immobilization of DNA to GEC-based transducers, schematically outlined in Figure 4.4.

The common method for the multisite physical adsorption of DNA and oligonucleotides on carbonaceous-based material can be classified in dry or wet adsorption (Figure 4.4a). Dry adsorption relies on leaving DNA to dry on the carbonaceous surface. DNA can adopt a variety of conformations depending on the degree of hydration. As an example, the most familiar double helix DNA—called B-DNA—can become the A-DNA form if it is strongly dehydrated. When the DNA solution is evaporated to dryness, the bases of DNA that have been dehydrated are exposed, thus the hydrophobic bases are strongly adsorbed flat on the electrode surfaces. Once it is adsorbed, DNA is difficult to rehydrate. Hence, DNA is not desorbed, no matter how long the adsorbed DNA is soaked in water, characteristic of irreversible adsorption. Once immobilized on GEC, DNA preserves its unique hybridization properties, which can be revealed using different strategies based on both enzymatic labeling (Figure 4.1c,d) (Pividori et al. 2003; Pividori and Alegret 2003) and the intrinsic signal coming from the DNA oxidation (Erdem et al. 2004) (Figure 4.1a). If the PCR product—or any other double-stranded DNA—is directly adsorbed on a GEC transducer, a denaturing alkaline procedure after the DNA dry-adsorption is mandatory to break the hydrogen bonds linking the complementary DNA strands in order to ensure the proper hybridization with the signaling probe (Pividori et al. 2003). This strategy is able to electrochemically detect the PCR amplicon coming from *Salmonella* spp. in a simple and inexpensive way (Pividori et al. 2003).

Besides the strategy in which the DNA target can be easily attached and detected by its complementary DNA signaling probe, a sandwich assay in which the DNA target is in solution can be easily performed by a double hybridization with a capture and a signaling probe, respectively (Pividori and Alegret 2005). This strategy was demonstrated to be useful for the detection a novel determinant of β-lactamase resistance in *S. aureus* using one- and two-step capture format. When compared with other reported designs, the genosensor based on dry adsorption is simpler and cheaper, showing detection limits of the same order of magnitude. The procedures based on previous activation/modification of the surface transducer and subsequent immobilization, as well as some blocking and washing steps that are tedious, expensive, and time-consuming, are avoided using GEC as electrochemical platform.

The direct adsorption of biomolecules on the GEC transducer has the main advantage of being simple and applicable to almost any type of bioreceptor. However, the adsorption procedure presents several limitations, mainly the absence of binding specificity as well as the lack of control in the orientation of molecules bound to the surface.

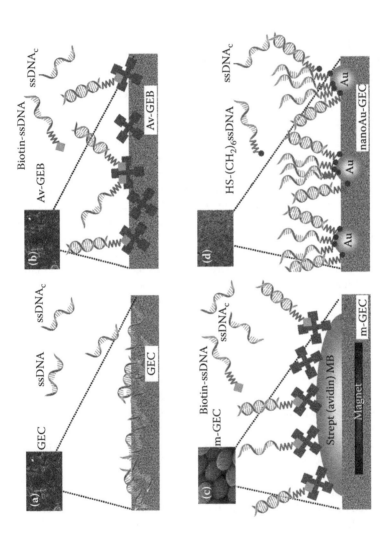

FIGURE 4.4 Schematic representation of the different strategies for DNA immobilization (related with pathogenic bacteria) in genosensing devices based on GEC composites, biocomposites, and nanocomposites: (a) dry or wet multisite adsorption on GEC; (b) avidin-biotin linkage on Av-GEB; (c) (strept)avidin-biotin linkage on magnetic beads captured on m-GEC; and (d) chemisorption on nanoAu-GEC.

A DNA probe can be easily immobilized in an oriented way on the surface of the avidin-modified transducer through the avidin–biotin reaction, since both nucleic acids as well as short oligonucleotides can be readily linked to biotin without serious effects on their biological, chemical, or physical properties. Biotinylated DNA can be firmly single-point attached in Av-GEB (Figure 4.4b). In this case, a capture format is used in which the immobilization of the biotinylated probe together with the hybridization with a digoxigenin (Dig) signaling probe is performed in a one-step procedure (Williams et al. 2003). The hybridization can be revealed using enzymatic labeling (Figure 4.1d). The utility of Av-GEB platform is demonstrated for the determination of the mecA DNA sequence related to methicillin-resistant *Staphylococcus aureus* (MRSA) (Williams et al. 2003).

We have designed a double-tagged PCR assay in order to achieve not only the amplification but also the labeling of the amplicon of pathogenic bacteria (Figure 4.5)

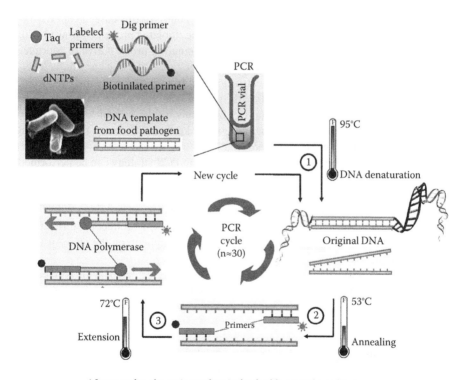

After n cycles, the main product is the double-tagged amplicon:

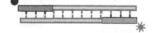

FIGURE 4.5 Schematic representation of the double-tagging PCR strategy, in order to obtain a double-tagged amplicon labeled with both biotin and digoxigenin tags, from a genome template of the *Salmonella* and *E. coli*, for food safety. The same approach can be performed with thiolated and digoxigenin tags for the amplification of the *Salmonella* genome.

for further immobilization and electrochemical detection. To achieve this task, a different set of primers can be used with different tags, such as -biotin, -digoxigenin, -SH.

The rapid electrochemical verification of the amplicon coming from the *E. coli* O157:H7 genome is performed by double tagging the amplicon during PCR with a set of two labeled PCR primers—one of them with biotin and the other one with digoxigenin (Lermo et al. 2008) (Figure 4.5). During PCR, not only the amplification of the *E. coli* is achieved but also the double tagging of the amplicon ends with (1) the biotinylated capture primer to achieve the immobilization on a biosensor based on an Av-GEB transducer (Figure 4.4b) and (2) the digoxigenin signaling primer to achieve the electrochemical detection (Figure 4.1d) (Lermo et al. 2008).

Thiolated DNA can be also easily attached on gold-based transducers, such as nanoAu-GEC, as schematically shown in Figure 4.4d. Hybridization efficiency is expected to be higher on the edging of the gold nanoparticles surrounded by nonreactive GEC (Figure 4.4d) (Oliveira Marques et al. 2009). In this case, a capture format can be used in which the immobilization of the thiolated probe is followed with the hybridization with a digoxigenin (or biotin) signaling probe (Oliveira Marques et al. 2009) (Figure 4.6b1). The hybridization can be revealed using enzymatic labeling (Figure 4.6c). Regarding other electrochemical transducers previously reported, such as an Av-GEB (Lermo et al. 2008), the main advantages of the inorganic nano-Au-GEC electrode compared with the biocomposite are the lack of loss of activity and that the latter requires the temperature to be kept at 4°C due to the biological nature of the modifier, the protein avidin.

Moreover, and for the first time, a double-tagging PCR strategy can be performed with a thiolated primer for the detection of *Salmonella* spp. (Oliveira Marques et al. 2009) (Figure 4.5). The rapid electrochemical verification of the amplicon coming from the pathogenic genome of *Salmonella* performed by PCR with a set of two labeled primers is an easy way for the thiolation of the PCR product (Figure 4.6b2). The thiolated end allowed the immobilization of the amplicon on the nanoAu-GEC electrode in an easy way.

With this strategy, as low as 200 fmol can be easily detected, with an electrochemical signal of almost 3 μA. This double-tagging PCR strategy opens new routes not only for immobilization purposes but also for ease in labeling with gold or quantum dots during PCR.

Instead of the direct modification of the electrode surface, the biological reactions (as immobilization, hybridization, enzymatic labeling) and the washing steps can be successfully performed on magnetic beads. After the modifications, the magnetic beads can be easily captured by magnetic forces onto the surface of m-GECs (Figure 4.4c).

Once immobilized on m-GEC, the hybridization performed on the magnetic beads can be electrochemically revealed using different strategies based on both enzymatic labeling (Figure 4.1d) and the intrinsic signal coming from the DNA oxidation (Figure 4.1a). A single-point immobilization procedure on the magnetic meads based on streptavidin–biotin interaction can be also performed. Biotinylated DNA can be firmly attached on streptavidin-modified magnetic beads in that way

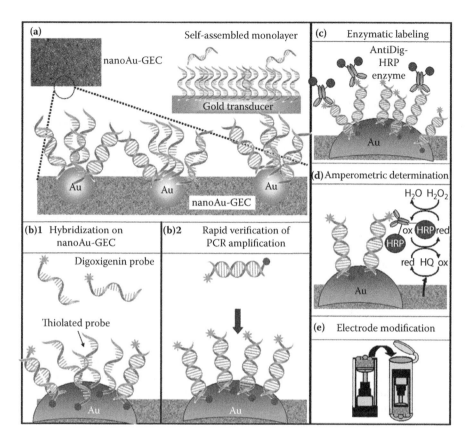

FIGURE 4.6 Schematic representation of (a) nanoAu-GEC material showing isolated gold nanoparticles able to produce "bioactive chemisorbing islands" instead of SAMs on a continuous layer of gold. (b1) Hybridization assay on the nanoAu-GEC electrode. (b2) Rapid electrochemical verification of thiolated and double-tagged amplicons on the nanoAu-GEC electrode. Parts (c–e) are common steps (electrode modification, enzymatic labeling, and amperometric determination) for both parts (b1) and (b2). (From Oliveira Marques, P.R.B., A. Lermo, S. Campoy et al., 2009, *Analytical Chemistry* 81:1332–1339. With permission.)

(Figure 4.4c) (Lermo et al. 2007). When the electrochemical detection is based on enzymatic activity determination, a capture format is used in which the immobilization of the biotinylated probe together with the hybridization with the digoxigenin signaling probe is performed in a one-step procedure (Lermo et al. 2007). The hybridization can be revealed using enzymatic labeling (Figure 4.1d).

The procedure for electrochemical DNA biosensing based on magnetic beads can be also used for the detection of *IS200* element specific for *Salmonella* spp.

This new electrochemical genomagnetic strategy using magneto electrodes in connection with magnetic particles offers many potential advantages compared with more traditional strategies for detecting DNA.

This strategy takes advantage of working with magnetic beads, such as improved and more effective biological reactions, washing steps, and magnetic separation after each step. This electrochemical genomagnetic assay provides more sensitive, rapid, and cheaper detection than other assays previously reported. This sensitivity of the GEC with respect to other electrochemical transducers and the selectivity conferred by the magnetic separation are also used for the detection of PCR amplicons coming from real samples.

The rapid electrochemical verification of the amplicon coming from the *Salmonella IS200* element (Lermo et al. 2007) as well as the eaeA gene, related with *E. coli* O157:H7 (Lermo et al. 2008), is performed by double labeling the amplicon during PCR with a set of two labeled PCR primers —one of them with biotin and the other one with digoxigenin, as schematically outlined in Figure 4.5. During PCR, not only the amplification of the bacteria genome is achieved but also the double labeling of the amplicon ends with (1) the biotinylated capture primer to achieve the immobilization on the streptavidin-modified magnetic bead (Figure 4.4c) and (2) the digoxigenin signaling primer to achieve the electrochemical detection (Figure 4.1d). Beside this double-labeling PCR strategy, a single-labeling PCR strategy with a further confirmation of the amplicon by its hybridization is achieved by performing the PCR with the biotin primer and a further hybridization step with a digoxigenin probe (Lermo et al. 2007). Moreover, a PCR reactor for real-time electrochemical detection is also reported (Lermo et al. 2007). In this case, the amplification and double labeling are performed directly on the streptavidin magnetic beads by using *magnetic bead primers* (Lermo et al. 2007). The rapid and sensitive verification of the PCR amplicon related with *Salmonella* can be achieved with 2.8 fmol of amplified product (Lermo et al. 2007). In the case of *E. coli*, the assay shows to be more sensitive compared with Q-PCR strategies based on fluorescent labels such as TaqMan probes (Lermo et al. 2007).

4.5 ELECTROCHEMICAL BIOSENSING APPROACHES COMBINING BOTH IMMUNOLOGICAL AND GENETIC INFORMATION FOR FOOD SAFETY

The double-tagged PCR strategy with electrochemical genosensing can be also combined with an immunoseparation step of the bacteria to improve the LOD for detecting pathogenic bacteria (Liébana et al. 2009a). The procedure consists briefly of the following steps, as schematically outlined in Figure 4.7: (1) immunomagnetic separation (iMS) of the bacteria from food samples (Figure 4.7a); (2) lysis of the bacteria and DNA separation (Figure 4.7b); (3) DNA amplification and double labeling of *Salmonella IS200* insertion sequence (Figure 4.7c, see also Figure 4.5 for details); (4) immobilization of the doubly labeled amplicon in which the biotin extreme of the dsDNA amplicon is immobilized on the streptavidin magnetic beads (Figure 4.7d); (5) enzymatic labeling using as enzyme label the antibody antiDIG-HRP capable of bonding the other labeled extreme of the dsDNA amplicon (Figure 4.7d); (6) magnetic capture of the modified magnetic particles (Figure 4.7d); and (7) amperometric determination (Figure 4.7d) (Liébana et al. 2009a).

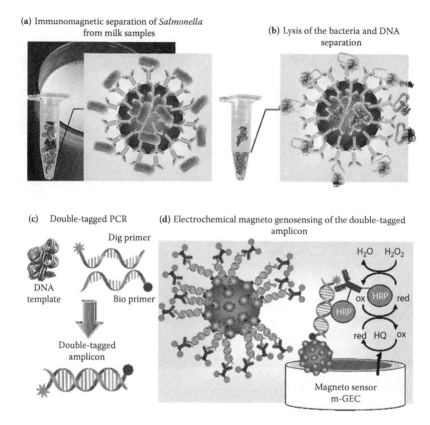

(a) Immunomagnetic separation of *Salmonella* from milk samples

(b) Lysis of the bacteria and DNA separation

(c) Double-tagged PCR

Dig primer

DNA template

Bio primer

Double-tagged amplicon

(d) Electrochemical magneto genosensing of the double-tagged amplicon

H_2O H_2O_2

ox HRP red

HRP

red HQ ox

Magneto sensor m-GEC

FIGURE 4.7 Schematic representation of the IMS/double-tagging PCR/m-GEC electrochemical genosensing approach. (From Liébana, S. et al., 2009a, *Analytical Chemistry* 81:5812-5820. With permission.)

In this approach, the bacteria are captured and preconcentrated from food samples with magnetic beads by immunological reaction with the specific antibody against *Salmonella*, as shown in the scanning electrode microscopy (SEM) images in Figure 4.8.

After the lysis of the captured bacteria, further amplification of the genetic material by PCR with a double-tagging set of primers is performed to confirm the identity of the bacteria. Both steps are rapid alternatives to the time-consuming classical selective enrichment and biochemical/serological tests. The double-tagged amplicon is then detected by electrochemical magneto genosensing using m-GEC electrodes. The "IMS/double-tagging PCR/m-GEC electrochemical genosensing" approach can be used for the sensitive detection of *Salmonella* artificially inoculated into skim milk samples. A limit of detection of 1 cfu mL^{-1} is obtained in 3.5 h without any pretreatment, in Luria-Bertani (LB) broth and in milk diluted 1/10 in LB. When the skim milk is preenriched for 6 h, the method is able to feasibly detect as low as 0.04 cfu mL^{-1} (1 cfu in 25 g of milk) with a signal-to-background ratio of 20 (Liébana et al. 2009a).

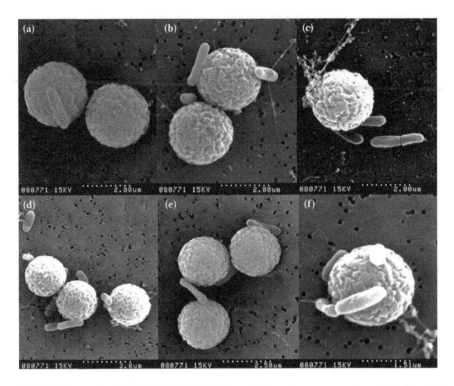

FIGURE 4.8 Evaluation of the IMS by SEM at a *Salmonella* concentration of 10^4 cfu mL^{-1}. The images show the *Salmonella* cells attached to the magnetic beads. In all cases, identical acceleration voltage (15 KV) is used. (From Liébana, S. et al., 2009a, *Analytical Chemistry* 81:5812-5820. With permission.)

Interestingly, the specificity of this approach is conferred by both the antibody in the IMS and the set of primer during the double-tagging PCR, in this case for detecting *Salmonella* spp. The same approach could be also designed for detecting different *Salmonella* or *E. coli* serotypes by selecting a specific pair of primers or antibody.

4.6 CONCLUSION

The converging of technologies such as nanotechnology and biotechnology are opening new horizons in electrochemical biosensors. The integration of micro and nanostructured materials in biosensing devices (such as graphite microparticles, bioreceptors, gold nanoparticles, magnetic micro- or nanoparticles) provides excellent analytical performances for the detection of contaminants affecting food safety. One of the key contributions of these technologies in the electrochemical biosensing field relies on the design of novel transducers, not only with enhanced transducing features but also with improved immobilization of biomolecules while retaining their biological activity. Future trends are focused on the integration of materials and procedures for the design of powerful nano-bio-magneto-electro-analytical systems.

REFERENCES

Akerstrom, B., T. Brodin, K. Reis, and L. Bjorck. 1985. Protein-G: A powerful tool for binding and detection of monoclonal and polyclonal antibodies. *Journal of Immunology* 135:2589–2592.

Alegret, S. 1996. Rigid carbon–polymer biocomposites for electrochemical sensing: A review. *Analyst* 121:1751–1758.

Alfonta, L., A.K. Singh, and I. Willner. 2001. Liposomes labelled with biotin and horseradish peroxidase: A probe for the enhanced amplification of antigen– antibody or oligonucleotide–DNA sensing processes by the precipitation of an insoluble product on electrodes. *Analytical Chemistry* 73:91–102.

Andreescu, S., and J.-L. Marty. 2006. Twenty years research in cholinesterase biosensors: From basic research to practical applications. *Biomolecular Engineering* 23:1–15.

Armistead, P.M., and H.H. Thorp. 2000. Modification of indium tin oxide electrodes with nucleic acids: Detection of attomole quantities of immobilized DNA by electrocatalysis. *Analytical Chemistry* 72:3764–770.

Au, A.M. 2003. Pesticides and herbicides: Types, uses, and determination of herbicides. In *Encyclopedia of food science and nutrition*, ed. B. Caballero, L. Trugo, and P.H. Fimglas, 4483–4487. New York: Academic Press.

Baeumner, A.J. 2003. Biosensors for environmental pollutants and food contaminants. *Analytical and Bioanalytical Chemistry* 377:434–445.

Barlen, B., S.D. Mazumdar, O. Lezrich, P. Kämpfer, and M. Keusgen. 2007. Detection of *Salmonella* by surface plasmon resonance. *Sensors* 7:1427–1446.

Bergwerff, A.A., and J. Schloesser. 2003. Antibiotics and drugs: Residue determination. In *Encyclopedia of food science and nutrition*, ed. B. Caballero, L. Trugo, and P.H. Fimglas, 254–261. New York: Academic Press.

Bowtell, D.D.L. 1999. Options available—From start to finish—For obtaining expression data by microarray. *Nature Genetic Supplement* 21:25–32.

Cacciatore, G., M. Petz, S. Rachid, R. Hakenbeck, and A. Bergwerff. 2004. Development of an optical biosensor assay for detection of β-lactam antibiotics in milk using the penicillin-binding protein 2x. *Analytica Chimica Acta* 520:105–115.

Cai, H., C. Xu, P. He, and Y. Fang. 2001. Colloid Au-enhanced DNA immobilization for the electrochemical detection of sequence-specific DNA. *Journal of Electroanalytical Chemistry* 510:78–85.

Carpini, G., F. Lucarelli, G. Marrazza, and M. Mascini. 2004. Oligonucleotide-modified screen-printed gold electrodes for enzyme-amplified sensing of nucleic acids. *Biosensors and Bioelectronics* 20:167–175.

Carter, M.T., M. Rodriguez, and A.L. Bard. 1989. Voltammetric studies of the interaction of metal chelates with DNA. 2: Tris-chelated complexes of cobalt(III) and iron(II) with 1,10-phenanthroline and 2,2%-bipyridine. *Journal of the American Chemical Society* 111:8901–8911.

Cassidy, J.F., A.P. Doherty, and J.G. Vos. 1998. Amperometric methods of detection. In *Principles of Chemical and Biological Sensors*, ed. D. Diamond, Chapter 3. New York: John Wiley & Sons.

Chu, F.S. 2003. Immunoassays: Radioimmunoassay and enzyme immunoassay. In *Encyclopedia of food science and nutrition*, ed. B. Caballero, L. Trugo, and P.H. Fimglas, 3248–3255. New York: Academic Press.

Compton, B.J., M. Lewis, F. Whigham, J.S. Gerald, and G.E. Countryman. 1989. Analytical potential of protein-A for affinity-chromatography of polyclonal and monoclonal-antibodies. *Analytical Chemistry* 61:1314–1317.

Croci, L., E. Delibato, G. Volpe, and G. Palleschi. 2001. A rapid electrochemical ELISA for the detection of *Salmonella* in meat samples. *Analytical Letters* 34:2597–2607.

Croci, L., E. Delibato, G. Volpe, D. De Medici, and G. Palleschi. 2004. Comparison of PCR, electrochemical enzyme-linked immunosorbent assays, and culture method for detecting *Salmonella* in meat products. *Applied and Environmental Microbiology* 70:1393–1396.

Da Silva, S., L. Grosjean, N. Ternan, P. Mailley, T. Livache, and S. Cosnier. 2004. Biotinylated polypyrrole films: An easy electrochemical approach for the reagentless immobilization of bacteria on electrode surfaces. *Bioelectrochemistry* 63:297–301.

Darain, F., S.-U. Park, and Y.-B. Shim. 2003. Disposable amperometric immunosensor system for rabbit IgG using a conducting polymer modified screen-printed electrode. *Biosensors and Bioelectronics* 18:773–780.

Deisingh, A.K., and M. Thompson. 2004. Biosensors for the detection of bacteria. *Canadian Journal of Microbiology* 50:69–77.

Dequaire, M., C. Degrand, and B. Limoges. 2000. An electrochemical metalloimmunoassay based on a colloidal gold label. *Analytical Chemistry* 72:5521–5528.

Erdem, A., K. Kerman, B. Meric, and M. Ozsoz. 2001. Methylene blue as a novel electrochemical hybridization indicator. *Electroanalysis* 13:219–223.

Erdem, A., M.I. Pividori, M. del Valle, and S. Alegret. 2004. Rigid carbon composites: A new transducing material for label-free electrochemical genosensing. *Journal of Electroanalytical Chemistry* 567:29–37.

Ferguson, J., A. Baxter, P. Young et al. 2005. Detection of chloramphenicol and chloramphenicol glucuronide residues in poultry muscle, honey, prawn and milk using a surface plasmon resonance biosensor and Qflex® kit chloramphenicol. *Analytica Chimica Acta* 529:109–113.

Fojta, M., and E. Paleček. 1997. Supercoiled DNA modified mercury electrode: A highly sensitive tool for the detection of DNA damage. *Analytica Chimica Acta* 342:1–12.

Gaudin, V., J. Fontaine, and P. Maris. 2001. Screening of penicillin residues in milk by a surface plasmon resonance-based biosensor assay: Comparison of chemical and enzymatic sample pre-treatment. *Analytica Chimica Acta* 436:191–198.

Gehring, A.G., C.G. Crawford, R.S. Mazenko, L.J. Van Houten., and J.D. Brewster. 1996. Enzyme-linked immunomagnetic electrochemical detection of *Salmonella typhimurium*. *Journal of Immunological Methods* 195:15–25.

González-García, M.B., C. Fernández-Sánchez, and A. Costa-García. 2000. Colloidal gold as an electrochemical label of streptavidin–biotin interaction. *Biosensors and Bioelectronics* 15:315–321.

Green, N.M. 1990. Avidin and streptavidin. *Methods in enzymology* 184:51–67.

Gustavsson, E., P. Bjurling, and A. Sternesjö. 2002. Biosensor analysis of penicillin G in milk based on the inhibition of carboxypeptidase activity. *Analytica Chimica Acta* 468:153–159.

Haasnoot, W., M. Bienenmann-Ploum, U. Lamminmäki, M. Swanenburg, and H. van Rhijn. 2005. Application of a multi-sulfonamide biosensor immunoassay for the detection of sulfadiazine and sulfamethoxazole residues in broiler serum and its use as a predictor of the levels in edible tissue. *Analytica Chimica Acta* 552:87–95.

Hashimoto, K., K. Ito, and Y. Ishimori. 1994. Novel DNA sensor for electrochemical gene detection. *Analytica Chimica Acta* 286:219–224.

Haukanes, B.I., and C. Kvam, 1993. Application of magnetic beads in bioassays. *Biotechnology* 11:60–63.

Humphrey, T., and P. Stephens. 2003. *Salmonella* setection. In *Encyclopedia of food science and nutrition*, ed. B. Caballero, L. Trugo, and P.H. Fimglas, 5079–5084. New York: Academic Press.

Ivnitski, D., I. Abdel-Hamid, P. Atanasov, and E. Wilkins. 1999. Biosensors for detection of pathogenic bacteria. *Biosensors and Bioelectronics* 14:599–624.

Ivnitski, D., I. Abdel-Hamid, P. Atanasov, E. Wilkins, and S. Stricker. 2000. Application of electrochemical biosensors for detection of food pathogenic bacteria. *Electroanalysis* 12:317–325.

Jelen, F., B. Yosypchuk, A. Kourilová, L. Novotný, and E. Paleček. 2002. Label-free determination of picogram quantities of DNA by stripping voltammetry with solid copper amalgam mercury in the presence of copper. *Analytical Chemistry* 74:4788–4793.

Jones, M.L., and G.P. Kurzban. 1995. Noncooperativity of biotin binding to tetrameric streptavidin. *Biochemistry* 34:11750–11756.

Jue, R., J.M. Lambert, L.R. Pierce, and R.R. Traut. 1978. Addition of sulfhydryl groups to *Escherichia coli* ribosomes by protein modification with 2-iminothiolane (methyl 4-mercaptobutyrimidate). *Biochemistry* 17:5399–5405.

Laube, T., S. Alegret, and M.I. Pividori. 2009. Gliadin for food safety of celiac patients. PhD diss., Universitat Autònoma de Barcelona.

Leonard, P., P. Hearty, J. Brennan et al. 2003. Advances in biosensors for detection of pathogens in food and water. *Enzyme and Microbial Technology* 32:3–13.

Lermo, A., S. Campoy, J. Barbé, S. Hernández S. Alegret, and M.I. Pividori. 2007. *In situ* DNA amplification with magnetic primers for the electrochemical detection of food pathogens. *Biosensors and Bioelectronics* 22:2010–2017.

Lermo, A., S. Fabiano, S. Hernández et al. 2009. Rapid electrochemical magneto immunosensing of folic acid in vitamin-fortified food products. *Biosensors and Bioelectronics* 24:2057–2063.

Lermo, A., E. Zacco, J. Barak et al. 2008. Towards Q-PCR of pathogenic bacteria with improved electrochemical double-tagged genosensing detection. *Biosensors and Bioelectronics* 23:1805–1811.

Liébana, S., A. Lermo, S. Campoy, J. Barbé, S. Alegret, and M.I. Pividori. 2009a. Magneto immunoseparation of pathogenic bacteria and electrochemical magneto genosensing of the double-tagged amplicon. *Analytical Chemistry* 81:5812–5820.

Liébana, S., A. Lermo, S. Campoy, M.P. Cortés, S. Alegret, and M.I. Pividori. 2009b. Rapid detection of *Salmonella* in milk by electrochemical magneto immunosensing. *Biosensors and Bioelectronics* 25:510–513.

Limoges, B., J.-M. Savéant, and D. Yazidi. 2003. Quantitative analysis of catalysis and inhibition at horseradish peroxidase monolayers immobilized on an electrode surface. *Journal of the American Chemical Society* 125:9192–9203.

Luk, J.M., U. Kongmuang, R.S.W. Tsang, and A.A. Lindberg. 1997. An enzyme-linked immunoadsorbent assay to detect PCR products of the rfbS gene from serogroup D salmonellae: A rapid screening prototype. *Journal of Clinical Microbiology* 35:714–718.

Mauriz, E., A. Calle, L.M. Lechuga, J. Quintana, A. Montoya, and J.J. Manclús. 2006. Real-time detection of chlorpyrifos at part per trillion levels in ground, surface and drinking water samples by a portable surface plasmon resonance immunosensor. *Analytica Chimica Acta* 561:40–47.

Mehervar, M., and M. Abdi. 2004. Recent development, characteristics, and potential applications of electrochemical biosensors. *Analytical Sciences* 20:1113–1126.

Mellgren, C., A. Sternesjo, P. Hammer, G. Suhren, L. Bjorck, and W. Heeschen. 1996. Comparison of biosensor, microbiological, immunochemical and physical methods for detection of sulfamethazine residues in raw milk. *Journal of Food Protection* 59:1223–1226.

Mello, L.D., and L.T. Kubota. 2002. Review of the use of biosensors as analytical tools in the food and drink industries. *Food Chemistry* 77:237–256.

Millan, K.M., and S.K. Mikkelsen. 1993. Sequence-selective biosensor for DNA based on electroactive hybridization indicators. *Analytical Chemistry* 65:2317–2323.

Millan, K.M., A. Saraullo, and S.K. Mikkelsen. 1994. Voltammetric DNA biosensor for cystic fibrosis based on a modified carbon paste electrode. *Analytical Chemistry* 66:2943–2948.

Moser, I., T. Schalkhammer, F. Pittner, and G. Urban. 1997. Surface techniques for an electrochemical DNA biosensor. *Biosensors and Bioelectronics* 12:729–37.

Mullett, W.M., E.P.C. Lai, and J.M. Yeung. 2000. Surface plasmon resonance-based immuno-assays. *Methods* 22:77–91.

Nakamura, H., and I. Karube. 2003. Current research activity in biosensors. *Analytical and Bioanalytical Chemistry* 377:146–168.

Nawaz, S. 2003. Pesticides and herbicides: Residue determination. In *Encyclopedia of food science and nutrition*, ed. B. Caballero, L. Trugo, and P.H. Fimglas, 4487–4493. New York: Academic Press.

Nuzzo, R.G., and D.L. Allara. 1983. Adsorption of bifunctional organic disulfides on gold surfaces. *Journal of the American Chemical Society* 105:4481–4483.

Oliveira Brett, A.M., S.H.P. Serrano, I. Gutz, M.A. La-Scalea, and M.L. Cruz. 1997. Voltammetric behavior of nitroimidazoles at a DNA-biosensor. *Electroanalysis* 9:1132–1137.

Oliveira Marques, P.R.B., A. Lermo, S. Campoy et al. 2009. Double-tagging polymerase chain reaction with a thiolated primer and electrochemical genosensing based on gold nano-composite sensor for food safety. *Analytical Chemistry* 81:1332–1339.

Paleček, E. 1958. Oscillographic polarography of nucleic acids and their buildingblocks, *Naturwiss* 45:186–187.

Paleček, E. 1960. Oscillographic polarography of highly polymerized deoxyribonucleic acid. *Nature* 188:656–657.

Paleček, E., R. Kizek, L. Havran, S. Billova, and M. Fotja. 2002. Electrochemical enzyme-linked immunoassay in a DNA hybridization sensor. *Analytica Chimica Acta* 469:73–83.

Pang, D.W., and H.D. Abruña. 1998. Micromethod for the investigation of the interactions between DNA and redox-active molecules. *Analytical Chemistry* 70:3162–3169.

Patel, P.D. 2002. (Bio)sensors for measurements of analytes implicated in food safety: A review. *TrAC Trends in Analytical Chemistry* 21:96–115.

Pingarrón, J.M., P. Yáñez-Sedeño, and A. González-Cortés. 2008. Gold nanoparticle-based electrochemical biosensors. *Electrochimica Acta* 53:5848–5866.

Pividori, M.I., and S. Alegret. 2003. Graphite-epoxy platforms for electrochemical genosensing. *Analytical Letters* 36:1669–1695.

Pividori, M.I., and S. Alegret. 2005. Electrochemical genosensing based on rigid carbon composites. *Analytical Letters* 38:2541–2565.

Pividori, M.I., A. Merkoçi, and S. Alegret. 2000. Electrochemical genosensor design: Immobilization of oligonucleotides onto transducer surfaces and detection methods. *Biosensors and Bioelectronics* 15:291–303.

Pividori, M.I., A. Merkoçi, and S. Alegret. 2001. Classical dot-blot format implemented as an amperometric hybridisation genosensor. *Biosensors and Bioelectronics* 16:1133–1142.

Pividori, M.I., A. Merkoçi, J. Barbé, and S. Alegret. 2003. PCR-genosensor rapid test for detecting *Salmonella*. *Electroanalysis* 15:1815–1823.

Rajesh, A.T., and D. Kumar. 2009. Recent progress in the development of nano-structured conducting polymers/nanocomposites for sensor applications. *Sensors and Actuators B: Chemical* 136:275–86.

Roberts, D. 2003. Food poisoning. In *Encyclopedia of food science and nutrition*, ed. B. Caballero, L. Trugo, and P.H. Fimglas, 2654–2658. New York: Academic Press.

Rooney, R., and P.G. Wall. 2003. Food safety. In *Encyclopedia of food science and nutrition*, ed. B. Caballero, L. Trugo, and P.H. Fimglas, 2682–2688. Academic Press.

Sanders, G.H.W., and A. Manz. 2000. Chip-based microsystems for genomic and proteomic analysis. *TrAC Trends in Analytical Chemistry* 19:364–378.

Shimomura, M., Y. Nomura, W. Zhang et al. 2001. Simple and rapid detection method using surface plasmon resonance for dioxins, polychlorinated biphenylx and atrazine. *Analytica Chimica Acta* 434:223–230.

Sjoquist, J., B. Meloun, and H. Hjelm. 1972b. Protein A isolated from *Staphylococcus aureus* after digestion with lysostaphin. *European Journal of Biochemistry* 29:572–578.

Sjoquist, J., J. Movitz, I.B. Johansson, and H. Hjelm. 1972a. Localization of protein-A in bacteria. *European Journal of Biochemistry* 30:190–194.

Snejdarkova, M., M. Rehak, and M. Otto. 1993. Design of a glucose minisensor based on streptavidin glucose-oxidase complex coupling with self-assembled biotinylated phospholipid membrane on solid support. *Analytical Chemistry* 65:665–668.

Sternesjo, A., C. Mellgren, and L. Bjorck. 1995. Determination of sulphamethazine residues in milk by a surface resonance based-biosensors assay. *Analytical Biochemistry* 226:175–181.

Sun, X., P. He, S. Liu, L. Ye, and Y. Fang. 1998. Immobilization of single stranded deoxyribonucleic acid on gold electrode with self-assembled aminoethanethiol monolayer for DNA electrochemical sensor applications. *Talanta* 47:487–495.

Teles, F.R.R., and L.P. Fonseca. 2008. Applications of polymers for biomolecule immobilization in electrochemical biosensors. *Materials Science and Engineering C* 28:1530–1543.

Terry, L.A., S.F. White, and L.J. Tigwell. 2005. The application of biosensors to fresh produce and the wider food industry. *Journal of Agricultural and Food Chemistry* 53:1309–1316.

Tietjen, M., and D.Y.C. Fung. 1995. *Salmonella* and food safety. *Critical Reviews in Microbiology* 21:53–83.

Todd, E. 2003. Contamination of food. In *Encyclopedia of food science and nutrition*, ed. B. Caballero, L. Trugo, and P.H. Fimglas, 1593–1600. New York: Academic Press.

Traut, R.R., A. Bollen, T.T. Sun, J.W. Hershey, J. Sundberg, and L.R. Pierce. 1973. Methyl 4-mercaptobutyrimidate as a cleavable cross-linking reagent and its application to the *Escherichia coli* 30S ribosome. *Biochemistry* 12:3266–3273.

Ulman, A. 1996. Formation and structure of self-assembled monolayers. *Chemical Reviews* 96:1533–1554.

Vakurov, A., C.E. Simpson, C.L. Daly, T.D. Gibson, and P.A. Millner. 2005. Acetylecholinesterase-based biosensor electrodes for organophosphate pesticide detection: II. Immobilization and stabilization of acetylecholinesterase. *Biosensors and Bioelectronics* 20:2324–2329.

Velasco-Garcia, M.N., and T. Mottram. 2003. Biosensors for detection of pathogenic bacteria. *Biosystems Engineering* 84:1–12.

Wan, J., K. King, H. Craven, C. Mcauley, S.E. Tan, and M.J. Coventry. 2000. Probelia™ PCR system for rapid detection of *Salmonella* in milk powder and ricotta cheese. *Letters in Applied Microbiology* 30:267–271.

Wang, J. 2005. Carbon-nanotube based electrochemical biosensors: A review. *Electroanalysis* 17:7–14.

Wang, J., and Y.H. Lin. 2008. Functionalized carbon nanotubes and nanofibers for biosensing applications. *Trends in Analytical Chemistry* 27:619–626.

Wang, J., R. Polsky, and D. Xu. 2001a. Silver-enhanced colloidal gold electrochemical stripping detection of DNA hybridization. *Langmuir* 17:5739–5741.

Wang, J., X. Cai, C. Jonsson, and M. Balakrishnan. 1996. Adsorptive stripping potentiometry of DNA at electrochemically pretreated carbon paste electrodes. *Electroanalysis* 8:20–24.

Wang, J., D. Xu, A.-N. Kawde, and R. Polsky. 2001b. Metal nanoparticle-based electrochemical stripping potentiometric detection of DNA hybridization. *Analytical Chemistry* 73:5576–5581.

Wilchek, M., and E.A. Bayer. 1988. The avidin biotin complex in bioanalytical applications. *Analytical Biochemistry* 171:1–32.

Wilchek, M., and E.A. Bayer. 1990. Applications of avidin-biotin technology: Literature survey. *Methods in Enzymology* 184:14–45.

Wilchek, M., E.A. Bayer, and O. Livnah. 2006. Essentials of biorecognition: The (strept) avidin-biotin system as a model for protein-protein and protein-ligand interaction. *Immunology Letters* 103:27–32.

Williams, E., M.I. Pividori, A. Merkoçi, R.J. Forster, and S. Alegret. 2003. Rapid electrochemical genosensor assay using a streptavidin carbon-polymer biocomposite electrode. *Biosensors and Bioelectronics* 19:165–175.

Zacco, E., M.I. Pividori, and S. Alegret. 2006a. Electrochemical biosensing based on universal affinity biocomposite platforms. *Biosensors and Bioelectronics* 21:1291–1301.

Zacco, E., R. Galve, M.-P. Marco, S. Alegret, and M.I. Pividori. 2007a. Electrochemical biosensing of pesticides residues based on affinity biocomposite platforms. *Biosensors and Bioelectronics* 22:1707–1715.

Zacco, E., M.I. Pividori, S. Alegret, R. Galve, and M.-P. Marco. 2006b. Electrochemical magnetoimmunosensing strategy for the detection of pesticides residues. *Analytical Chemistry* 78:1780–1788.

Zacco, E., M. I. Pividori, X. Llopis, M. del Valle, and S. Alegret. 2004. Renewable Protein A modified graphite-epoxy composite for electrochemical immunosensing. *Journal of Immunological Methods* 286:35–46.

Zacco, E., J. Adrian, R. Galve, M.-P. Marco, S. Alegret, and M.I. Pividori. 2007b. Electrochemical magneto immunosensing of antibiotic residues in milk. *Biosensors and Bioelectronics* 22:2184–2191.

5 Electrochemical DNA Biosensors in Food Safety

Pınar Kara, Ozan Kılıçkaya,
and Mehmet Şengün Özsöz

CONTENTS

5.1 INTRODUCTION

Food safety is a global health goal. Foodborne pathogenic microorganisms take an important role in many diseases. The analyses of biological and chemical contaminants of foods have great importance for food safety and quality. Over the past decade, the control of food safety has been mainly carried out through product testing rather than process control.

Conventional detections for pathogenic microorganisms are based on traditional techniques that rely on specific microbiologicals to isolate and detect viable bacterial cells in food (Ricci et al. 2007). These methods are expensive and require several days to generate results.

The development of analytical methods based on low cost, ease to operate, and portable instrumentation is of great importance for the food industry because a large number of small producers dominate this sector. An adequate control requires real-time, simultaneous, and selective quantification of several key compounds that establish the food quality and safety (Rogers and Mascini 1998; Rodriguez-Mozaz et al. 2005).

Biosensors hold great promise in food safety and quality analysis due to their high specificity and sensitivity that allow rapid detection of a wide spectrum of samples with minimum sample pretreatment (Draisci et al. 1998; Panfili et al. 2000).

In this chapter, electrochemical DNA biosensor techniques for the detection of food safety and quality are presented dealing with past and novel developments including biosensor types. For this purpose, particular emphasis will be given to the most important approaches for electrochemical genosensing.

5.2 BIOSENSORS

A biosensor is defined by the International Union of Pure and Applied Chemistry (IUPAC) as a "device that uses specific biochemical reactions mediated by isolated enzymes, immunosystems, tissues, organelles or whole cells to detect chemical compounds usually by electrical, thermal or optical signals" (McNaughty and Wilkinson 1997).

A biosensor is capable of selective quantitative or semiquantitative detection using a biological recognition element. The main advantages of biosensors are short times of analysis, low cost of assays, portable equipment, real-time measurements, and use as remote devices (Farré et al. 2009).

The aim of the biosensor technologies is monitoring the desired analytes in both *in vivo* and *in vitro* applications. The biosensor applications cover the environmental analysis for monitoring drugs and foods for ensuring safety and clinical analysis (Maria-Pilar and Damia 1996; Schulze et al. 2002; Hildebrandt et al. 2008). The term *biosensor* has generally been referred to as the monitoring and analyzing of biological interactions by a sensor device.

The first biosensor device was produced in 1956 for the detection of dissolved oxygen in blood (Clark 1959). After the discovery, in 1962 an enzyme electrode was produced for the detection of glucose based on oxygen consumption (Clark Jr. and Lyons 1962). The first potentiometric enzyme biosensor was used to observe the urea level (Guilbault and Montalvo 1969). In 1975 Divis proposed a microorganism electrode for determining the alcohol level in a solution (Divis 1975). Also, the same year, the first glucose biosensor was produced commercially by Yellow Springs Instruments.

The first biosensor samples were based on the detection of the alteration in the current or potential. Since the 1980s, different sensors have been produced for different applications and different analytes.

The basic scheme of a biosensor device is based on an interaction surface, a transducer, a signal amplifier, and a data analysis device. When performing an analysis, biological samples that specifically interact with their substrates on the surface are detected by the recognition surface. The results of the interaction should form changes, which can be physical or chemical. After recognition, the detection signals are converted to another signal by the transducer that can be analyzed easily. The transformed signal is amplified and processed for user analysis.

Biosensors can be classified according to the biochemical reaction element (enzyme, tissue, cell, nucleic acid, etc.) and physicochemical transducer (electrochemical, optical, piezoelectrical, etc.).

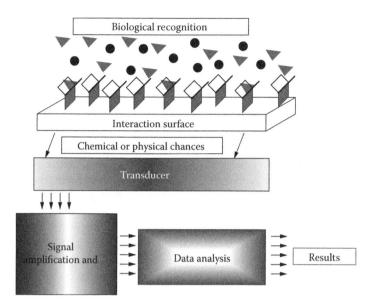

FIGURE 5.1 A basic scheme of a biosensor devices.

5.2.1 BIOSENSORS ACCORDING TO BIORECOGNITION ELEMENTS

Over the last decade, affinity biosensors have received considerable attention because they provide information about binding of enzyme–substrate (Lotierzo et al. 2004), cell receptors to ligands (Andreescu and Sadik 2005), antibodies to antigens (immunosensors) (Tsekenis et al. 2008), nucleic acids to complementary sequences (genosensors) (Masařík et al. 2007), or aptamers (aptasensors) to specific proteins (Tombelli et al. 2005).

5.2.1.1 Enzyme Biosensors

Enzyme-based biosensor samples are based on monitoring glucose in blood (Tothill 2001; D'Orazio 2003). Enzyme biosensors are the first generation of biosensor development due to their commercial availability or ease of isolation and purification. Among various oxidoreductases, glucose oxidase, horseradish peroxidase, and alkaline phosphatase have been employed in most biosensor studies (Laschi et al. 2000).

Enzyme biosensors have been widely used for detection of several biological or chemical compounds in food analysis such as biogenic amines, alcohol, glucose, lactate, vitamins, antibiotics, and so forth (Basu et al. 2007; Carelli et al. 2007; Lupu et al. 2007; McGlinchey et al. 2008; Asav and Akyilmaz 2010).

5.2.1.2 Immunosensors

When antibodies or antibody fragments are used as a molecular recognition element for specific analytes (antigens) to form a stable complex, the device is called an immunosensor. Based on the high selectivity and stability of the antibody–antigen reaction, the development of immunosensors has received considerable attention for several disease detections and environmental and food analysis monitoring (Luppa et al. 2001). Immunosensors for the detection of foodborne pathogens are generally thermometric, electrochemical, optical, or piezoelectrical based according to their signal transduction (Rasooly and Rasooly 1999; Waswa et al. 2007; Pal and Alocilja 2009).

5.2.1.3 Nucleic Acid Biosensors

DNA is an antiparallel and double-stranded biological polymer. Inside the cell it possesses the whole genetic information and controls the all metabolic and catabolic activities. Besides, DNA acts as a transporting and protecting center of the genetic information. It controls the replication of information and transports it to daughter cells in cell division and also protects the information by packaging (Watson and Crick 1953a).

DNA molecules are composed of nucleotides. These nucleotides are covalently linked to each other with ester bonds. All nucleotides have a phosphate group, a deoxyribose sugar, and an organic base. The phosphate group gives the negatively charged character to the double helix backbone and is located on the outer side of the helix. The deoxyribose sugar is pentose sugar and lacks an oxygen molecule in its second carbon position on the sugar molecule. The organic bases can be either a purine or a pyrimidine. DNA molecules carry adenine and guanine as purine molecules, and thymine and cytosine as pyrimidine molecules. Each adenine makes two hydrogen bonds with thymine; and each guanine makes three hydrogen bonds with cytosine on the complementary strand (Watson and Crick 1953b).

In DNA polymerization the nucleotides are added to the newly synthesized chains on its 3' ends. The phosphate group of the nucleotide makes an ester bond with third position of the deoxyribose sugar. Therefore, the direction of polymerization should be 5' to 3' of the molecule (Watson et al. 2008).

In the last decades there has been a considerable interest in DNA biosensors due to their significant analytical properties. The most popular application of DNA biosensors is based on nucleic acid hybridization detection of specific DNA sequences (Kelley et al. 1999).

Recent advances in nucleic acid recognition have enhanced the power of DNA biosensors (genosensors) and DNA chips. DNA-based sensors have potential applications that ranges from genomic sequencing to mutation detection, pathogen identification, and GMO analysis (Marrazza et al. 2001; Ozkan et al. 2002; Farabullini et al. 2007; Tichoniuk 2008).

DNA sensing techniques have been reported for simultaneous detection of foodborne pathogens including *Staphylococcus*, *Escherichia coli*, *Camphylobacter*, *Listeria* genes, and their virulence factors (Sergeev et al. 2004; Karamollaoglu et al. 2009).

5.2.2 BIOSENSORS ACCORDING TO TRANSDUCTION TECHNOLOGY

The transducer converts the biochemical interactions into a measurable analytical signal (Turner 1987; Coulet 1991). Electrochemical, optical, piezoelectrical acoustical, and mechanical transducers are among the many types found in biosensors.

5.2.2.1 Optical Biosensors

Optical sensors employ optical fibers or planar wave guides to direct light to the sensing film. The measured optical signals often include absorbance, fluorescence, chemiluminescence, surface plasmon resonance (to probe refractive index), or changes in light reflectivity. Optical biosensors are preferable for screening a large number of samples simultaneously; however, they cannot be easily miniaturized for insertion into the bloodstream. Most optical methods of transduction still require a spectrophotometer to detect any changes in signal (Luong et al. 2008).

Several kinds of optical transducers can form the basis of biosensors, including fluorescence (Piunno et al. 1995), surface plasmon resonance (Wood 1993; Wang et al. 2004; Dudak and Boyaci 2007), and raman spectroscope for the detection of foodborne pathogens.

5.2.2.2 Piezoelectrical Biosensors

Piezoelectric biosensors are the mass sensitive biosensors that can produce a signal based on the mass of chemicals that interact with the sensing film. Quartz crystal microbalance (QCM) sensors are operated by applying an oscillating voltage at the resonant frequency of the crystal, and measuring the change in resonant frequency when the target analyte interacts with the sensing surface (Spichiger-Keller 2007).

5.2.2.3 Electrochemical Biosensors

Electrochemical biosensors measure the electrochemical changes that occur when the biochemical element interacts with a sensing surface of the detecting electrode. The electrical changes can be based on a change in the measured voltage between the electrodes (potentiometric), a change in the measured current at a given applied voltage (amperometric), or a change in the ability of the sensing material to transport charge (conductometric). Electrochemical biosensors appear more suited for field monitoring applications (e.g., hand-held) and miniaturization toward the fabrication of an implantable biosensor. Based on their high sensitivity, simplicity, and cost competitiveness, more than half of the biosensors reported in the literature are based on electrochemical transducers (Meadows 1996; Palecek 1996).

In an electrochemical cell, the potential of the current passing through electrodes across the electrode interface depends on the chemical components of the solution. Therefore, the chemical reaction and components change can be detected by measuring the current or potential (Mao et al. 2008).

5.2.3 Electrochemical DNA Biosensors in Food Analysis

Electrochemical DNA biosensors based on electrochemical transduction of hybridization have a great promise for detection of foodborne pathogenic microorganisms and specific GMOs. Electrochemical detection of specific DNA sequences has an advantage in reducing the size of the total detection system (Wang 1999).

Electrochemical DNA sensors based on the voltammetric transduction of the formation of double-stranded DNA (ds DNA) have been reported by direct electrical response of the DNA bases (Wang and Kawde 2001), and the electrical response of the double-stranded assembly of the transition metal complexes (Erdem et al. 2002) or dyes (Kobayashi et al. 2001) that are intercalated or electrostatically attracted to the double-stranded assembly. The electrochemical response of these labels or indicators changes upon DNA hybridization, when the hybridization process occurs due to change of the indicator concentration at the electrode surface (Erdem and Ozsoz 2002). The oxidation signals of guanine and adenine were observed at low concentrations of DNA and peptide nucleic acid (PNA) by applying chronopotentiometry and voltammetry (Wang and Kawde 2002). Due to their electroactivity, label-free genosensing techniques have a great importance for sequence-specific detection in clinical, environmental forensic science, and food quality analysis.

The ultimate goal is to design DNA microarray systems allowing early diagnosis of microorganisms and polymorphisms in food safety and quality. For this purpose several techniques have been investigated based on recognition of DNA hybridization, by using electroactive labels, dye molecules, nanoparticles, or label-free methods. The coupling of DNA biosensors with PCR amplification reaction is a recent field of research and provides genetic identification of selected species (Ariksoysal et al. 2005; Pänke et al. 2007; Kara et al. 2008). Detection of specific base sequences in different matrices is becoming increasingly important in several areas ranging from the detection of GMO to forensic and environmental research.

5.2.3.1 Detection of Genetically Modified Organisms (GMOs)

GMOs are artificial organisms whose genetic information has been altered and changed beyond natural selection and evolution. For this purpose several artificial genes have been cloned and expressed from different species. The artificial recombination of genetic information of an organism can be used for enhancement, for example, increasing food nutrition ingredients or producing more resistant plants against insects and environmental conditions (Bertheau et al. 2002). Therefore, the detection of these GMOs in the food sector is crucial because of the unknown side effects of these products.

Although GMO suggests an alternative way of fighting against starvation or improving food quality, there are still several questions about its benefits or usage. Therefore, detecting GMOs has become an important and emerging area in bioanalysis. Several methods have been used to detect the GMO content of food; however, electrochemical DNA biosensors provide a dependable alternative for GMO detection in food (Shehata 2005; Ahmed et al. 2008).

GMOs can be defined as organisms that possess an artificial gene that can produce artificial proteins. These recombinant gene cassettes have control elements such as promoter and terminator genes. Vectors for expression of recombinant proteins contain some control elements. These elements are cauliflower mosaic virus (CaMV) 35S promoter sequence and *Agrobacterium tumafaciens* nopaline synthase gene termination sequence (TNOS) (Holden et al. 2010).

Several works have proposed to detect the 35S promoter and TNOS termination sequence electrochemically. All works have used short oligonucleotides for probing the desired sequence. Minunni et al. (2001) produced biotins labeled 25-mer probe DNA sequences and have used daunomycin as an intercalating agent. Meric et al. (2004) have designed a disposable genosensor for detection of NOS terminator by using methylene blue as a hybridization indicator. Voltammetric detection of hybridization between 25 mer probe sequence and complementary sequence of NOS–terminator gene was accomplished at screen-printed electrodes. Nica et al. (2004) used streptavidin/alkaline phosphatase reaction as a hybridization indicator by using biotinylated DNA probesat screen-printed electrodes surfaces.

Zhu et al. (2008) suggested a biosensor design based on methylene blue ss/dsDNA interaction. They reported that the electrochemical detection has been performed in both synthetic and real samples. Sun et al. (2008) also offered a lead sulphide (PbS) modified gold electrode for the detection of 35S promoter sequence.

A few works have been done for the electrochemical detection of GMOs. Kalogianni et al. (2006) investigated a nanoparticle-based genosensor for detection of PCR-amplified NOS terminator and 35S promoter sequences of real samples. In this study biotin-labeled primers were used to obtain biotinylated PCR products. Hybridization detection was enhanced by using gold nanoparticles at the detection limit of 0.16 nM.

Wang et al. (2008) developed a modified electrodeposited Pt-electrode system for the detection of 35S promoter gene, which generally locates in the beginning of the recombinant genes. The system has been performed using genetically modified soybean, and the detection limit has been 1.07×10^{-7} M.

Lien's groups used multiwalled carbon nanotubes (MWCN) for label-free detection of GMOs. Consequently they suggested a new approach for nanobiosensing in food quality.

5.2.3.2 Detection of Foodborne Pathogenic Microorganisms

Food contaminations by pathogenic bacteria such as *E. coli*, *Staphylococcus*, *Salmonella*, *Listeria*, and so forth cause major diseases. The demands of rapid, sensitive, and selective detection of these pathogenes are increasing. For this purpose several methods have been investigated to reduce expense and time, and to enhance sensitivity and selectivity.

Electrochemical hybridization biosensors offer a fast, reliable, and cost-effective diagnosis. In monitoring foodborne bacteria, nucleic acid hybridization biosensors have a high potential, and there are many reports in this field (Arora et al. 2006).

Mascini's group developed a genosensor for multiple detection of food pathogens (Farabullini et al. 2007). A sandwich assay model was designed for rapid detection of PCR products of *Salmonella* and *Listeria* species. Biotin-labeled signaling probes were conjugated streptavidin-alkaline phosphatase complex after hybridization occurred at disposable sensor surfaces. The reduction signal of napthol was used as an analytical response.

A DNA microdevice was designed for electrical detection of *E. coli* by Berganza et al. (2007). A sensor was patterned as a two-electrode configuration, a working electrode and a counter-electrode in the microdevice. The self-assembled monolayer (SAM) technique was used in the presence of C6 alkanethiol group for probe immobilization. After hybridization with *E. coli* strands, a cyclic voltammeter was used as a transducer in the presence of $K_4[Fe(CN)_6]$ solution.

Recent advances for bacterial food contamination analysis are based on electrochemical detection of rRNA/DNA hybridization recognition. Elsholz et al. (2006)

and Gabig-Ciminska et al. (2004) used streptavidin–alkaline phosphatase or streptavidin–horseradish peroxidase conjugates as reporter enzymes, which bind to biotin-labeled oligonucleotides and convert the nucleic acid hybridization event to an electrochemical signal. Pöhlmann et al. (2009) used thermostable reporter enzyme, esterase 2 (EST2) from *Alicyclobacillus acidocaldarius* for one-step detection of *E. coli*. rRNA/DNA complex recognition was accomplished by using 16srRNA of the bacteria and EST2/probe DNA reporter conjugate.

Recent methods are mostly based on nanocomposites in monitoring sequence-specific food pathogens. Lermo et al. (2007) have developed a genomagnetic assay for detection of food pathogens by using magnetic primers. A sandwich assay was applied with biotinylated capture probe and digoxigenin-modified signaling probe that binds to streptavidin-coated magnetic beads. An amperometric detection of *Salmonella* was performed coupled with PCR products at the detection limit of 21.5 pmol. A gold nanocomposite sensor for detection of *Salmonella* spp. was performed by the same group (Brasil de Oliveira Marques et al. 2009). A thiolated capture DNA was immobilized onto a gold nanoparticles-covered surface, and hybridization occurred with digoxigenin complementary oligonucleotide. Anti-DIG HRP was used as an electrochemical label. Sun et al. (2009) produced a chitosan/nanoV_2O_5/MWNTs composite film sensor and performed the detection of the loop-mediated isothermal amplification (LAMP) product *Yersinia enterocolitica*. This composite film sensor was used as a working electrode. Detection of hybridization between probe and target sequences was accomplished by using methylene blue (MB) as an electrochemical label. Differential pulse voltammetric responses of MB reduction signals were evaluated for *Y. enterocolitica* analysis.

5.3 CONCLUSION

Through this chapter, we demonstrated recent advances in food safety and quality analysis based on electrochemical DNA biosensors. As indicated above, electrochemical DNA biosensors have incredible potential for achieving reliable, rapid, sensitive, and selective detection methods in food safety and quality. New challenges and requirements for the design of ideal electrochemical genosensors include high-sensitivity, high-selectivity methods that provide relatively short times and lower detection limits.

REFERENCES

Ahmed, M.U., M.M. Hossain, and E. Tamiya. 2008. Electrochemical biosensors for medical and food applications. *Electroanalysis* 20:616–626.
Andreescu, S., and O.A. Sadik. 2005. Advanced electrochemical sensors for cell cancer monitoring. *Methods* 37:84–93.
Ariksoysal, D.O., H. Karadeniz, A. Erdem, A. Sengonul, A.A. Sayiner, and M. Ozsoz. 2005. Label-free electrochemical hybridization genosensor for the detection of Hepatitis B virus genotype on the development of lamivudine resistance. *Analytical Chemistry* 77:4908–4917.
Arora, K., S. Chand, and B.D. Malhotra. 2006. Recent developments in bio-molecular electronics techniques for food pathogens. *Analytica Chimica Acta* 568:259–274.

Asav, E., and E. Akyilmaz. 2010. Preparation and optimization of a bienzymic biosensor based on self-assembled monolayer modified gold electrode for alcohol and glucose detection. *Biosensors and Bioelectronics* 25:1014–1018.

Basu, A.K., P. Chattopadhyay, U. Roychoudhuri, and R. Chakraborty. 2007. Development of cholesterol biosensor based on immobilized cholesterol esterase and cholesterol oxidase on oxygen electrode for the determination of total cholesterol in food samples. *Bioelectrochemistry* 70:375–379.

Berganza, J., G. Olabarria, R. García, D. Verdoy, A. Rebollo, and S. Arana. 2007. DNA microdevice for electrochemical detection of *Escherichia coli* 0157:H7 molecular markers. *Biosensors and Bioelectronics* 22:2132–2137.

Bertheau, Y., A. Diolez, A. Kobilinsky, and K. Magin. 2002. Detection methods and performance criteria for genetically modified organisms. *Journal of AOAC International* 85:801–808.

Brasil de Oliveira Marques, P.R., A. Lermo, S. Campoy et al. 2009. Double-tagging polymerase chain reaction with a thiolated primer and electrochemical genosensing based on gold nanocomposite sensor for food safety. *Analytical Chemistry* 81:1332–1339.

Carelli, D., D. Centonze, C. Palermo, M. Quinto, and T. Rotunno. 2007. An interference free amperometric biosensor for the detection of biogenic amines in food products. *Biosensors and Bioelectronics* 23:640–647.

Clark, L.C. 1959. Monitor and control of blood and tissue oxygen tension. *Journal of the American Society for Artificial Internal Organs* 2:41–48.

Clark Jr., L.C., and C. Lyons. 1962. Electrode systems for continuous monitoring in cardiovascular surgery. *Annals of the New York Academy of Sciences* 102:29–45.

Coulet, P.R. 1991. What is a biosensor? In *Biosensor principles and applications*, ed. L.J. Blum and P.R. Coulet, 1–6. New York: Marcel Dekker Inc.

Divis, C. 1975. Remarques sur l'oxydation de l'ethanol par une electrode micro-bienned' acetobacter zylinum. *Annals of Microbiology* 126A:175–86.

D'Orazio, P. 2003. Biosensors in clinical chemistry. *Clinica Chimica Acta* 334:41–69.

Draisci, R., G. Volpe, L. Lucentini, A. Cecilia, R. Federico, and G. Palleschi. 1998. Determination of biogenic amines with an electrochemical biosensor and its application to salted anchovies. *Food Chemistry* 62:225–232.

Dudak, F.C., and I.H. Boyaci. 2007. Development of an immunosensor based on surface plasmon resonance for enumeration of *Escherichia coli* in water samples. *Food Research International* 40:803–807.

Elsholz, B., R. Worl, L. Blohm et al. 2006. Automated detection and quantitation of bacterial RNA by using electrical microarrays. *Analytical Chemistry* 78:4794–4802.

Erdem, A., and M. Ozsoz. 2002. Electrochemical DNA biosensors based on DNA-drug interactions. *Electroanalysis* 14:965–74.

Erdem, A., K. Kerman, B. Meric, D. Ozkan, P. Kara, and M. Ozsoz. 2002. DNA biosensor for *Microcystis* spp. sequence detection by using methylene blue and ruthenium complex as electrochemical hybridization labels. *Turkish Journal of Chemistry* 26:851–862.

Farabullini, F., F. Lucarelli, I. Palchetti, G. Marrazza, and M. Mascini. 2007. Disposable electrochemical genosensor for the simultaneous analysis of different bacterial food contaminants. *Biosensors and Bioelectronics* 22:1544–1549.

Farré, M., L. Kantiani, S. Pérez, and D. Barceló. 2009. Sensors and biosensors in support of EU directives. *TrAC Trends in Analytical Chemistry* 28: 170–185.

Gabig-Ciminska, M., A. Holmgren, H. Andresen, et al. 2004. Electric chips for rapid detection and quantification of nucleic acids. *Biosensors and Bioelectronics* 19:537–546.

Guilbault, G.G., and J.G. Montalvo. 1969. Urea-specific enzyme electrode. *Journal of the American Chemical Society* 91:2164–2165.

Hildebrandt, A., R. Bragós, S. Lacorte, and J.L. Marty. 2008. Performance of a portable biosensor for the analysis of organophosphorus and carbamate insecticides in water and food. *Sensors and Actuators B: Chemical* 133:195–201.

Holden, M., M. Levine, T. Scholdberg, R. Haynes, and G. Jenkins. 2010. The use of 35S and *Tnos* expression elements in the measurement of genetically engineered plant materials. *Analytical and Bioanalytical Chemistry* 396:2175–2187.

Kalogianni, D.P., T. Koraki, T.K. Christopoulos, and P.C. Ioannou. 2006. Nanoparticle-based DNA biosensor for visual detection of genetically modified organisms. *Biosensors and Bioelectronics* 21:1069–1076.

Kara, P., B. Meric, and M. Ozsoz. 2008. Application of impedimetric and voltammetric genosensor for detection of a biological warfare: Anthrax. *Electroanalysis* 20:2629–634.

Karamollaoglu, I., Oktem, H.A., and M. Mutlu. 2009. QCM-based DNA biosensor for detection of genetically modified organisms (GMOs). *Biochemical Engineering Journal* 44:142–150.

Kelley, S., E. Boon, J. Barton, N. Jackson, and M. Hill. 1999. Single-base mismatch detection based on charge transduction through DNA. *Nucleic Acids Research* 27:4830–4837.

Kobayashi, M., T. Mizukami, Y. Morita, Y. Murikami, K. Yokohama, and E. Tamiya. 2001. Electrochemical gene detection using microelectrode array on a DNA chip. *Electrochemistry* 69:1013–1016.

Laschi, S., M. Fránek, and M. Mascini. 2000. Screen-printed electrochemical immunosensors for PCB detection. *Electroanalysis* 12:1293–1298.

Lermo, A., S. Campoy, J. Barbé, S. Hernández, S. Alegret, and M.I. Pividori. 2007. *In situ* DNA amplification with magnetic primers for the electrochemical detection of food pathogens. *Biosensors and Bioelectronics* 22:2010–2017.

Lien, T.T.N., T.D. Lam, V.T.H. An et al. 2010. Multi-wall carbon nanotubes (MWCNTs)-doped polypyrrole DNA biosensor for label-free detection of genetically modified organisms by QCM and EIS. *Talanta* 80:1164–1169.

Lotierzo, M., O.Y.F. Henry, S. Piletsky et al. 2004. Surface plasmon resonance sensor for domoic acid based on grafted imprinted polymer. *Biosensors and Bioelectronics* 20:145–152.

Luong, J.H.T., K.B. Male, and J.D. Glennon. 2008. Biosensor technology: Technology push versus market pull. *Biotechnology Advances* 26:492–500.

Luppa, P.B., L.J. Sokoll, and D.W. Chan. 2001. Immunosensors—Principles and applications to clinical chemistry. *Clinica Chimica Acta* 314:1–26.

Lupu, A., A. Valsesia, F. Bretagnol, P. Colpo, and F. Rossi. 2007. Development of a potentiometric biosensor based on nanostructured surface for lactate determination. *Sensors and Actuators B: Chemical* 127:606–612.

Mao, X.-L., J. Wu, and Y.-B. Ying. 2008. Application of electrochemical biosensors in fermentation. *Chinese Journal of Analytical Chemistry* 36:1749–1755.

Maria-Pilar, M., and B. Damia. 1996. Environmental applications of analytical biosensors. *Measurement Science and Technology* 7:1547–1562.

Marrazza, G., S. Tombelli, M. Mascini, and A. Manzoni. 2001. Detection of human apolipoprotein E genotypes by DNA biosensors coupled with PCR. *Clinica Chimica Acta* 307:241–248.

Masařík, M., K. Cahová, R. Kizek, E. Paleček, and M. Fojta. 2007. Label-free voltammetric detection of single-nucleotide mismatches recognized by the protein MutS. *Analytical and Bioanalytical Chemistry* 388:259–270.

McGlinchey, T.A., P.A. Rafter, F. Regan, and G.P. McMahon. 2008. A review of analytical methods for the determination of aminoglycoside and macrolide residues in food matrices. *Analytica Chimica Acta* 624:1–15.

McNaughty, A.D., and A. Wilkinson. 1997. *Compendium of chemical terminology*. 2nd ed. Oxford: Blackwell Scientific Publications.

Meadows, D. 1996. Recent developments with biosensing technology and applications in the pharmaceutical industry. *Advanced Drug Delivery Reviews* 21:179–189.

Meric, B., K. Kerman, G. Marrazza, I. Palchetti, M. Mascini, and M. Ozsoz. 2004. Disposable genosensor, a new tool for the detection of NOS-terminator, a genetic element present in GMOs. *Food Control* 15:621–626.

Minunni, M., S. Tombelli, E. Mariotti, and M. Mascini. 2001. Biosensors as new analytical tool for detection of genetically modified organisms (GMOs). *Fresenius' Journal of Analytical Chemistry* 369:589–593.

Nica, A.C., M. Mascini, and A.A. Ciucu. 2004. DNA-based biosensor for detection of genetically modified organisms. *Chimie Annual* 23:84–94.

Ozkan, D., A. Erdem, P. Kara et al. 2002. Allele-specific genotype detection of factor V leiden mutation from polymerase chain reaction amplicons based on label-free electrochemical genosensor. *Analytical Chemistry* 74:5931–5936.

Pal, S., and E.C. Alocilja. 2009. Electrically active polyaniline coated magnetic (EAPM) nanoparticle as novel transducer in biosensor for detection of *Bacillus anthracis* spores in food samples. *Biosensors and Bioelectronics* 24:1437–1444.

Palecek, E. 1996. From polarography of DNA to microanalysis with nucleic acid-modified electrodes. *Electroanalysis* 8:7–14.

Panfili, G., P. Manzi, D. Compagnone, L. Scarciglia, and G. Palleschi. 2000. Rapid assay of choline in foods using microwave hydrolysis and a choline biosensor. *Journal of Agricultural and Food Chemistry* 48:3403–407.

Pänke, O., A. Kirbs, and F. Lisdat. 2007. Voltammetric detection of single base-pair mismatches and quantification of label-free target ssDNA using a competitive binding assay. *Biosensors and Bioelectronics* 22:2656–662.

Piunno, P.A.E., U.J. Krull, R.H.E. Hudson, M.J. Damha, and H. Cohen. 1995. Fiber-optic DNA sensor for fluorometric nucleic acid determination. *Analytical Chemistry* 67:2635–2643.

Pöhlmann, C., Y. Wang, M. Humenik, B. Heidenreich, M. Gareis, and M. Sprinzl. 2009. Rapid, specific and sensitive electrochemical detection of foodborne bacteria. *Biosensors and Bioelectronics* 24:2766–2771.

Rasooly, L., and A. Rasooly. 1999. Real time biosensor analysis of Staphylococcal enterotoxin A in food. *International Journal of Food Microbiology* 49:119–127.

Ricci, F., G. Volpe, L. Micheli, and G. Palleschi. 2007. A review on novel developments and applications of immunosensors in food analysis. *Analytica Chimica Acta* 605:111–129.

Rodriguez-Mozaz, S., M.J.L. Alda, M.-P. Marco, and D. Barceló. 2005. Biosensors for environmental monitoring: A global perspective. *Talanta* 65:291–297.

Rogers, K.R., and M. Mascini. 1998. Biosensors for field analytical monitoring. *Field Analytical Chemistry & Technology* 2:317–331.

Schulze, H., E. Scherbaum, M. Anastassiades, S. Vorlová, R.D. Schmid, and T.T. Bachmann. 2002. Development, validation, and application of an acetylcholinesterase-biosensor test for the direct detection of insecticide residues in infant food. *Biosensors and Bioelectronics* 17:1095–1105.

Sergeev, N., M. Distler, S. Courtney et al. 2004. Multipathogen oligonucleotide microarray for environmental and biodefense applications. *Biosensors and Bioelectronics* 20:684–698.

Shehata, M.M. 2005. Genetically modified organisms (GMOs) food and feed: Current status and detection. *International Journal of Food, Agriculture and Environment* 3:43–55.

Spichiger-Keller, U.E. 2007. *Chemical sensors and biosensors for medical and biological applications*. Weinheim, Germany: Wiley-VCH Verlag.

Sun, W., J. Zhong, P. Qin, and K. Jiao. 2008. Electrochemical biosensor for the detection of cauliflower mosaic virus 35 S gene sequences using lead sulfide nanoparticles as oligonucleotide labels. *Analytical Biochemistry* 377:115–119.

Sun, W., P. Qin, H. Gao, G. Li, and K. Jiao. 2009. Electrochemical DNA biosensor based on chitosan/nano-V2O5/MWCNTs composite film modified carbon ionic liquid electrode and its application to the LAMP product of *Yersinia enterocolitica* gene sequence. *Biosensors and Bioelectronics* 25:1264–1270.

Tichoniuk, M., M. Ligaj, and M. Filipiak. 2008. Application of DNA hybridization biosensor as a screening method for the detection of genetically modified food components. *Sensors* 8:2118–2135.

Tombelli, S., M. Minunni, and M. Mascini. 2005. Analytical applications of aptamers. *Biosensors and Bioelectronics* 20:2424–2434.

Tothill, I.E. 2001. Biosensors developments and potential applications in the agricultural diagnosis sector. *Computers and Electronics in Agriculture* 30:205–218.

Tsekenis, G., G.-Z. Garifallou, F. Davis, P.A. Millner, T.D. Gibson, and S.P.J. Higson. 2008. Label-less immunosensor assay for myelin basic protein based upon an ac impedance protocol. *Analytical Chemistry* 80:2058–2062.

Turner, A.P.F., I. Karube, and G.S. Wilson. 1987. *Biosensors: Fundamentals and applications*. Oxford: Oxford University Press.

Wang, J. 1999. Towards genoelectronics: Electrochemical biosensing of DNA hybridization. *Chemistry—A European Journal* 5:1681–1685.

Wang, J., and A.-N. Kawde. 2001. Pencil-based renewable biosensor for label-free electrochemical detection of DNA hybridization. *Analytica Chimica Acta* 431:219–224.

Wang, J., and A.N. Kawde. 2002. Amplified label-free electrical detection of DNA hybridization. *Analyst* 127:383–386.

Wang, M.-Q., X.-Y. Du, L.-Y. Liu, Q. Sun, and X.-C. Jiang. 2008. DNA biosensor prepared by electrodeposited Pt-nanoparticles for the detection of specific deoxyribonucleic acid sequence in genetically modified soybean. *Chinese Journal of Analytical Chemistry* 36:890–894.

Wang, R., S. Tombelli, M. Minunni, M.M. Spiriti, and M. Mascini. 2004. Immobilisation of DNA probes for the development of SPR-based sensing. *Biosensors and Bioelectronics* 20:967–974.

Waswa, J., J. Irudayaraj, and C. DebRoy. 2007. Direct detection of *E. coli* O157:H7 in selected food systems by a surface plasmon resonance biosensor. *LWT—Food Science and Technology* 40:187–192.

Watson, J.D., and F.H.C. Crick. 1953a. Molecular structure of nucleic acids: A structure for deoxyribose nucleic acid. *Nature* 171:737–738.

Watson, J.D., and F.H.C. Crick. 1953b. Genetic implications of the structure of deoxyribonucleic acid. *Nature* 171:964–967.

Watson, J.D., T.A. Baker, S.P. Bell, A. Gann, M. Levine, and R. Losick. 2008. *Molecular biology of the gene*. 6th ed. Cold Spring Harbor Laboratory Press.

Wood, S.J. 1993. DNA-DNA hybridization in real time using BIAcore. *Microchemical Journal* 47:330–337.

Zhu, L., R. Zhao, K. Wang, H. Xiang, Z. Shang, and W. Sun. 2008. Electrochemical behaviors of methylene blue on DNA modified electrode and its application to the detection of PCR product from NOS sequence. *Sensors* 8:5649–5460.

6 Biosensors for the Assessment of Natural Toxins in Food

Beatriz Prieto-Simón, Thierry Noguer, and Mònica Campàs

CONTENTS

6.1 INTRODUCTION

Among natural chemicals present in foods, toxins appear as one of the most harmful threats to human health. Some of these natural toxins are synthesized by plants as a means to protect themselves from insects or to create resistance to certain diseases, and thus may be present in fruits and vegetables. Other natural toxins are produced as a consequence of the damage to a plant or the growth of molds on it. Another kind of natural toxin is those produced by microalgae in the sea or in fresh water. In the former case, bivalves, due to their condition from filter-feeding mollusks, are more likely to contain these toxins than fish. In the latter case, cyanotoxins such as microcystins, which are hepatotoxins produced by cyanobacteria, can be present in water for agriculture, cattle breeding, or human consumption (Carmichael et al. 2001). Microbial toxins produced by microorganisms such as bacteria may also contaminate foodstuffs. Some examples are *botulinum neurotoxin*, the most potent toxin known, which induces botulism, a potentially fatal paralysis (Singh 2000); *cholera toxin*, which is responsible for the harmful effects of cholera infection; and *enterotoxin*, which causes the death of the cells of the intestinal wall.

Due to the ubiquity of toxins in nature, there is an increasing concern for the control of their widespread presence in foodstuff. Humans have some detoxification mechanisms that allow for the assimilation of certain toxin amounts without deleterious consequences, but when consumed in large quantities and/or over a long period of time, toxins can be considered harmful compounds. Apart from the amount and frequency of the toxin ingested, the toxic effects may vary according to other factors such as environmental conditions, synergistic effects induced by other nutrients ingested and/or adsorbed by the organism, the organism, and the health status of the individual. Therefore, it is important to set maximum permitted levels and to develop reliable methods to determine the toxin amounts in food samples.

Experts in pollutant-risk assessment consider toxins produced by filamentous fungi, known as mycotoxins, to be the most important chronic dietary risk factor, above other natural toxins or synthetic compounds (Kuiper-Goodman 1998). The first reported case of adverse health effects caused by these secondary metabolites was in the 10th century, when a chronicler described the "St. Anthony's fire" disease, today called ergotism. Nevertheless, it was not until the second part of the 20th century when significant research began to study the causes of mycotoxicosis.

Phycotoxins, produced by some marine microalgae, are also causing an important health concern. These secondary metabolites enter into the food chain through the phytoplankton ingested mainly by shellfish, although other fish species can also act as vectors for transmitting them to humans. Seafood contaminated by phycotoxins tastes and smells like noncontaminated foodstuff, thus causing problems in the exploitation of marine resources (Shumway 1990). Cases of illnesses linked to the consumption of raw shellfish were first reported at the end of the 19th century. This was a result of the presence of domoic acid (DA) and the cause of the gastrointestinal and neurological symptoms, characteristics of the amnesic shellfish poisoning (ASP).

Regulations have been established in many countries to protect consumers from the deleterious effects of natural toxins. Due to the high-risk level related to contamination of foodstuffs with mycotoxins and phycotoxins, this chapter will focus on the impact of these toxins on food safety and on the tools that are currently available and those that are in development to determine the toxin content in food samples.

6.1.1 Mycotoxins

Mycotoxins can be found in a wide range of foodstuffs depending on the environmental conditions during production and storage (Prieto-Simón et al. 2007). They show both acute and chronic toxicological effects, some of which include carcinogenics, teratogenics, nephrotoxics, hepatotoxics, or immunotoxics. Some mycotoxins, such as the case of the genotoxic aflatoxins, are so harmful that there is not a threshold below which it is completely safe to eat the contaminated food. Nevertheless, maximum levels must be set with a safety margin that is able to cover the uncertainty associated with toxicological evaluation, as well as the synergistic effects produced in the presence of other compounds in food samples, and the variation in uptake by different consumers, with special attention on infants. The European Union aims to standardize the legislation related to mycotoxins within the different countries by the adoption of a directive or regulation set by the Steering Committee for Food.

TABLE 6.1
Contaminated Food Products and Health Effects of the Main Mycotoxins

Toxin	Source Organism	Contaminated Food	Health Effects on Humans (IARC* Classification)
Aflatoxins (B$_1$, B$_2$, G$_1$, G$_2$, M$_1$, M$_2$)	*Aspergillus flavus* *Aspergillus nomius* *Aspergillus parasiticus*	Groundnuts, other edible nuts, peanut butter, figs, spices, corn, rice, cassava, tobacco, seeds, milk, yogurt, cheese	Hepatomegaly, jaundice, hepatitis, cirrhosis, liver cancer, Reye's syndrome, kwashiorkor, immunosuppression (Group 1: Carcinogenic)
Ochratoxin A	*Aspergillus alliaceus* *Aspergillus auricomus* *Aspergillus carbonarius* *Aspergillus glaucus* *Aspergillus melleus* *Aspergillus niger* *Aspergillus ochraceus* *Penicillium cyclopium* *Penicillium verrucosum* *Penicillium viridicatum*	Cereals and derived products, rice, dried vine fruit, nuts, coffee, cocoa, wine, beer, spices, pork, cheese	Kidney damage (i.e., Balkan endemic nephropathy) (Group 2B: Possible carcinogen)
Patulin	*Aspergillus* spp. *Byssochlamys* spp. *Penicillium expansum* *Penicillium griseofulvum*	Fruits, vegetables, cereals, pies, jam, apple juice, animal silage	Stomach disturbances, nausea, vomiting, gastrointestinal hyperemia, ulceration, hemorrhages (Group 3: Carcinogenic properties not classifiable)
Fumonisins (B$_1$, B$_2$)	*Aspergillus alternata* *Fusarium anthophilum* *Fusarium dlamini* *Fusarium moniliforme* *Fusarium napiformeI* *Fusarium nygama* *Fusarium proliferatum* *Fusarium verticillioides*	Cereals and derived products, rice, beans, asparagus, beer	Esophageal cancer (Group 2B: Possible carcinogen)
Zearalenone, deoxynivalenol, nivalenol, fusarenone X	*Fusarium culmorum* *Fusarium crookwellense* *Fusarium graminearum*	Cereals and derived products, rice, walnuts, milk, beer, meet, animal-feed products	Acutely toxic to humans, sickness, gastrointestinal and abdominal pains, vomiting, diarrhea, cervical cancer (Group 3: Carcinogenic properties not classifiable)

Source: European Mycotoxins Awareness Network (http://www.mycotoxins.org/).
*IARC: International Agency for Research on Cancer

Maximum permitted levels for mycotoxins are regulated in the European Union by the Commission Regulation (EC) No. 1881/2006 and No. 1126-2007. These levels depend on the mycotoxin type as well as on the contaminated product.

Validated analytical methods to detect mycotoxins at the limits set by legislation are required. Since mycotoxins are heterogeneously distributed in samples, the sampling step plays an important role in the determination, and thus, legislation establishes several methods for each mycotoxin according to the final purpose and the detection method (European Mycotoxins Awareness Network) (see Table 6.1). These methods involve expensive and/or time-consuming steps such as solid-phase extraction (SPE) with organic solvents from complex matrices, sample cleanup to remove interferents, preconcentration, and analyte derivatization. The use of immunoaffinity columns (IACs) has improved the sample extraction and purification steps, but protocols are still laborious and require trained personnel. Traditional detection methods are mainly based on high-performance liquid chromatography (HPLC), thin-layer chromatography (TLC), and gas chromatography (GC) coupled with ultraviolet-visible (UV-Vis) or fluorescence spectroscopy, mass spectrometry (MS), tandem MS (MS/MS), sequential MS (MSn), or electrochemical detection (ECD) (Pittet 2005). Enzyme-linked immunosorbent assays (ELISAs) have also been widely applied to mycotoxin detection. They allow parallel analysis of multiple samples, being a powerful tool for rapid screening. Although most ELISA protocols do not require sample cleanup other than filtration and dilution, they still involve several washing steps.

6.1.2 Phycotoxins

Microalgae produce phycotoxins that may be ingested by shellfish and accumulated in their digestive glands without apparent toxic effects. However, marine mammals and humans develop several illnesses after ingestion of contaminated shellfish. As a result, phycotoxins pose a serious threat to human health, as well as to environmental protection and the fishing industry (Food and Agriculture Organization 2004) (see Table 6.2). Depending on the symptoms, phycotoxins can be classified into three main groups:

• Diarrheic shellfish poisoning (DSP) toxins
• Paralytic shellfish poisoning (PSP) toxins
• Amnesic shellfish poisoning (ASP) toxins

Nowadays, the *DSP toxins* terminology is being replaced with *lipophilic toxins*, since it has been observed that not all toxins included in this group are involved in diarrhoea episodes.

Legislation on the quality and safety of seafood products sets maximum permitted levels for certain phycotoxins in edible tissue. These levels are usually given in μg of toxin equivalents/g, since not all the toxins belonging to the same group exhibit the same toxicity. Another frequently used concentration unit is mouse units/g (MU/g), defined as the amount of intraperitoneally injected toxin causing the mouse death within 24 h. The European Commission set a maximum level of PSP toxins of 80 μg saxitoxin (STX) eq/100 g shellfish flesh through Directive 91/492/EEC and an upper safe limit of 20 μg/g for the total ASP toxins content in the edible parts of mollusks

TABLE 6.2
Contaminated Seafood Products and Health Effects of the Main Phycotoxins

Toxin	Source Organism	Contaminated Seafood	Health Effects on Humans
Diarrheic shellfish poisoning (DSP) toxins	*Dinophysis acuminata* *Dinophysis acuta* *Dinophysis caudata* *Dinophysis fortii* *Dinophysis hastata* *Dinophysis mitra* *Dinophysis norvegica* *Dinophysis rotundata* *Dinophysis sacculus* *Dinophysis tripos* *Gonyaulax polyhedra* *Phalacroma rotundatum* *Protoceratium reticulatum* *Protoperidinium oceanicum* *Protoperidinium pellucidum* *Prorocentrum arenarium* *Prorocentrum belizeanum* *Prorocentrum concavum* *Prorocentrum lima* *Prorocentrum redfieldi*	Clams, mussels, oysters, scallops	Okadaic acid (OA) and dynophysistoxins: Gastrointestinal disorders, nausea, abdominal pain, vomiting, diarrhea Potent tumor promoters and possible mutagenic and immunotoxic agents Pectenotoxins (PTXs): Liver damage Yessotoxins (YTXs): Cardiac muscle and liver damage
Paralytic shellfish poisoning (PSP) toxins	*Alexandrium andersonii* *Alexandrium catenella* *Alexandrium cohorticula* *Alexandrium fraterculus* *Alexandrium fundyense* *Alexandrium minutum* *Alexandrium tamarensis* *Aphanazomenon flos-aquae* *Gymnodinium catenatum* *Pyrodinium bahamense* *Spondylus butler* *Zosimous acnus*	Clams, crabs, cockles, cods, copepods, gastropods, herrings, lobsters, mackerels, mussels, ormers, oysters, puffer fish, salmon, scallops, starfish, whales, whelks	Saxitoxin (STX): Neurological and gastrointestinal (nausea, vomiting and diarrhea) symptoms After severe intoxication, general muscular incoordination, dysmetria, respiratory distress and potentially death
Amnesic shellfish poisoning (ASP) toxins	*Alsidium corallinum* *Amphora coffaeformis* *Chondria armata* *Chondria baileyana* *Nitzschia navis-varingica* *Pseudo-nitzschia australis* *Pseudo-nitzschia fraudulenta* *Pseudo-nitzschia multiseries* *Pseudo-nitzschia multistriata* *Pseudo-nitzschia pseudodelicatissima* *Pseudo-nitzschia pungens* *Pseudo-nitzschia seriata* *Pseudo-nitzschia turgidula*	Anchovies, clams, crabs, gastropods, lobsters, mackerels, mussels, oysters, scallops	Domoic acid (DA) and its derivatives: Potent neurotoxin with neurological (memory loss and disorientation) and gastrointestinal (abdominal cramps, nausea, vomiting and diarrhea) symptoms Optical problems (disconjugate gaze, diplopia and ophthalmoplegia) After severe intoxication, important neurological deficits (confusion, mutism, seizures, autonomic dysfunction, lack of response to painful stimuli and uncontrolled crying or aggressiveness) and potentially coma and death

Sources: Food and Agriculture Organization (FAO), Food and Nutrition Paper 80; Rome 2004; Campàs et al. 2007.

through Commission Decision 2002/226/EC. Additionally, in 2002 it was agreed to set tolerable levels for okadaic acid (OA), dynophysistoxins (DTXs), and pectenotoxins (PTXs) present at the same time in edible tissue of 160 μg OA eq/kg shellfish (equivalent to 40 MU/kg), and a maximum level of yessotoxins (YTXs) of 1 μg YTX eq/g shellfish (Commission Decision 2002/225/EC). Since some regulations have been adjusted to the limitations imposed by the detection methods, it is obvious that there is a need for new, highly sensitive, and reliable methods to monitor and control contamination of seafood by phycotoxins.

Traditionally used as the method of reference, the mouse bioassay is useful as a preliminary tool for toxicity screening. However, in addition to ethical implications, it is poor in selectivity and accuracy, and time-consuming. Chromatographic techniques represent an important improvement of selectivity and sensitivity. In fact, liquid chromatography (LC) coupled to UV detection is the Association of Official Analytical Chemists International (AOAC) official method for DA, and LC coupled to fluorimetric (FLD), MS or MS/MS detection is gaining increasing success. Another emerging analysis method is capillary electrophoresis usually coupled to a UV or fluorescence detector. Immunoassays in the format of ELISAs have been also shown as promising tools for routine detection and quantification of shellfish toxins, due to the high sample throughput and relative low cost (Garthwaite 2000; Campàs et al. 2007).

Despite the existence of both classical and emergent analyses, there is still a need to advance toward new screening and confirmatory tests directed at replacing the bioassay with definitive and reliable quantitative detection. In this direction, the European Union Commission Regulation (EC) No. 2074/2005 recognizes a series of methods, such as HPLC-FLD, LC-MS, and immunoassays and functional assays, such as the protein phosphatase inhibition assay, as potential alternative or supplementary methodologies to the biological testing methods for the determination of lipophilic toxins, provided that these methods are not less effective and that their implementation provides an equivalent level of public health protection.

6.2 BIOSENSORS FOR NATURAL TOXINS DETECTION IN FOOD

In looking for rapid, robust, specific, and sensitive analytical methods for the detection of natural toxins, biosensors appear as highly performing devices that additionally, under the proper conditions, can deal with complex matrices and with toxin mixtures. Biosensors have been shown as powerful screening tools for many analytes, including natural toxins such as mycotoxins and phycotoxins. Those devices have been developed using different biorecognition elements, although immunosensors, based on the high-affinity interactions between antigens and antibodies, have till now, been the ones that show better performance. This chapter will focus on electrochemical biosensors, because apart from being highly sensitive and enabling low limits of detection, they can be easily miniaturized. Furthermore, the possibility to be portable and automated and the subsequent ease of use by nonskilled personnel are interesting features for *in situ* monitoring of food samples.

Currently, the biosensors developed for the detection of natural toxins in food have to be considered as tools for the preliminary screening of the toxicity, and positive samples are still subject to further analysis with complementary analytical techniques to provide an accurate toxin determination. The main limitations of biosensors come from the application to food samples. Complex matrices with potentially interfering compounds, as well as the presence of other toxins that can show synergistic effects with the target toxin, have limited the final application of the developed devices. In fact, although in principle biosensors are able to deal with such samples, much work should be done to validate their applicability. Thus, this final step of biosensor development is crucial.

6.2.1 Electrochemical Biosensors for Natural Toxin Detection and Their Applicability to Food Samples

Although the development of electrochemical biosensors for the detection of natural toxins is an interesting trend in toxin analysis, only a few devices have demonstrated their suitability to determine toxin content in food samples (Table 6.3). Important issues when using natural samples are the presence of organic solvents used in the toxin extraction processes and of interfering compounds from the matrix. Ben Rejeb et al. (2009) developed a biosensor for detecting aflatoxin B1 (AFB_1) in olive oil, based on the inhibition caused by this toxin to acetylcholinesterase (AChE). First, they should investigate the percentage of organic solvent used in the extraction protocol, in this case methanol, that the biosensor can afford without compromising its performance. Experiments showed that 1% of methanol does not interfere in the measurement, and thus, after extraction, olive oil samples were evaporated and subsequently solubilized in phosphate buffer with 1% methanol. Matrix effects mainly caused by phenolic compounds and pesticides were almost negligible after the optimization of the conditions for sample treatment and electrochemical measurement. Another interesting parameter to prove the reliability of the determination of toxin content in food samples is the recovery percentage. In this case, the authors found a recovery percentage of 78±9% for 10 ppb of AFB_1, demonstrating the suitability of the whole method. Nevertheless, a preconcentration step was necessary in order to attain the 2 ppb level set by the European Commission. Also devoted to the detection of AFB_1, Tan et al. (2009) developed an immunosensor exhibiting almost no matrix effects and good recoveries (from 88% to 112%) for spiked rice samples, after a one-step extraction procedure with 80% of methanol.

Another example is the analysis of aflatoxin M1 (AFM_1) in milk. Milk is a complex matrix that contains many interfering compounds. Vig et al. (2009) developed an impedimetric immunosensor for AFM_1, showing that even after centrifugation for defatting milk and dilution, different proteins are still present in the sample. These proteins can be potentially adsorbed onto the electrode surface, affecting the measured signal. Parker and Tothill (2009) performed an extensive study of the interferences to the electroanalytical signal during the analysis of milk using a biosensor based on chronoamperometric measurements. They conclude that apart from the centrifugation step, a simple sample pretreatment based on the addition of calcium

TABLE 6.3
Electrochemical Biosensors for the Detection of Natural Toxins in Food Samples

Toxin	Principle	Electrochemical Detection	Food Samples	Sample Treatment	References
Aflatoxin B$_1$ (AFB$_1$)	Enzymatic biosensor based on inhibition of acetylcholinesterase (AChE)	Amperometry	Olive oil	Extractions with aqueous solution of methanol, hexane and chloroform; evaporation; reconstitution in PBS	Ben Rejeb et al. 2009
Aflatoxin B$_1$ (AFB$_1$)	Immunosensor based on competitive ELISA	Linear Sweep Voltammetry	Rice	Grinding; extraction with 80% methanol; centrifugation; supernatant dilution in PBS	Tan et al. 2009
Aflatoxin M$_1$ (AFM$_1$)	Immunosensor based on competitive ELISA	Impedance / Linear Sweep Voltammetry	Milk	Defatting-centrifugation; pH adjustment; centrifugation; addition of thrichloroacetic acid; centrifugation; addition of calcium chloride during analysis	Vig et al. 2009
Aflatoxin M$_1$ (AFM$_1$)	Immunosensor based on competitive ELISA	Chronoamperometry	Milk	Defatting-centrifugation; addition of calcium chloride; supernatant dilution in PBS	Parker and Tothill 2009
Aflatoxin M$_1$ (AFM$_1$)	Immunosensor based on competitive ELISA	Chronoamperometry	Milk	Defatting-centrifugation	Micheli et al. 2005
Ochratoxin A (OTA)	Immunosensor based on competitive ELISA	Differential Pulse Voltammetry	Wheat	Extraction with aqueous acetonitrile	Alarcón et al. 2006
Ochratoxin A (OTA)	Immunosensor based on competitive ELISA	Chronoamperometry/ Differential Pulse Voltammetry	Wine	Treatment with PVP; filtration; pH adjustment	Prieto-Simón et al. 2008
Okadaic acid (OA)	Immunosensor based on competitive ELISA	Chronoamperometry	Mussels, oysters	Homogenization; extraction with aqueous solution of methanol; centrifugation; filtration; evaporation; reconstitution in PBS	Campàs et al. 2008

chloride is necessary to stabilize whey proteins in milk samples and to eliminate the interference caused by their adsorption on the electrode. On the contrary, the AFM_1 biosensor developed by Micheli et al. (2005), also based on chronoamperometric measurements, did not show matrix effects when directly measuring simply centrifuged milk, without dilution or other pretreatment steps. A possible explanation could be that they used a lower working potential than that used by Parker and Tothill (2009). All these works show that the electrochemical technique used in the measurement step as well as the operating parameters may be crucial in the applicability of biosensors to natural samples.

Electrochemical biosensors for ochratoxin A (OTA) determination in wheat and wine have also been developed. In the first case, Alarcón et al. (2006) used a one-step extraction procedure with aqueous acetonitrile without any cleanup of the extract. Since this simple sample treatment did not avoid the matrix effects, they performed a calibration curve for OTA in the presence of wheat extract, which allowed them to accurately determine toxin contents in this food. A good correlation was observed with the results obtained using HPLC, demonstrating the applicability of the strategy. For the analysis of OTA in wine, Prieto-Simón et al. (2008) removed interfering polyphenols by pretreating wine samples with the binding agent poly(vinylpyrrolidone) (PVP) and then adjusting the pH to optimum conditions for antigen–antibody binding. The assay was performed using two different enzyme labels, horseradish peroxidase (HRP) and alkaline phosphatase (ALP), involving different working potentials for the amperometric detection. While a slope deviation of 25% was found when using ALP, this percentage was lower than 4% when using HRP. This difference can be explained by the oxidation of the electroactive interfering compounds found in wine, such as tartaric or malic acids, which may be adsorbed onto the electrode surface during the competition step. Whereas these compounds may be oxidized at +0.25 V (working potential applied for ALP detection), they are probably not at −0.2 V (working potential applied for HRP detection).

Similarly, Campàs et al. (2008) developed an immunosensor for OA detection in oyster and mussel samples. In this case, the device provided underestimated OA contents in relation to other techniques, probably due to the adsorption of some matrix compounds on the electrode surface, which can be directly electrooxidized or even cause the subsequent nonspecific binding of the antibodies on them. Despite these underestimated contents, the immunosensor was able to detect and quantify much lower toxin contents than LC-MS/MS, due to the higher sensitivity inherent in both the biorecognition event and the electrochemical detection. In what concerns the applicability of the immunosensor for OA, despite the appropriate limits of detection obtained, the underestimation should be corrected. The optimization of the extraction protocols and the eventual addition of a purification step prior to the analysis could, in principle, minimize the matrix effects.

6.3 CONCLUSION

The European Union is encouraging the use of new methodologies for the determination of natural phycotoxins, alternative or supplementary to the traditional biological testing methods. In the case of mycotoxins, although the official analysis methods are not biological, new approaches with high-throughput analysis abilities are gaining interest. In this direction, biosensors appear as promising tools for the preliminary screening of natural toxins in food samples. Although the research of biosensors has already derived appropriate devices for the determination of toxin content at the laboratory bench, much effort should be done on validation to demonstrate their applicability and possible integration into food quality monitoring programs. These studies should include the optimization of the extraction protocols, the assessment of matrix effects from different foods, the application to samples with multitoxin profiles, the validation with an exhaustive number of samples, and the analysis of multiple samples simultaneously. Nevertheless, the first step toward their real applicability has already been covered, as demonstrated by the high-performance analytical devices described in this chapter.

ACKNOWLEDGMENTS

This research was partially funded by INIA through the Project RTA2008-00084-00-00. Dr. Campàs gratefully acknowledges financial support from the Departament d'Educació i Universitats de la Generalitat de Catalunya, through the Beatriu de Pinós program.

REFERENCES

Alarcón, S.H., G. Palleschi, D. Compagnone, M. Pascale, A. Visconti, and I. Barna-Vetró. 2006. Monoclonal antibody based electrochemical immunosensor for the determination of ochratoxin A in wheat. *Talanta* 69:1031–1037.
Ben Rejeb, I., F. Arduini, A. Arvinte, et al. 2009. Development of a bio-electrochemical assay for AFB$_1$ detection in olive oil. *Biosensors and Bioelectronics* 24:1962–1968.
Campàs, M., B. Prieto-Simón, and J.-L. Marty. 2007. Biosensors to detect marine toxins: Assessing seafood safety. *Talanta* 72:884–895.
Campàs, M., P. de la Iglesia, M. Le Berre, M. Kane, J. Diogène, and J.-L. Marty. 2008. Enzymatic-recycling-based amperometric immunosensor for the ultrasensitive detection of okadaic acid in shellfish. *Biosensors and Bioelectronics* 24:716–722.
Carmichael, W.W., S.M.F.O. Azevedo, J.S. An, et al. 2001. Human fatalities from cyanobacteria: Chemical and biological evidence for cyanotoxines. *Environmental Health Perspectives* 109:663–668.
Commission Regulation (EC) No. 2074/2005. European Union. Laying down implementing measures for certain products under Regulation (EC) No. 853/2004. http://eur-lex.europa.eu/LexUriServ/LexUriServ.do?uri=OJ:L:2005:338:0027:0059:EN:PDF (accessed September 20, 2010).
Commission Regulation (EC) No. 1881/2006. European Union. Setting maximum levels for certain contaminants in foodstuffs. http://eur-lex.europa.eu/LexUriServ/site/en/oj/2006/l_364/l_36420061220en00050024.pdf (accessed September 20, 2010).
Commission Regulation (EC) No. 1126/2007. European Union. Amending Regulation (EC) No 1881/2006. Setting maximum levels for certain contaminants in foodstuffs as regards *Fusarium* toxins in maize and maize products. http://eur-lex.europa.eu/LexUriServ/site/en/oj/2007/l_255/l_25520070929en00140017.pdf (accessed September 20, 2010).
European Mycotoxins Awareness Network, http://www.mycotoxins.org.

European Union Commission Decision (2002/225/EC). The maximum levels and the methods of analysis of certain marine biotoxins in vbivlave mollusks, echinoderms, tunicates and marine gastropods. http://eur-lex.europa.eu/LexUriserv/LexUriserv. co?uri=CELEX:32002d0225:EN: HTML (accessed September 20, 2010).

European Union Commission Decision (2002/226/EC). Establishing special health checks for the harvesting and processing of certain bivalve mollusks with a level of amnesic shellfish poison (ASP) exceeding the limit laid down by Council Directive 91/4921 EEC. http://eur-lex.europa.eu/LexUriserv/LexUriserv.do?uri=oj:L:2002:075:0065:0066:EN: PDF (accessed September 20, 2010).

Food and Agriculture Organization (FAO). Food and Nutrition Paper 80. Room 2004. http://www.fao.org/docrep/007/y5486e00.HTM (accessed September 20, 2010).

Garthwaite, I. 2000. Keeping shellfish safe to eat: A brief review of shellfish toxins, and methods for their detection. *Trends in Food Science and Technology* 11:235–244.

Kuiper-Goodman, T. 1998. Food safety: Mycotoxins and phycotoxins in perspective. In *Mycotoxins and phycotoxins: Developments in chemistry, toxicology and food safety*, ed. M. Miraglia, H. van Edmond, C. Brera and J. Gilbert, 25–48. Fort Collins, CO: Alaken Inc.

Micheli, L., R. Grecco, M. Badea, D. Moscone, and G. Palleschi. 2005. An electrochemical immunosensor for aflatoxin M_1 determination in milk using screen-printed electrodes. *Biosensors and Bioelectronics* 21:588–596.

Parker, C.O., and I.E. Tothill. 2009. Development of an electrochemical immunosensor for aflatoxin M_1 in milk with focus on matrix interference. *Biosensors and Bioelectronics* 24:2452–2457.

Pittet, A. 2005. Modern methods and trends in mycotoxin analysis. *Mitteilungen aus Lebensmitteluntersuchung und Hygiene* 96:424–444.

Prieto-Simón, B., T. Noguer, and M. Campàs. 2007. Emerging biotools for assessment of mycotoxins in the past decade. *TrAC Trends in Analytical Chemistry* 26:689–702.

Prieto-Simón, B., M. Campàs, J.-L. Marty, and T. Noguer. 2008. Novel highly-performing immunosensor-based strategy for ochratoxin A detection in wine samples. *Biosensors and Bioelectronics* 23:995–1002.

Shumway, S.E. 1990. A review of the effects of algal blooms on shellfish and aquaculture. *Journal of the World Aquaculture Society* 21:65–104.

Singh, B.R. 2000. Intimate details of the most poisonous poison. *Nature Structural Biology* 7:617–619.

Tan, Y., X. Chu, G.-L. Shen, and R.-Q. Yu. 2009. A signal-amplified electrochemical immunosensor for aflatoxin B_1 determination in rice. *Analytical Biochemistry* 387:82–86.

Vig, A., A. Radoi, X. Muñoz-Berbel, G. Gyemant, and J.-L. Marty. 2009. Impedimetric aflatoxin M_1 immunosensor based on colloidal gold and silver electrodeposition. *Sensors and Actuators B: Chemical* 138:214–220.

7 Biosensors for Pesticides and Foodborne Pathogens

Munna S. Thakur, Raghuraj S. Chouhan,
and Aaydha C. Vinayaka

CONTENTS

7.1   INTRODUCTION

Quality assurance and safety of food during the manufacturing process are the essential requirements in the food industry. These should be carried out during the food processing at appropriate points to get an immediate quality status of the product being manufactured. Quality- and safety-related issues of the food basically include physical, organoleptic, chemical, microbiological, and toxicological characteristics. Food manufacturers must ensure that no product is dispatched until all the specified tests for the quality and safety have been satisfactorily achieved (Grunert 2005). Thus there is a need to develop fast, reliable, sensitive, and robust analytical systems.

Food contamination may be classified into two groups: (1) chemicals, and (2) biologicals.

A *pesticide* is a substance or mixture of substances intended for preventing, destroying, repelling, or mitigating any pest (Environmental Protection Agency, EPA). Although, there are benefits to the use of pesticides for preharvest and postharvest protection of food grains, there are also drawbacks, such as potential toxicity to humans and other animals. In this category different chemical classes such as organophosphates, organometallics (tin/mercury), carbamates, pyrethroids, and phenolic/phenoxy compounds have become popular for the following reasons:

1. They are *broad-spectrum* chemicals; that is, in terms of activity, they are toxic to a wide range of insects/pests, yet appear to have low toxicity to mammals.
2. They are *effective* in very low quantities.
3. They are *persistent* (do not break down rapidly in the environment by the action of soil microbes) so that they don't have to be applied often.

The use of pesticides is indispensable for crop protection, which is creating problems in the food chain due to their persistence for long periods and high toxicity at trace levels. Monitoring of these pesticides has become essential (Soler et al. 2007). Recently, it was reported that the concentration levels of pesticides detected were high as compared with maximum residue levels (MRLs) for drinking water and soft drinks. The detected pesticides were mainly those applied for postharvest treatments, and some of them contain chlorine atoms in their structure. The level of pesticides gets accumulated in the food chain and is increasing day by day, which leads to the depletion of normal health of the flora and fauna on the earth. Because of its high toxicity and risks of exposure to agricultural workers and to birds, and in response to manufacturers' requests, parathion application was canceled by the EPA in January 1992 on fruit, nut, and vegetable crops. The only uses retained are those on alfalfa, barley, corn, cotton, sorghum, soybeans, sunflowers, and wheat. Further, to reduce exposure to agricultural workers, parathion may be applied to these crops only by commercially certified aerial applicators, and treated crops may not be harvested by hand.

Analytical methods commonly used for pesticide analysis include gas chromatography (GC), high-performance liquid chromatography (HPLC), and spectrophotometric methods. However, these demand extensive sample preparation, skilled manpower, and huge investments. Hence, simple alternative methods like biosensors are preferred as they make the assay easier, quicker, and accurate with minimal sample pretreatment (Jiang et al. 2008).

7.2 BIOSENSORS

In brief, a *biosensor* is an instrument or device containing a sensing element of biological origin, which is either integrated within or is in intimate contact with a physicochemical transducer designed for quantitative or semiquantitative analysis of an analyte (Thakur and Karanth 2003). The biocomponent is the responsive element, which provides specificity and selectivity to the biosensor system. The biological components used in the biosensor construction can be divided into two categories: those where the primary sensing event results from catalysis (e.g., enzymes, microorganisms, cells, and tissues) and those that depend on an essentially irreversible binding of the target molecule (e.g., antibodies, receptors, nucleic acids) (Dennison and Turner 1995). Depending upon the physical/chemical change brought about by the biological element, different physicochemical transducers that are compatible with the change in signals can be used (Figure 7.1). These include electrochemical, optical, mechanical, and thermometric transducers (Turner 1992).

Electrochemical transducers include amperometric, potentiometric, and colorimetric transducers that are based on the measurement of change in electric current, total potential difference, and total charge transferred on the working electrode as compared with the reference electrode, respectively. Optical transducers are based on the principles of emission spectroscopy (e.g., fluorescence, phosphorescence, chemiluminescence, bioluminescence) reflectance spectroscopy (e.g., surface plasmon resonance [SPR], resonant mirror [RM] wave guide transducers). Mechanical transducers are based on the piezoelectric effect whereby the change in frequency of a quartz crystal oscillating in an applied electric field changes on the binding of a ligand to the immobilized biomolecular layer (e.g., quartz crystal microbalance [QCM] and surface acoustic

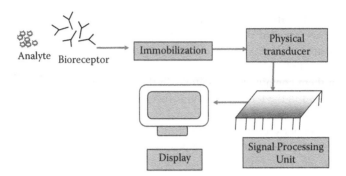

FIGURE 7.1   Schematic diagram of a biosensor unit.

wave [SAW] sensors) (Namsoo et al. 2007; Lange et al. 2008). Biosensors based on the change in temperature due to interaction of ligand/substrate with the biological molecules have traditionally used microcalorimeters, thermistors, peltiers, and other macro devices. Temperature-sensitive polymer films have been successfully used in conjunction with the optical techniques (Johansson et al. 1976; Dessy et al. 1990; Patra et al. 2008). An ideal biosensor should meet the following criteria:

1. *Selectivity*: The device should be highly selective for the target analyte; no cross-reactivity with moieties having similar chemical structure.
2. *Sensitivity*: The device should be able to measure in the range of interest for a given target analyte with minimum additional steps such as precleaning and preconcentration of the samples.
3. *Linearity and reproducibility of signal response*: The linear response range of the system should cover the concentration range over which the target analyte is to be measured.
4. *Response time and recovery time*: The sensor response should be rapid so that real-time monitoring of the target analyte can be done efficiently. The recovery time should be short for reusability of the system.
5. *Stability and operating life*: The biological element used for sensing should be active enough for a long time to get a reproducible response.

7.3   PESTICIDE DETECTION USING BIOSENSORS

Pesticides have been detected using enzyme- and antibody-based biosensor systems.

7.3.1   ENZYME-BASED BIOSENSORS FOR ORGANOPHOSPHOROUS PESTICIDES

Exposure to organophosphorous (OP) pesticides even at trace levels poses health hazards as OPs affect neurotransmission. Therefore, efficient and reliable monitoring of OP pesticides is important.

There are reports on both single- as well as multienzyme-based systems for OP pesticide detection (Gouda et al. 1997). Single-enzyme-based biosensors use either acetylcholinesterase (AChE) or butyrylcholinesterase (BuChE) as the biological component, and thiocholine production is monitored amperometrically (Rekha et al. 2000a) or acid production is monitored potentiometrically. Multienzyme-based biosensor systems use cholinesterase in conjunction with choline oxidase and measure hydrogen peroxide production (Rekha et al. 2000b) or oxygen consumption (Gouda et al. 2001a,b). There are also reports of using acid phosphatase (Gouda et al. 1997) and alkaline phosphatase (Ayyagari et al. 1995) for OP pesticide determination.

Multienzyme-based biosensors for OP pesticide detection in which one of the enzyme activities is inhibited by the pesticide in conjunction with the indicator enzyme that uses the product of the first enzymatic reaction as the substrate with concomitant consumption of oxygen which can be readily monitored using an oxygen electrode. Even though the pesticide inhibits only one of the two enzymes, monitoring of this inhibition is not possible without coupling it with a second enzyme. This disadvantage can be overcome by using single-enzyme systems like the one described by Rekha et al. (2000a,b). The use of an oxidase enzyme that is readily inhibited by the pesticide avoids the need for coupling a second enzymatic reaction merely for the purpose of monitoring the inhibition reaction. It is known that organophosphates exhibit their pesticidal property through a strong inhibition of AChE activity. This inhibition principle has been used to develop a biosensor for detection of OP pesticides (Gulla et al. 2002)

The principle involved in an AChE-based biosensor is the hydrolysis of acetylthiocholine to thiocholine and acetic acid catalyzed by AChE as given below:

$$H_2O + (CH)_3 - N - (CH_2)_2 - S - C(CH_3) = O \xrightarrow{Ach}$$

Acetyl Thiocholine

$$(CH_3)_3 - N - (CH_2)_2 - SH + CH_3 - COOH$$

$$\downarrow$$

(+410mV Anodicoxidation)

$$\downarrow$$

$$S - (CH_2)_2 \quad -N - (CH_3)_3$$
$$|$$
$$S - (CH_2)_2 \quad -N - (CH_3)_3 \quad +2H^+ \quad +2e^-$$

Thiocholine oxidizes at the electrode surface at +410 mV, thereby resulting in an increase in current output. OP pesticide inhibits AChE, decreasing thiocholine production, and in turn causes a decrease in current output, which is correlated with the pesticide concentration.

Using this principle, a laboratory biosensor has been constructed at the Central Food Technological Research Institute (CFTRI) (Mysore, India), for paraoxon with a sensitivity of 0.5 ppm.

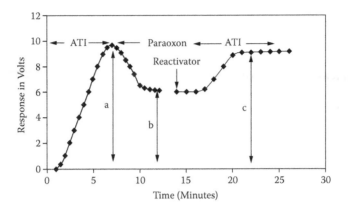

FIGURE 7.2 Typical voltage response of enzyme electrode for substrate (ATI) followed by the OP pesticide paraoxon. (Reprinted from *Biochimica et Biophysica Acta (BBA)—Protein Structure and Molecular Enzymology* 1597, Gulla, K.C., M.D. Gouda, M.S. Thakur, and N.G. Karanth, Reactivation of Immobilized Acetyl Cholinesterase in an Amperometric Biosensor for Organophosphorus Pesticide, 133–139. Copyright 2002, with permission from Elsevier.)

7.3.1.1 Enzyme Inactivation Problems with AchE

Biosensors based on AchE inhibition have been known for monitoring of OP pesticides in food and water samples. However, strong inhibition of the enzyme is a major drawback in practical application of the biosensor; this can be at least partially overcome by reactivation of the enzyme for repeated use. A study on the enzyme reactivation by oximes was explored (Gulla et al. 2002). Two oximes, namely, 1,1'-trimethylene-bis-4-formylpyridinium bromide dioxime (TMB-4) and pyridine 2-aldoxime methiodide (2-PAM) were compared for the reactivation of the immobilized AChE (Figure 7.2). TMB-4 was found to be a more efficient reactivator under repeated use, retaining more than 60% of initial activity.

7.3.1.2 Acid Phosphatase Inhibition-Based Detection

An amperometry-based biosensor has also been developed to analyze the OP pesticide using the dual enzyme acid phosphatase and glucose oxidase (GOD) (Gouda et al. 1997). The biochemical reactions occurring in this dual enzyme system are

$$\text{Glucose 6-phosphate} + H_2O \xrightarrow{\text{acid phosphatase}} \text{glucose} + HPO_4^{2-}$$

$$\text{Glucose} + O_2 \xrightarrow{\text{GOD}} \text{gluconic acid} + H_2O_2$$

Glucose 6-phosphate is converted to glucose in the presence of acid phosphatase and further oxidized to gluconic acid catalyzed by GOD. Activity of the acid phosphatase is inhibited by the OP pesticide, resulting in a reduction in the second reaction, monitored in terms of a change in output voltage of the biosensor, which is proportional to the pesticide concentration.

Using the above system, an amperometric biosensor consisting of a potato layer rich in acid phosphatase and an immobilized GOD membrane, when operated in conjunction with a Clark-type electrode, detected the pesticide. A notable advantage of this biosensor is that the inhibition of acid phosphatase by the pesiticide is reversible and thereby eliminates the serious problem of enzyme inactivation and the necessity for its reactivation, which is not efficient.

7.3.1.3   Ascorbate Oxidase-Based Biosensors

Biochemical studies have indicated that a number of enzymes including ascorbate oxidase are inhibited by OP pesticides (Rekha et al. 2000a). A biosensor for the quantification of OP pesticides by making use of ascorbic acid oxidase enzyme inhibition has been developed at CFTRI (Rekha et al. 2000a). In this system, cucumber (*Cucumis sativus*) tissue, rich in ascorbic acid oxidase, was used for the detection of OP pesticide ethyl paraoxon, which inhibited the ascorbic acid oxidase activity. The optimal concentration of ascorbic acid used as substrate was found to be 5.67 mM. The biosensor response was found to reach steady state within 2 minutes. A linear relationship was obtained between the percentage inhibition of the enzyme substrate reaction and the pesticide (ethyl paraoxon) concentration in the range 1–10 ppm.

The biochemical oxidation of L-ascorbate catalyzed by ascorbate oxidase is given below.

$$2 \text{ L-ascorbate} + O_2 \xrightarrow{\text{Ascorbate oxidase}} 2 \text{ Dehydroascorbate} + 2H_2O$$

Many biosensors that are used for pesticide detection are based on the inhibition reaction or catalytic activity of several enzymes in the presence of pesticides (Anh et al. 2004; Deo et al. 2005; Laschi et al. 2007). Since many pesticides have a similar mode of action affecting the activity of the same enzyme, most enzyme-based biosensors are used for screening purposes and are unspecific for individual pesticides. Hence they can only detect total pesticide content and do not provide specific information about a particular pesticide (Jiang et al. 2008).

7.3.1.4   Biosensors for Organochlorine Pesticides

The most commonly used organochlorine pesticides are dichlorodiphenyltrichloroethane (DDT) and hexachlorohexane (HCH). It is reported that most water bodies and soils are getting contaminated with these chemicals and their degradation products (Thakur and Karanth 2003). DDT persists with a half-life of about 10 years. Toxic effects have been attributed to these pesticides including neurotoxic, nephrotoxic, mutagenic, carcinogenic, teratogenic, and immunosuppressiveness (Karamus et al. 2002). Due to these toxic effects, these pesticides have been either banned or allowed for restricted applications in agriculture and public health programs. The persistence of these pesticide residues in the environment and their entry into the food chain have made it mandatory to develop new, simple, specific, rapid, sensitive, and portable techniques for their detection. A work on the immobilized dehydrohalogenase-based potentiometric biosensor is being done for the detection of DDT and HCH in water samples at CFTRI, Mysore. The chloride ion released as a result of dehalogenation by immobilized dehydrohalogenase during the degradation of DDT and HCH respectively was detected by using an ion-selective electrode.

7.3.2 Immunosensors

Immunoassays are highly selective due to their inherent selectivity between antibodies and antigens. Antibodies (Abs) are serum glycoproteins of the immunoglobulin (Ig) class and are produced by the vertebrate immune system against foreign materials called antigens (Ags). Antibodies bind to their antigens with very high specificity. This specificity is exploited in the development of immunoassays (IAs). The result of the binding reaction between the Ab and an analyte is usually made visible by means of enzymatic, chemiluminescent, fluorescent, or radioactive markers. According to the label used, IAs can be classified into enzyme immunoassays (EIAs), radioimmunoassays (RIAs), fluorescence immunoassays (FIAs), or chemiluminescent immunoassays (CLIAs) (Gosling 1990; Chouhan et al. 2006). Immunochemical assays have gained the interest of the scientific community as a potential tool that can overcome difficulties encountered during pesticide detection and analysis (Mullet et al. 2000; Suri et al. 2002; Thakur and Karanth 2003; Farre et al. 2007).

7.3.2.1 Advantages of Immunosensor Methods over Conventional Methods

1. High specificity to target analyte
2. Rapid response and simplicity of operation
3. Require no or minimal pretreatment of samples
4. Can operate in turbid/colored solutions
5. Can be used without addition of reagents
6. Easier for miniaturization, portability, and economy
7. Amenability for continuous operation
8. Can be easily interfaced with microprocessor or computer for data processing and control

Due to strong and specific binding of an antigen with antibody, an immunosensor needs harsh conditions like low pH and higher ionic concentrations to dissociate the antigens and regenerate the binding sites of the antibody (Kandimalla et al. 2004). Under these harsh conditions, the antibody is washed away or denaturation occurs, resulting in a decrease in the number of reuses and therefore overall operating life (Hermanson et al. 1992; Kandimalla et al. 2004).

Biosensor technology has received an impetus because of the generation of specific antibodies against target pesticide molecules (Jiang et al. 2008). Immunosensors are similar to immunoassays; the only difference is that an immunoassay transducer is not an integral part of the analytical system. In the immunosensor, the transduction is achieved by the use of specific markers/labels, where the principle is very similar to the immunoassays. Based on the marker/labels used for detection, immunosensors can be divided in to two groups:

1. Labeled Formats
2. Label-Free Formats

7.3.2.2 Labeled Formats

This method involves a label to quantify the amount of antibodies or analyte bound during the incubation. The most commonly used labels are enzymes such as HRP, alkaline

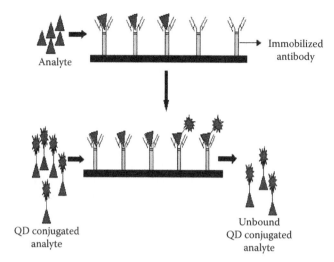

FIGURE 7.3 Direct competitive immunosensor/assay. (With kind permission from Springer Science & Business Media: *Analytical and Bioanalytical Chemistry*, Focus on Quantum Dots as Potential Fluorescent Probes for Monitoring Food Toxicants and Foodborne Pathogens, Vol. 397, 2010, 1445–1455, Vinayaka, A.C. and M.S. Thakur.)

phosphatase, glucose oxidase, β-galactosidase, and so forth. Fluorescence markers such as fluorescien isothiocyanate (FITC) and quantum dots (QDs) are reported in the literature for the detection of pesticides (Mulchandani et al. 2001; Vinayaka et al. 2009). More recently, nanoparticle-based labeling is becoming more efficient (Lisa et al. 2009) for detection of pesticides in food samples. Commonly, two different formats adopted for labeled immunosensors are available: (1) sandwich assays and (2) competitive assays. A sandwich assay consists of two recognition steps. In the first step, the antibodies are immobilized on a transducer surface, allowing it to capture the analyte of interest. In the second step, labeled secondary antibodies are used for binding with the previously captured analyte. The immunocomplexes are formed, and the signals from labels increase in proportion to the analyte concentration (Sadik and Van Emon 1996).

In competitive assays, the analyte competes with labeled analyte for a limited number of antibody binding sites (Figure 7.3). As the analyte concentration increases, more labeled analyte is displaced, giving a decrease in signal if antibody-bound labeled analyte is detected (Bange et al. 2005; Vinakyaka et al. 2009). Although the labeled format is usually more sensitive, labeled immunosensors are not capable of real-time monitoring of the Ab–Ag reaction and increase both development and operation costs compared with label-free immunosensors. The amount of target analyte can be inferred from the amount of labels that bind to the interface.

7.3.2.3 Label-Free Formats

This procedure detects the binding of pesticide and the antibodies on a transducer surface without any labels. There are also two basic types in this format: direct and indirect (Jiang et al. 2008). In the first type, the response is directly proportional to the amount of pesticides present. The vital advantage of these direct immunosensors is the simple, single-stage reagentless operation. However, such direct immunosensors are often inadequate to generate a highly sensitive signal resulting from antibody–antigen binding interactions, and it is still difficult to meet the demand of sensitive detection. The second type, also based on competitive formats, is carried out as a binding inhibition test. The antigen (pesticide–protein conjugate) is first immobilized onto the surface of a transducer, and then pesticide–antibody mixtures are preincubated in solution. After being injected on the sensor surface, the antibody binding to the immobilized conjugate is inhibited by the presence of target pesticide.

The biorecognition element determines the degree of selectivity or specificity of the biosensor, whereas the sensitivity of the biosensor is greatly influenced by the transducer. According to the transduction mechanism, immunosensors can be further classified into electrochemical, optical, piezoelectric, and nanomechanic immunosensors.

7.3.3 ELECTROCHEMICAL IMMUNOSENSORS

Electrochemical transducers are commonly used for biosensor application. The principle is based on the electrical properties of the electrode that is affected by Ab–Ag interactions. Determination of the level of pesticides can be done by measuring the change of potential, current, conductance, or impedance caused by the immunoreaction. These electrochemical-based devices are not affected by sample turbidity, quenching, or interference from absorbing and fluorescing compounds commonly found in biological samples.

7.3.3.1 Potentiometric Methods

Potentiometric immunosensors are based on measuring the changes in potential induced by the label used, which occur after the specific binding of the Ab–Ag. During these reactions measurement of the potential across an electrochemical cell containing the Ab or Ag is monitored. The measured potential is given by the Nernst equation:

$$E = cons \tan t = \frac{RT}{nF \ln a}$$

where E is the potential to be measured; R, T, and F are constants; n is the electron transfer number; and a is the relative activity of the ion of interest. Using this principle 2,4-dichlorophenoxyacetic acid (2,4-D) and 2,4,5-trichlorophenoxyacetic acid (2,4,5- T) have been reported (Dzantiev et al. 1996). The assay monitored the competitive binding of free pesticide and pesticide-peroxidase conjugate with Ab immobilized on a graphite electrode by potentiometric measurement of peroxidase activity in the immune complexes on the electrode surface. Detection limits were 40 ng/mL for 2,4-D and 50 ng/mL for 2,4,5-T. The life of the electrode extended to 60 sequential measurements. The assay optimization steps (choice of electrode, substrate mixture, regeneration regime, etc.) were characterized in another paper for herbicide simazine monitoring (Yulaev et al. 2001). The limit of simazine detection was 3 ng/mL. The operating lifetime of the sensor was 15 days.

An immunosensor based on ion-selective field effect transistors (ISFETs) was developed for the quick determination of the herbicide simazine (Starodub et al. 2000). The activity of bound peroxidase was measured by a basic pH shift of ascorbic acid solution after addition of hydrogen peroxide. The limit of simazine detection by competitive method was 1.25 ng/mL and the linearity was observed in the range of 5–175 ng/mL.

7.3.3.2 Amperometric Methods

Amperometric sensors are a subclass of voltametric sensors. Amperometric immunosensors detect the concentration-dependent current generated when an electroactive species is either oxidized or reduced at the electrode surface to which Ab–Ag binds specifically (Patel 2002). It is held at a fixed electrical potential. The current is directly proportional to specific Ab–Ag binding. The current and bulk concentration of the detecting species can be approximated as

$$i = ZFkmC^*$$

where i is the current to be measured, Z and F are constants, km is the mass transfer coefficient, and C^* is the bulk concentration of the detection species.

While amperometric methods offer good resolution, they cannot discriminate different electroactive species undergoing an electrochemical reaction at the electrode (Bond 1980). This is important for avoiding interference that could be present in the sample solution.

For the simultaneous analysis of several samples using a device, Skládal and Kaláb (1995) developed a multichannel immunosensor for detection of 2,4-D. The 2,4-D molecule conjugated to horseradish peroxidase was used as a tracer, which was determined amperometrically using hydrogen peroxide and hydroquinone as substrates. The detection limit for free 2,4-D in water was 0.1 ng/mL. A similar immunosensor coupled with an enzyme-linked immunosorbent assay (ELISA) microtiter plate was also reported (Deng and Yang 2007), and the detection limit of 0.072 ng/mL was achieved. The advantages of the electrochemical detector were high stability and sample throughput, lower detection limit, and the ability to be repeatedly used.

In another amperometric immunosensor (Baumner and Schmid 1998), hapten-tagged liposomes entrapping ascorbic acid as a marker molecule were chosen for the generation and amplification of the signal, and the sensitivity of measurements in tap water was below 1 ng/mL of atrazine and terbutylazine.

A separation-free electrochemical technique was developed for the determination of atrazine (Keay and McNeil 1998) and chlorsulfuron (Dzantiev et al. 2004) in water samples. Signal detection was carried out by a screen-printed electrode, incorporating horseradish peroxidase in the carbon ink working electrode to determine peroxidase activity. Grennan et al. (2003) described an amperometric immunosensor for the analysis of atrazine using recombinant single-chain antibody (scAb) fragments. The sensor was based on carbon paste screen-printed electrodes incorporating the conducting polymer, which enables direct mediatorless coupling to take place between the redox centers of antigen-labeled HRP and the electrode surface (Figure 7.4). The system was capable of measuring atrazine to a detection limit of 0.1 ng/mL.

Killard et al. (2001) described an amperometric immunosensor for real-time environmental monitoring. It was possible to detect atrazine as low as 0.13 μM using

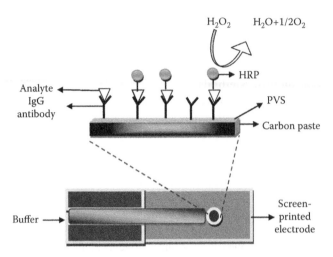

FIGURE 7.4   Schematic diagram of the electrochemical real-time sensing process.

equilibrium incubation with an analytical range of 0.1–10 μM. Since an amperometric sensor can operate in other modes, cyclic voltammetry was performed in the analysis of the pesticide (Killard et al. 2000). An electrochemical immunosensor with paraoxon Ab loaded on the gold nanoparticles to monitor the concentration of paraoxon in aqueous samples was reported by Hu et al. (2001), and the detection limit of 12 ng/mL was achieved. To a large extent, the immunosensor can eliminate the interferences from other coexisting electroactive pollutants.

7.3.3.3 Capacitance/Conductance/Impedance Methods

The concepts of impedance, conductance, capacitance, and resistance are different ways of monitoring the test system and are all interrelated (Milner et al. 1998). Impedance immunosensors measure changes of an electrical field. Those changes could be overall electrical conductivity of the solution and capacity alteration due to the Ab–Ag interaction on the electrode surface, which also can be reflected in impedimetric response.

Electrochemical impedance spectroscopy (EIS) is a sensitive technique that detects the electrical response of the system studied after application of a periodic small-amplitude alternating current (AC) signal. The electrochemical impedance immunosensors have attracted extensive interest in the sensing of Ag–Ab interaction (Sadik et al. 2002; Darain et al. 2004; Yang and Li 2005; Chen et al. 2006). EIS in connection with immunochemical methods was tested for the direct determination of the herbicide 2,4-D (Navartilova and Skladal 2004). Formation of immunocomplex resulted in the change of impedance parameters on the sensing surfaces that were detected for analyte concentration. It was possible to estimate 2,4-D in a concentration range from 45 nM to 450 μM (Hleli et al. 2006).

7.3.4 Optical Immunosensors

A general optical sensor system consists of a complex optical bench having a light source, a number of optical components with specific characteristics that direct this

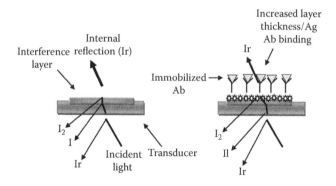

FIGURE 7.5 Schematic diagram of the RIfS detection system.

light to a modulating agent, a modified sensing head, and a photodetector. Different techniques can be used for creating an optical change, for example, reflectometric interference spectroscopy (RIfS), interferometry, optical wave-guide light mode spectroscopy (OWLS), total internal reflection fluorescence (TIRF), scattering, and SPR.

Optical immunosensors offer advantages of compactness, flexibility, resistance to electrical noise, and a small probe size. A label-free approach of detection is preferred. In addition, the use of very low amounts of reagents makes optical immunosensors advantageous to use.

7.3.5 REFLECTOMETRIC INTERFERENCE SPECTROSCOPY

RIfS is one of the reflectometric techniques that have been applied to direct immunosensing. A white incident light passing the interface between different refractive indexes will be reflected in part. These reflected beams superimpose and build a characteristic interference spectrum. The binding of Ag on the immobilized Ab to the surface changes the thickness of the layer, which causes a change in the reflectance spectrum (Figure 7.5). Thus, the interaction process between the Ab and the hapten derivative on the surface can be detected (Proll et al. 2004).

The optical thickness of the transducer changes during binding events onto the surface. Brecht et al. (1995) investigated a direct optical immunosensor based on RIfS for detection of atrazine with surface-immobilized antigen. They made comparison to other optical transducers and the feasibility of this direct optical immunosensor in pesticide detection. The exiting RIfS transducer technology had not reached the European Union limit of 0.1 ng/mL or below. However, several other optical transducers could reach the required limits of detection. Sample handling and physical constraints limited the device performance under practical conditions, such as the change in refractive index of the sample with temperature.

7.3.6 INTERFEROMETRY

When an immunoreaction takes place on the wave-guide surface, it brings about a shift in the refractive-index profile within the evanescent field; consequently, the effective refractive index of the wave-guide system is changed. With the Mach–Zehnder inter-ferometry (MZI), an optical wave guide is split into two arms and after a certain distance they are recombined again. The sensor arm will be exposed to a variation of the refractive index due to immunoreactions in the sensor channel. During this time, light traveling in the sensing arm will experience a phase shift in comparison to reference wave-guided light (Prieto et al. 2003).

Lechuga et al. (1995) discussed the feasibility of an evanescent wave interferome-ter immunosensor for pesticide detection. The interferometer immunosensor showed a high resolution (2×10^{-3} nm). An increase of both the receptor layer binding capac-ity and net response rate should result in the required increased sensitivity of the MZI, which made the direct detection of pesticides.

7.3.7   OPTICAL WAVE-GUIDE LIGHT MODE SPECTROSCOPY

The OWLS technique is a new sensing technique using evanescent field for the *in situ* and label-free study of surface processes at molecular levels. It is based on the precise measurement of the resonance angle of linearly polarized laser light, dif-fracted by a grating and incoupled into a thin wave-guide layer (Adanyi et al. 2006). The incoupling is a resonance phenomenon that occurs at a defined angle of inci-dence that depends on the refractive index of the medium covering the surface of the wave guide. In the wave-guide layer, light is guided by total internal reflection to the edges where it is detected by photodiodes. By varying the angle of incidence of the light, the mode spectrum is obtained from which the effective refractive indexes are calculated for both electrical and magnetic field waves (Luppa et al. 2001). This technique has been applied for the detection of the herbicide trifluralin (Székács et al. 2003). Within the immobilized Ab-based immunosensors, this method allowed the detection of trifluralin only above 100 ng/mL.

7.3.8   TOTAL INTERNAL REFLECTION FLUORESCENCE

TIRF is based on utilization of the evanescent wave of an electromagnetic field, which extends out from the interface of two different media into the medium with lower refractive index. This field is created when light from a specific angle is totally reflected due to differences in refractive index of the two different media. Molecules with fluorescence characteristics are excited in the evanescent field of the wave guide, creating a fluorescent evanescent wave. This evanescent wave allows quanti-tative measurements of adsorbed molecules on the surface without influence of any moderate bulk concentration of the same analyte (Engstrom et al. 2006).

The advantages of the format are its robustness, availability, versatility, low cost, and portability, which allow it to be used in the field, while its main drawback is the need for labels (Gonzalez-Martinez et al. 2006). A TIRF immunosensor was shown to allow the detection of analytes in one single test cycle (Klotz et al. 1998). Calibration curves obtained for 2,4-D and simazine had detection limits of 0.035 and 0.026 ng/mL, respectively. One limiting factor on the ability to perform more than one assay simultaneously on the same transducer was the availability of low cross-reactant Ab combined with high affinity between the antibody and the analyte.

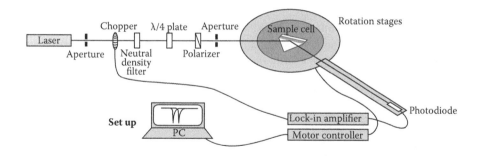

Surface Plasmon Resonance

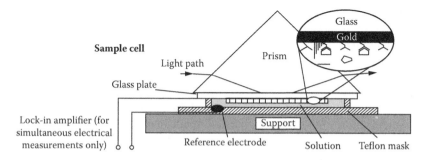

FIGURE 7.6   Schematic diagram of an SPR device.

Prototype immunosensors for multianalyte applications, such as the European River Analyzer (RIANA) based on a TIRF that takes place on an optical transducer chip, were developed in recent years and applied to monitor pesticides in water samples (Mallat et al. 2001a,b; Rodriguez-Mozaz et al. 2004). Detection limits of several analytes were at parts per trillion level. Individual assays were completed within 15 minutes including total regeneration of the transducer.

7.3.9   SURFACE PLASMON RESONANCE

SPR immunosensing involves immobilizing Ab (or Ag) by a coupling matrix to the thin gold surface deposited on the reflecting surface of a glass prism (Figure 7.6). Interaction of Ag and Ab on the surface will elicit a change in the refractive index as variations in light intensity (Dutra and Kubot 2007). The detection principle relies on detecting changes in the refractive index of the solution close to the surface of the sensor chip due to shifts in mass occurring after biomolecule binding. This is in turn directly related to the concentration of solute in the surface layer (Hock 1997; Homola 2003).

  SPR has an inherent advantage over other types of biosensors in its versatility and capability of monitoring binding interactions without the need for labeling of the biomolecules. It is versatile owing to its outstanding attributes of miniaturization, reliable portable instrumentation, and automation.

Monitoring of the pesticide chlorpyrifos in water samples was performed using SPR immunosensors (Mauriz et al. 2006a). The chlorpyrifos derivative was immobilized onto the gold-coated sensing surface and competed with free chlorpyrifos for binding to the Ab, and, as a result, increasing concentrations of chlorpyrifos will reduce the SPR signal. Other examples of single and multianalyte assays for simultaneous detection of different pesticides by SPR were reported by the same research group (Mauriz et al. 2006b,c,d). This portable immunosensor based on SPR technology could provide a highly sensitive detection of pesticide analytes at nanogram per liter levels. The regeneration of the immunosurface was accomplished throughout more than 200 assay cycles without degradation of the covalently immobilized molecule. The stability of the immunosensor was proved by performing 15 series of daily measurements. Another sensitive and reusable SPR-based immunosensor was developed for the determination of 2,4-D (Gobi et al. 2005). The SPR sensor was capable of detecting parts-per-billion levels of 2,4-D in 20 min and the regeneration ability enabled the achievement of as many as 20 measurement cycles.

SPR immunosensors may suffer from disturbances caused by changes in the refractive index or temperature. The use of a reference surface makes it possible to separate signals related to binding events from signals caused by differences in refractive index between a sample and running buffer (Mullett et al. 2000).

Schmid et al. (2006) have reported a method for the site-directed immobilization of antibodies on gold substrates for SPR applications (Figure 7.6). Protein A was used to modify gold surface on a glass slide via homobifunctional cross-linker dithiobissuccinimide propionate (DSP) to achieve uniform, stable, and sterically accessible antibody coating. The modified gold surface was stable for several weeks, and the reproducibility was satisfactory.

7.3.10 FLUORESCENCE/LUMINESCENCE

Fluorescence occurs when a valence electron is excited from its ground state to an excited singlet state. The excitation is produced by the absorption of light of sufficient energy (Lazcka et al. 2007). The common principle of luminescence immunosensors is that an indicator or chemical reagent placed inside or on an immmunoreactor is used as a mediator to produce an observable optical signal. Typically, conventional techniques, such as spectrometers, are employed to measure changes in the optical signal. This approach has been reviewed by Schobel et al. (2000).

A fluorometric immunosensor system was developed based on the principle of heterogeneous competitive enzyme assay using mouse monoclonal anticarbaryl Ab either in solution or immobilized for the determination of pesticides (Gonzalez-Martinez et al. 1997). In the direct format, the limit of detection was 26 ng/L (11 min/ assay), and the useful life of the sensor was 60–70 cycles. In the indirect format, the limit of detection was 284 ng/L (17 min/assay), and the useful life of the sensor was 160–200 cycles. Another portable miniaturized flow-injection immunosensor was designed for field analysis of environmental pollutants (Ciumasu et al. 2005).

In an optical immunosensor, indium tin oxide (ITO) could make a good platform using fluorescent nanoparticle labels in a competitive assay format for small pesticide molecule detection. However, when used in combination with fluorescent labels,

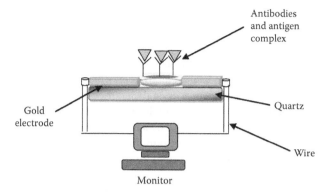

FIGURE 7.7 Schematic diagram of a piezoelectric immunocrystal.

a highly sensitive excitation/detection system was needed. A disposable multiband optical plastic capillary fluoroimmunosensor using Ab labeled with fluorescein was exploited for the simultaneous determination of four different pesticides in the same sample (Mastichiadis et al. 2002). The detection limits of this immunosensor were below or equal to 0.1 ng/mL.

Fluorescence polarization immunoassay (FPIA) is a type of homogeneous competitive fluorescence immunoassay. In this technique antigen from the specimen and antigen-fluorescein (AgF)-labeled reagent compete for the binding sites on the antibody. As a homogeneous immunoassay, the reaction is carried out in a single reaction solution, and the bound antibody-AgF complex does not require a wash step to separate it from the "free" labeled AgF. FPIA uses the three principles of fluorescence, rotation of molecules in solution, and polarized light (Farre et al. 2006).

FPIA is utilized to provide accurate, sensitive measurement of small analytes such as therapeutic drugs and some hormones. An FPIA based on monoclonal antibody (MAb) for the detection of flouresin-labeled parathion-methyl and its derivatives has been developed with a linear range of 0.025–10 mg/L. The limit of detection was 15 µg/L. Recovery in spiked samples averaged between 85% and 100%. This method showed high specificity and reproducibility (Kolosova et al. 2003).

7.3.11 PIEZOELECTRIC IMMUNOSENSORS

Piezoelectric immunosensors are devices based on materials such as quartz crystals with Ag or Ab immobilized on their surface, which resonate on application of an external alternating electric field (Figure 7.7).

The resonant frequency has a proportional relation to the mass changes of the quartz crystal. The biospecific reaction between the two interactive molecules, one immobilized on the surface and the other free in solution or gas phase, can be followed in real time. The potential of QCM devices in chemical sensor applications was realized after Sauerbrey derived the frequency-to-mass relationship

$$\Delta F = (-2.3 \times 10^6) f_o^2 \frac{\Delta M}{A}$$

where ΔF (Hz) is the change in the fundamental frequency of the coat crystal, f_0 is its resonant frequency, A (cm^2) is the area coated, and ΔM (g) is the mass deposited. This Sauerbrey equation, however, holds only for the case of rigid coated material. The frequency responses are also influenced by many factors, such as effective viscosity, conductivity, dielectric constant, electrode morphology, density and temperature of the liquid, and ionic status of a crystal/electrode interface to a water/buffer. Thus, when used for analysis, operating conditions should be evaluated.

In many cases, piezoelectric immunosensors can be designed for detecting pesticides without the need for expensive or hazardous labels. They offer advantages such as real-time output, high sensitivity, simplicity of use, and cost effectiveness. The main problem associated with these immunosensors is a change from a sample with low ionic strength to a sample with high ionic strength and vice versa. Moreover, a long time is needed to get a stable baseline due to changes in viscosity and other parameters close to the QCM surface.

Improved direct piezoelectric immunosensors operating in liquid solution for the competitive label-free assay of 2,4-D were developed (Horacek and Skladal 1997). Development of piezoelectric immunosensors for competitive and direct determination of atrazine are reported (Pribyl et al. 2003) and 2,4-D (Halamek et al. 2001). In the competitive format, the mixture of object pesticide and specific Ab was preincubated in solution and passed through the flow cell with the piezoelectric crystal modified with hapten-pesticide. In the direct method, an immunosensor was constructed by oriented immobilization of Ab to activated sensing surface, followed by direct binding of analyte present in solution. The result showed that the detected concentration of a direct assay was low, while the competitive assay provided sensitive detection.

7.3.12 Micronanomechanics Immunosensors

Micronanomechanics biosensors are based on the surface stress and consequently the bending of microfabricated silicon cantilevers, caused by the adsorption of molecules onto the sensor surface (Moulin et al. 2000; Raiteri et al. 2001). This technique was applied for detection of DDT by performing a competitive assay (Alvarez et al. 2003). A synthetic hapten of the DDT conjugated with bovine serum albumin (BSA) was covalently immobilized on the gold-coated side of the cantilever by using thiol self-assembled monolayers, and the cantilever was exposed to a mixed solution of the monoclonal Ab and DDT. It allowed pesticide detection below the nanomolar range without the need of labels. Compared with other label-free biosensors such as SPR and QCM biosensors, the reaction area was tiny (100 μm^2). Cantilever miniaturization combined with micronanofluidics could lead to femto molar sensitivity with low reagent consumption. Kaur et al. (2004) used an atomic force microscope (AFM) to directly evaluate specific interactions between pesticides and Ab on a biosensor surface. Oriented immobilization of Ab against 2,4-D and atrazine on gold was carried out to create the active immunosensor surfaces.

The study indicated that AFM could be utilized as a convenient immunosensing tool for confirming the presence and also assessing the strength of Ab–hapten interactions on biosensor surfaces (Figure 7.8). Future trends in nanomechanical biosensors will mainly address new fabrication methods to enhance cantilever response.

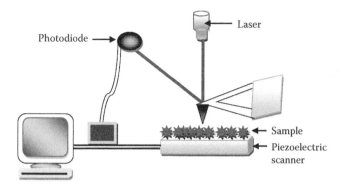

FIGURE 7.8 Schematic representation of the components of an atomic force microscope.

7.4 CHEMILUMINESCENCE

Chemiluminescence (CL) is the generation of electromagnetic radiation as light by the release of energy from a chemical reaction. While the light, in principle, is emitted in the ultraviolet, visible, or infrared region, those emitting visible light are the most common (Figure 7.9). Light emission occurring as a result of chemical processes is referred to as CL and is the most interesting and useful for several application in biosensor development.

CL reaction is summarized as follows:

$$HRP + H_2O_2 \rightarrow Complex\,1 + H_2O$$

$$Complex\,1 + Luminol(LH_2) \rightarrow Luminol\;radical(LH^-)$$

$$Complex\,2 + Luminol\;radical(LH^-) \rightarrow HRP + Oxidized\;Luminol$$

FIGURE 7.9 Chemiluminescence image.

In the presence of hydrogen peroxide (H_2O_2), HRP catalyzes the oxidation of luminol. Immediately following the oxidation of luminol, it reaches to an excited state and decays to the ground state by emitting light. Using this method it is possible to detect the pesticide (Chouhan et al. 2006).

A large number of molecules are capable of undergoing CL reactions; all the reactions studied to date being of an oxidative nature. The best-known CL reactions biochemical interests are those of luminol and its derivatives.

Recently, CL has had an impact on biochemical analysis, on cell biology, and on medicine. CL compounds have also had a recent impact on biotechnology, as replacements for radioactive labels such as I^{125} and P^{32}. The chemical mechanism of light emission in several groups of luminous organisms, "Living Light," has also been established.

7.4.1 LUMINOL-BASED CHEMILUMINESCENCE

These glowing reactions are generally oxidations, and a good example is the oxidation of 5-aminophthalhydrazide, or luminol, which produces a brilliant blue-green light.

Luminol is widely used as a CL reagent. The CL emitter is a "direct descendent" of the oxidation of luminol (or an isomer like isoluminol) by an oxidant in basic aqueous solution. Probably the most useful oxidant is hydrogen peroxide; however, other oxidants have been used such as perborate, permanganate, hypochlorite, and iodine. If the fuel is luminol, the emitting species is 3-amminophthalate; however, luminol-derivatized analytes allow for determination of compounds that would not normally show CL in the system and presumably have slightly different emitters.

CL and bioluminescent reactions usually involve the cleavage or fragmentation of the O-O bond of an organic peroxide compound. Peroxides, especially cyclic peroxides, are prevalent in light-emitting reactions because the relatively weak peroxide bond is easily cleaved and the resulting molecular reorganization liberates a large amount of energy. In order to achieve the highest levels of sensitivity, a CL reaction must be as efficient as possible in generating photons. Each CL compound or group can produce no more than one photon.

It is possible to increase the yield of CL when the emitter is poorly fluorescent. A highly fluorescent acceptor is used in these cases in order to transfer the excitation energy from the primary excited state compound to the fluorescent acceptor/emitter.

An appreciation of some of these fundamental principles of CL reactions will help in understanding how to design CL assays.

Chemiluminescent reactions can be grouped into three types:

1. Chemical reactions using synthetic compounds and usually involving a highly oxidized species such as a peroxide are commonly termed *chemiluminescent reactions.*
2. Light-emitting reactions arising from a living organism, such as the firefly or jellyfish, are commonly termed *bioluminescent reactions.*
3. Light-emitting reactions that take place by the use of electrical current are designated *electrochemiluminescent reactions.*

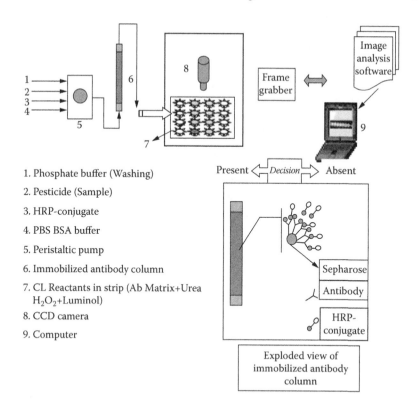

1. Phosphate buffer (Washing)

2. Pesticide (Sample)

3. HRP-conjugate

4. PBS BSA buffer

5. Peristaltic pump

6. Immobilized antibody column

7. CL Reactants in strip (Ab Matrix+Urea H$_2$O$_2$+Luminol)

8. CCD camera

9. Computer

FIGURE 7.10   Flow chart of qualitative determination of methyl parathion. (From Chouhan, R.S., Babu, K.V., Kumar M.A. et al., 2006, *Biosensors and Bioelectronics* 21:1264–1271. With permission.)

7.4.2   CHEMILUMINESCENT IMMUNOASSAYS

CL compounds can also be used to label analytes. A CL label produces light when combined with a trigger reagent. CL detection generally provides higher sensitivity than the chromogenic assay.

Samsonova et al. (1999) combined five different antibodies in a CL immunoassay to simultaneously identify and estimate the amounts of three s-triazine herbicides in one single assay. An immuno-CL assay was developed by Chouhan et al. (2006) for the qualitative detection of methyl parathion (MP). MP antibodies raised in poultry were used for biorecognition of MP in the samples. MP antibodies were immobilized on sepharose CL-4B matrix by the periodate oxidation method and packed in a glass capillary immune-reactor column. MP was made to react with the immobilized MP-antibodies. Then HRP-conjugated antibodies were added to the column. The immobilized pesticide and the MP–HRP conjugate were taken out using a phosphate buffer saline-BSA solution and then analyzed after addition of the CL substrate. Image analysis was done using a charged-coupled device (CCD) (Figure 7.10). The light intensity produced during this process was directly proportional to the pesticide concentration.

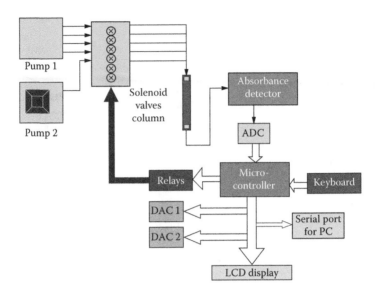

FIGURE 7.11   Automated system for flow of ELISA. (From Kumar, M.A. et al., 2006, *Analytica Chimica Acta* 560:30–34. With permission.)

It can be explained based on the principle that when the amount of pesticide in the sample is less, more conjugates bind to immobilized antibodies, and therefore more MP–HRP would be available for reaction with H_2O_2, leading to its higher degradation. Hence, less H_2O_2 is available for $K_3(FeCN)_6$, thereby resulting in a lesser amount of light. Similarly, when there is more pesticide, more light is produced. As $K_3(FeCN)_6$ is an electron mediator, it is used to increase the light intensity to a desired level to be grabbed by the camera. It is possible to detect the presence or absence of pesticide in the range of 10 ppt to 1000 ppb (Chouhan et al. 2006).

Kumar et al. (2006) and Chouhan et al. (2010a) describe the flow injection analyzer technique, which provide a very high sensitivity with high throughput of analyses. Automation of this analysis scheme (Figures 7.11 and 7.12) ensures precise detection with high accuracy. It employs 8952 microcontrollers for precise flow of reagents, samples, substrate, and conjugates used for analysis to be passed through an immobilized antibody column at a predetermined time. With the sequence and flow control of buffers used, it also provides the option for reuse of the immobilized antibody column. The system is flexible to accommodate multiple sequences up to a maximum of 99 steps. It can control up to 8 solenoid valves (DC 24V) and two peristaltic pumps and has one 12-bit analog channel for data acquisition. With the serial interface port, the system provides convenient means for data acquisition into the computer (Figure 7.11). The microcontroller program was devised to program up to 99 sequence steps. At each sequence step, the On/Off switching times of valves and flow rate of pumps could be prescribed. This could facilitate flow of reagents in a precise manner to carry out the flow of ELISA. Figure 7.12 shows the schematic diagram of the flow ELISA system. The system software controls the flow ELISA sequence of operations in a phased manner. The amounts of reagents and buffers

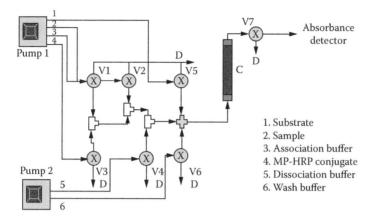

FIGURE 7.12 Schematic of flow of an ELISA system for pesticide analysis. (From Kumar, M.A. et al., 2006, *Analytica Chimica Acta* 560:30–34. With permission.)

that can flow through the valves were regulated by the flow rate of the two peristaltic pumps (P1 and P2). The flow injection analyzer was a more sensitive method for the detection of MP up to 50 ppt.

With the flow injection analyzer technique the task was much simpler as compared with plate ELISA. Optimization of the flow ELISA requires less time with automation. On the other hand, plate ELISA requires more skill, precision, and accuracy. The added advantage of automated flow ELISA is that the possible sources of human error on the part of the analyst are eliminated. For precise flow and timing control of reagents used, the use of a microcontroller made the task easier, and accurate determination of pesticide at ppt levels was possible.

More recently a sensitive CL-based immunoassay technique based on both dipstick and flow injection analytical formats has been reported by Chouhan et al. (2010a) for the detection of atrazine. In the dipstick-based immunoassay technique, antibody (antiatrazine) was first immobilized on the nitrocellulose (NC) membranes. The dipstick was then treated with atrazine and atrazine-horseradish peroxidase conjugate (atra-HRP) to facilitate the competitive binding. The dipstick was further treated with urea-hydrogen peroxide (U-H_2O_2) and luminol to generate photons. The number of photons generated was inversely proportional to the atrazine concentration.

7.5 RADIOIMMUNOASSAYS

In this technique, the antigen is made radioactive by labeling with radioactive isotopes of certain elements. The radiolabeled antigen is then allowed to react with the known quantity of antibody, and then the food sample with the pesticide residue is allowed to competitively bind with the radiolabeled antigens.

Radioactivity of the free antigen remaining is measured, and hence the amount of pesticide present may be quantitatively estimated. Five different isotopes, namely, ^3H, ^{14}C, ^{32}P, ^{35}S, and ^{125}I, are commonly being used. ^3H and ^{14}C are most commonly used because of their longer half-lives and low-energy beta emission. The chemistry of pesticides is different; hence, incorporating radioisotopes into these compounds without altering their structures requires intensive chemical synthesis, and sometimes satisfactory quality is not obtained (Knopp et al. 1986).

Hall et al. (1989) used conjugated [3H]-glycine for quantitative detection of two different auxinic herbicides, picloram and 2,4-Dinitrophenol (DNP). The conjugation method provided a convenient, simple, and inexpensive method for radiolabeling haptens. These RIAs had detection limits for picloram and DNP of 100 and 50 ppb, respectively.

7.6 NANOPARTICLE-BASED IMMUNOASSAYS AND IMMUNOSENSORS

7.6.1 Gold Nanoparticle-Based Immunosensors and Immunoassays

Gold nanoparticle-based immunosensors and immunoassays are accurate in terms of specificity, simplicity, sensitivity, and reliability. Many different types of detection systems have been designed for the efficient and speedy detection of pesticides.

Lisa et al. (2009) have developed a one-step gold nanoparticle-based dipstick competitive immunoassay for the rapid qualitative detection of organochlorine pesticides such as DDT at nanogram level (ppb).

In this method, gold nanoparticles of a definite size and shape conjugated to anti-DDT antibodies (IgY) acted as detecting agents. An immunocomplex was formed by a reaction of these anti-DDT antibodies that are conjugated gold nanoparticles with different concentrations of free DDT ranging from 0.7 ng/mL to 1000 ng/mL. A nitrocellulose membrane–containing strip was used for the immobilization of 2,2-bis-(p-chlorophenyl) acetic acid (DDA)-BSA conjugate (antigen). When this strip was dipped in the immunocomplex solution, the free gold nanoparticles conjugated anti-DDT antibodies in the immunocomplex solution competitively bind to the immobilized (Figure 7.13). Red color was developed in the detection zone of the NC membrane–containing strip, depending upon the concentration of free DDT in the sample. The intensity of the color developed was inversely proportional to the DDT concentration, and maximum intensity of color was developed at zero DDT concentrations. The lowest detection limit of DDT was found to be 27 ng/mL.

Lisha et al. (2009) have reported the detection of chlorpyrifos and malathion adsorbed on gold nanoparticles. The color change as a result of pesticide adsorption on gold nanoparticle (GNP)-enhanced by sodium sulphate was monitored. The presence of chlorpyrifos up to 20 ppb and malathion up to 100 ppb can be estimated without any sample preparation and preconcentration (Lisha et al. 2009).

DNP can be detected using a colorimetric nanobiosensor. In this technique, first the gold nanoparticles were functionalized with DNP–BSA, which is a toxin analog for DNP (Park et al. 2003).

Amino-derived microsphere dispersion was conjugated with anti-DNP antibodies in an Eppendorf tube. The DNP–BSA conjugated gold nanoparticles and anti-DNP antibody latex microsphere complex were obtained by the antigen–antibody interaction between these two solutions. This immunocomplex solution was then reacted with model toxin, DNP–glycin (DNP–GLY) to check the effectiveness

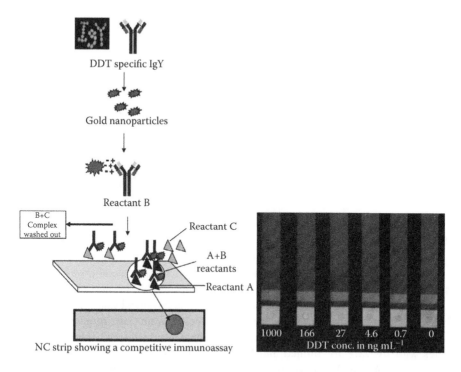

FIGURE 7.13 Detection of DDT using gold nanoparticles-based immunoassay. (From Lisa, M. et al., 2009, *Biosensors and Bioelectronics* 25:224–227. With permission.) Reactant A—mobilized DDA-BSA conjugate; Reactant B—GNP conjugated IgY; Reactant C—free DDT; Reactant A+B—DDA-BSA immobilized on NC strip + GNP conjugated IgY. Reactants B+C—GNP conjugated IgY + free DDT.

of this nanobiosensor. DNP–BSA conjugated gold nanoparticles were displaced due to the competition for the binding pocket in the anti-DNP antibody (Figure 7.14). An ultraviolet (UV)-spectrophotometer was used to quantify the extent of displacement of DNP–BSA conjugated gold nanoparticles by DNP–GLY molecules (Ko et al. 2010).

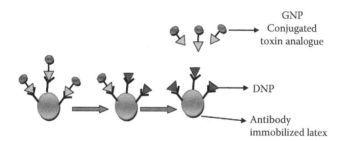

FIGURE 7.14 Schematic diagram of GNP-based immunodetector model.

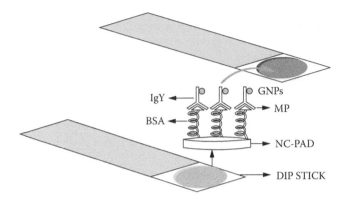

FIGURE 7.15   Schematic diagram of an immunochromatographic dipstick assay for MP detection.

Extensive work has been carried out at CFTRI to develop a lateral-flow-based dipstick immunoassay for the rapid screening of the herbicide MP in water (Figure 7.15).

A lateral-flow–based dipstick was developed using NC membrane. Anti-MP antibody was immobilized at the test line of the detection zone. The immunoassay was based on the competitive inhibition where the hapten-protein-gold conjugate competes with free MP for limited anti-MP antibody binding sites. Here hapten is a derivative of MP, and BSA is used as a carrier protein. The limit of detection was found to be up to 0.1 ppb.

A rapid immunochromatography one-step strip test for the detection of atrazine was developed by Kaur et al. 2007. NC membrane strips were used to produce the one-step strip kit (Figure 7.16).

A BSA-atrazine antigen was immobilized on the NC membrane. Water samples with atrazine were tested using the NC membrane strips. The antiatrazine IgG

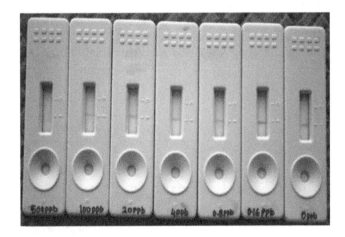

FIGURE 7.16   Immunochromatographic detection of atrazine. (From Kaur, J., Singh, K.V., Boro, R. et al., 2007, *Environmental Science and Technology* 41:5028–5036. With permission.)

antibodies labeled with colloidal gold particles reacted with the free atrazine by competitive inhibition. Zhou et al. (2004) developed an immunoassay for the detection of the N-Methylcarbamate pesticide carbofuran. Nanocolloidal gold particles were prepared and labeled to an anticarbofuran MAb. This conjugate was dispensed on the conjugated pad of a porous glass fiber. Ovalbumin (OVA)-carbofuran and goat antimouse IgG were dispensed on the NC membrane and served as the test line and control line, respectively. The carbofuran-containing sample migrated to the NC membrane and reacted with the anticarbofuran MAb labeled with the colloidal gold. The mixture diffused along the membrane and passed through the OVA-carbofuran in the test line via capillary action.

The more analyte present in the sample, the more effectively it will compete with the carbofuran immobilized on the test line for binding to the limited amount of antibody labeled with colloidal gold. An adequate amount of carbofuran could prevent attachment of the colored conjugate to the test line. The presence or absence of a colored band on the test line could indicate a negative or positive result, respectively. The limit of detection was found to be 0.25 mg/L of carbofuran.

Gold immunochromatography-based assay for the simultaneous detection of carbofuran and triazophos in water samples was developed by Guo et al. (2009). Two gold-based lateral-flow strips (strip A and strip B) were investigated for simultaneous detection of carbofuran and triazophos. For the strip A format, a bispecific monoclonal antibody (BsMAb) against both carbofuran and triazophos was employed to prepare the immunogold probe. For the strip B format, anticarbofuran MAb and antitriazophos MAb separately labeled with colloidal gold were combined as detector reagents.

By comparison of visual results from standard pesticide tests between the two formats, the strip B assay manifested higher sensitivities for both pesticides. Analysis of spiked water samples by the preferable strip indicated that the detection limits for carbofuran and triazophos were 32 µg/L and 4 µg/L, respectively (Guo et al. 2009).

Gold-coated chips were covered with a capture layer consisting of a protein derivative of the herbicide atrazine covalently bound to a self-assembled monolayer (SAM) containing a carboxy-terminated thiolate. Successive binding of antiatrazine antibody and secondary antirabbit IgG antibody resulted in a change of the infrared (IR) absorption properties of the organic film at the sensor surface (Salmaina et al. 2008)

The two prominent amide I and II bands observed on the surface IR spectra were taken for semiquantitative analysis of the adsorbed protein amount. The presence of increasing amounts of atrazine resulted in the progressive inhibition of antibodies binding to the sensors, yielding a relative lower increase of the IR signals. The test midpoint (IC_{50}) and the limit of detection (IC_{80}) were found to be in the nanomolar range (Salmiana et al. 2008).

A conductimetric immunosensor for atrazine detection has been designed and developed by Valera et al. (2008). This immunosensor uses antibodies labeled with gold nanoparticles. The immunosensor consists of an array of two coplanar nonpassivated interdigitated metallic µ-electrodes (IDµE) and immunoreagents specifically developed to detect this pesticide. The chemical recognition layer was covalently immobilized on the interdigital space.

Immunochemical detection of the concentration of atrazine is achieved by a competitive reaction that occurs before the inclusion of the labeled antibodies. It is shown that the gold nanoparticles provide an amplification of the conductive signal and hence make it possible to detect atrazine by means of simple direct current (DC) measurements. The immunosensor developed detects atrazine with limits of detection in the order of 0.1–1 µg/L (Valera et al. 2008).

Dasary et al. (2008) have developed a gold-nanoparticles-based surface-enhanced fluorescence for rapid and sensitive screening of OP agents. In this method Eu^{+3} ions are bound within the electromagnetic field of gold nanoparticles. In the presence of OP agents, these Eu^{+3} ions are released from the gold nanoparticle surface and thus gave a distinct fluorescence signal change (Dasary et al. 2008).

Hu et al. (2001) have developed a novel electrochemical immunosensor for the direct determination of paraoxon, based on the biocomposites of gold nanoparticles loaded with paraoxon antibodies. The biocomposites are immobilized on the glassy carbon electrode (GCE) using Nafion membrane. On the immunosensor, prepared paraoxon shows well-shaped cyclic voltammetry with reduction and oxidation peaks located –0.08 and –0.03 mV versus a saturated calomel electrode (SCE), respectively. The immunosensor has been employed for monitoring the concentrations of paraoxon in aqueous samples up to 1920 µg/ L with a detection limit of 12 µg/ L.

7.6.2 Quantum Dot-Based Fluorescence Immunoassays

Quantum dots (QDs) are semiconductor nanoparticles made up of elements from II–VI, III–V, or IV–VI groups of the periodic table with narrow, specific, stable emission spectra. They have attracted much interest in biological research because of their unique spectral properties such as the long photostability, broad absorption, and narrow emission bands. Thus, they are superior to conventional organic fluorescent dyes (Vinayaka et al. 2009; Chouhan et al. 2010b). Therefore, the bioconjugation of these QDs for biological fluorescent labeling is of interest for the sensitive detection of target analytes. From the viewpoint of ensuring food safety, there is a need to develop rapid, sensitive, and specific detection techniques to monitor food toxicants in food and environmental samples. Vinayaka and Thakur (2010) have described the application of water-soluble bioconjugated QDs for the detection of food contaminants such as pesticides, foodborne pathogens, and their toxins such as botulinum toxins and enterotoxins produced by *Staphylococcus aureus* and *Escherichia coli*. They also described a possible application of resonance energy transfer phenomenon resulting from nanobiomolecular interactions obtained through the bioconjugation of QDs with biomolecules.

Quantum dots can be used as suitable labels for the immunoassays for the detection of small biomolecules (biomarkers of pesticide exposure) because their photostability permits prolonged excitation for image observation and optimization. The spectral properties of QDs allow the multianalyte detection of the targets (Zhelev et al. 2006; Vinayaka et al. 2009).

Chouhan et al. (2010b) has reported the application of these intrinsic properties of QDs for sensitive detection of MP at picogram levels. Cadmium telluride quantum dot (CdTe QD) was bioconjugated to MP–BSA (Figure 7.17), and competitive

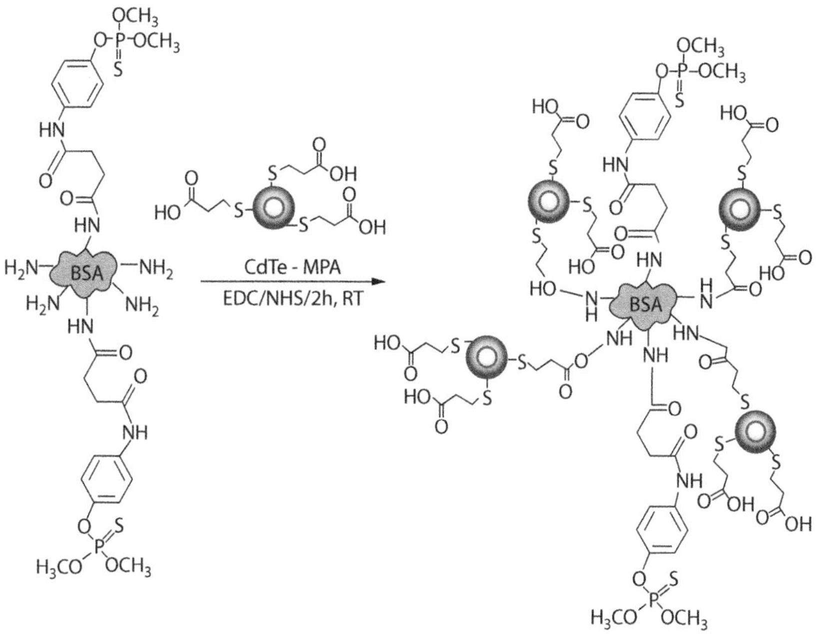

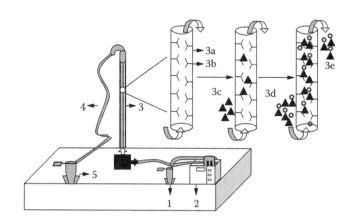

FIGURE 7.17 Bioconjugation of CdTe QD with MP-BSA. (With kind permission from Springer Science & Business Media: *Analytical and Bioanalytical Chemistry*, Thiol-Stabilized Luminescent CdTe Quantum Dot as Biological Fluorescent Probe for Sensitive Detection of Methyl Parathion by a Fluoroimmunochromatographic Technique, Vol. 397, 2010b, 1395–1633, Chouhan, R.S., A.C. Vinayaka, and M.S. Thakur.)

FIGURE 7.18 Immunoreactor column showing competitive binding for detection of 2,4-D. (1) Sample inlet, (2) peristaltic pump, (3) Immunoreactor column, (3a) glass capillary column, (3b) Immobilized anti-2,4-D antibodies, (3c) 2,4-D, (3d) 2,4-D-ALP-CdTeconjugate, (3e) unbound 2,4-D-ALP-CdTe conjugate, (4) polystyrene tubing, (5) Sample collection point. (From Vinayaka, A.C., S. Basheer, and M.S. Thakur, 2009, *Biosensors and Bioelectronics* 24:1615–1620. With permission.)

binding between free MP and CdTe QD bioconjugated MP (MP-BSA-CdTe) with immobilized anti-MP IgY antibodies was monitored in a flow-injection system. MP was detected in a linear range of 0.1–1 ng mL−1 with a regression coefficient $R^2 =$ 0.9905 with this method.

Nichkova et al. (2007) has developed multiplexed immunoassays for the detection of 3-phenoxybenzoic acid (PBA) and atrazine-mercapturate (AM), using quantum dots as the fluorescence labels.

PBA and AM are biomarkers of exposure to pyrethroid insecticides and the herbicide atrazine, respectively. First, the antigen BSA–PBA is microcontact printed and antigen BSA–AM is immobilized on a cover glass as solid substrate. Immunoreaction between the anti-PBA antibody and bionylated anti-AM antibody with the respective antibodies is allowed. Then they are incubated with secondary immunoreagents antirabbit QDot 580 and streptavidin-QDot 620 to obtain the fluorescence signal (Nichkova et al. 2007).

A reliable and rapid method for analysis and detection of herbicide 2,4-D has been developed (Vinayaka et al. 2009) using CdTe QD nanoparticle. Fluoroimmunoassay based on the fluorescent property of quantum dot was used along with immunoassay to detect 2,4-D for the development of this detection technique. Initially, 2,4-D was conjugated to alkaline phosphatase which were further conjugated to CdTe capped with mercaptopropionic acid. The conjugation was carried out using N-(3-dimethylaminopropyl)-N-ethylcarbodiimide hydrochloride and a coupling reagent like N-hydroxysuccinimide. Anti-2,4-D IgG antibodies were immobilized in an immunoreactor column using Sepharose CL-4B as an inert matrix. The detection of 2,4-D was carried out by fluoroimmunoassay-based biosensor using competitive binding between conjugated 2,4-D–ALP–CdTe and free 2,4-D with immobilized anti-2,4-D antibodies in an immunoreactor column (Figure 7.18). It was possible to detect 2,4-D up to 250 pg/mL (Vinayaka et al. 2009).

7.7 BIOSENSORS FOR PATHOGEN DETECTION

Detection, identification, and quantification of microbial pathogens are crucial for public health protection. Clinical diagnosis, water and environmental analysis, food safety, and defense are some of the areas where detection of microbial pathogens plays a crucial role. The traditional methods used for their detection are polymerase chain reaction (PCR) (Burtscher and Wuertz 2003), culture and colony counting (Allen et al. 2004), immunological techniques (Van Dyck et al. 2001), and fluorescence-based assays using organic dye molecules (Regnault et al. 2000). Although these methods have been found to be highly accurate, sensitive, and error-proof, they are laborious and time-consuming, require skilled personnel and sophisticated equipment, and sometimes do not give the required detectability and specificity toward the target. Also they require preenrichment or amplification, and hence, for all these reasons, the analysis cost goes very high.

As the microorganisms evolve rapidly and can grow and produce toxins at a very fast rate even under stress conditions, there is a need for the development of rapid detection techniques that are reliable, fast, user-friendly, inexpensive, and that can be used online or in the field.

Recent research and development of nanoparticles and biosensor techniques for application in bioanalysis have helped greatly in the design of sensitive and label-free detection of foodborne pathogens and their toxins in unprocessed and complex food samples. The specific recognition of the target organism is done with the help of nanobiorecognition technology, in which the nanoparticles are biofunctionalized by conjugation with a biomolecule (Yang et al. 2008).

Though immunosensors are more effective than other biosensors because of their specificity, versatility, flexibility, and ability for real-time detection, immunoassays and microarrays are the preferred detection techniques because they are less expensive and more cost effective and also because large throughputs can be obtained.

7.7.1 IMMUNOASSAYS

Immunoassays are popular because they can be used for the wide variety of analytes ranging from molecules to intact cells. They are also useful in point-of-care applications because the surface epitopes of the pathogens can be targeted; hence, the sample extraction that is required for the nucleic-acid-based detection can be avoided (Skottrup et al. 2008). Immunoassays are often based on two different biochips: nanoarrays and nanofluidic devices.

Nanoparticle-based assays have been used as an alternative to the planar surface nanoarrays. These assays are more sensitive and speedy because of increased surface for reaction. Nanoparticles like gold, quantum dots, plastic with magnetic core, silica nanoparticles, nanoshells, and others have been used (Santra et al. 2004; Kim and Park 2005; Medintz et al. 2005). Antibody-conjugated silica fluorescent nanoparticles have been used for detection of bacteria. This immunoassay is very fast and sensitive and can detect even a single bacterium. This assay gives a better fluorescent signal than an organic dye because thousands of antibody-conjugated nanoparticles are used to label a single bacterium as compared with a single molecule of the organic dye. Also this assay doesn't require any amplification or enrichment (Zhao et al. 2004).

The label-dependent antibody-based immunoassays such as dipstick methods and lateral flow devices are the most convenient and fast techniques for the online detection of pathogens and their toxins. The primary screening technique for *Listeria monocytogenes* with an immunochromatography strip test has been developed by Shim et al. (2007). The detection limit of the strip test was found to be 10^5 cfu/mL.

7.7.2 GOLD NANOPARTICLE-BASED DETECTION OF PATHOGENS AND THEIR TOXINS

Detection of pathogens and their toxins with the help of gold nanoparticles is highly sensitive and specific. Some examples of the detection systems designed for the detection of pathogens are discussed below.

An immunochromatographic assay based on gold nanoparticles for rapid detection of *S. aureus* was developed by Huang (2006, 2007).

Staphylococcal protein A is a cell wall protein of *S. aureus*, which is a virulence factor in *S. aureus* infections as it interacts with several host components. A sandwich assay is constructed by using antiprotein A IgG with two distinct specificities. According to the method given by Huang (2007), 20-nm gold nanoparticles were conjugated with antiprotein A IgG.

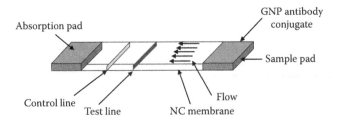

FIGURE 7.19   Immunochromatographic strip for the detection of *S. aureus*.

The assay was constructed in the form of sandwich by using antiprotein A IgG with two distinct specificities. One antiprotein A IgG was immobilized in a defined detection zone on a porous NC membrane, while the other antiprotein A IgG was conjugated with gold nanoparticles. The sample flows along the porous membrane by capillary action past the antiprotein A IgG in the detection zone, binding the particles that surface protein A was already bound to the surface of, yielding a red color (Figure 7.19). As a result, the sensitivity in the immunochromatographic test was 0.9, 1.2, 2.4, and 6 CFU/g, respectively, for *S. aureus* (Huang 2007).

Yang et al. (2008) have used gold-nanoparticles-based enhanced CL immunosensor for the detection of Staphylococcal enterotoxin B (SEB) in food. Gold nanoparticles were partially immobilized on a polycarbonate surface and then conjugated with rabbit anti-SEB affinity purified IgG. The gold nanoparticle-antibody immunosensor was then incubated with different concentrations of SEB for 45 minutes. This was then exposed to HRP-conjugated anti-SEB IgG.

Enhanced CL was measured after adding H_2O_2 with a custom built point-of-care charged-coupled device. Here, gold nanoparticles increase the sensitivity of the assay because their increased surface area enables better immobilization of the primary antibody. It is also believed that in the close proximity of the nanoparticles, the antibodies give better luminescence/fluorescence, and there is better orientation of the immobilized antibodies to the sensor surface. The limit of detection was found to be 0.01 ng/mL (Yang et al. 2009).

Salmonella typhi can be detected with the help of a highly sensitive electrochemical amplification immunoassay based on copper-enhanced gold label (Dungchai et al. 2008). Monoclonal rabbit antibodies for polysaccharides of *S. typhi* were immobilized in the polystyrene wells and acted as capture antibodies. Polyclonal rabbit antibodies for polysaccharides of *S. typhi* 0901 were conjugated with gold nanoparticles acting as reporter antibodies. After immobilization of the primary antibody, the wells were treated with *S. typhi* and incubated, followed by exposure to the secondary antibody conjugated with gold nanoparticles. A copper-enhanced solution containing ascorbic acid and copper (II) sulphate was added to the wells. The ascorbic acid reduces the Cu^{2+} ions to Cu, which deposits on the gold nanoparticles. This copper is dissolved using nitric acid, and released copper ions are detected by anodic stripping voltammetry.

The amount of copper deposited gave the amount of gold nanoparticle tags present and that subsequently gave the concentration of *S. typhi* cells present in the sample. The limit of detection was found to be as low as 98 cfu/mL.

A lateral-flow assay device has been developed for the detection of SEB. Colloidal gold probes of 25 nm were conjugated with anti-SEB IgG. Goat antirabbit IgG and rabbit anti-SEB IgG were sprayed on an NC membrane at the control and the test lines, respectively. These strips were then treated with the food samples with SEB, and red color was obtained when the SEB interacted with the rabbit anti-SEB IgG on the test line because of the gold-conjugated anti-SEB IgG. The limit of detection was found to be 1ng/mL (Rong-Hwa et al. 2010).

Label-free electrochemical impedance immunosensors for the detection and quantification of *E. coli* using SAMs modified gold screen-printed electrodes (AuSPEs) were developed by Escamilla-Gomez et al. (2009). Two different immunosensor configurations were tested and compared. In the first one, the immunosensing design was based on the covalent immobilization of anti–*E. coli* at AuSPEs, using the cross-linker 3,3'-dithiobis[sulfosuccinimidylpropionate] (DTSSP), which is homobifunctional in nature. The other one was based on the immobilization of the thiolated antibody onto the electrode surface.

In both cases, the evaluation of the developed immunosensors' performance was accomplished through the monitoring of the electron-transfer resistance detected by electrochemical impedance spectroscopy (EIS) in the presence of [Fe(CN)$_6$ $^{3-}$] / [Fe(CN)$_6$$^{4-}$] as a redox probe. The configuration using the thiolated antibodies gave rise to a better analytical performance, exhibiting a linear relationship between the increment in the electron-transfer resistance (ΔR_{et}) and the logarithmic value of the *E. coli* concentration in the 108 cfu/mL range. The limit of detection achieved without preconcentration or preenrichment steps was 3 cfu/mL. The developed immunosensors showed a high selectivity against *S.aureus* and *Salmonella choleraesuis* (Escamilla-Gomez et al. 2009).

Zhang et al. (2009a) have developed a sensitive electrochemical immunoassay for rapid detection of *E. coli* by anodic stripping voltammetry (ASV) based on core–shell CuAu nanoparticles as anti-*E. coli* antibody labels. CuAu-labeled antibodies were made to react with the immobilized *E. coli* on a polystyrene-modified ITO chip. CuAu nanoparticles were dissolved by oxidation to the metal ionic forms, and the released Cu^{2+} ions were determined at GC/Nafion/Hg–modified electrode by ASV.

The utilization of a GC/Nafion/Hg–modified electrode could enhance the sensitivity for Cu^{2+} detection with a concentration as low as 9.0×10^{-12} mol/L. Since CuAu nanoparticle labels were only present when an antibody reacted with *E. coli*, the amount of Cu^{2+} directly reflected the number of *E. coli*. The technique could detect *E. coli* with a detection limit of 30 cfu/mL, and the overall analysis could be completed in 2 h. By introducing a preenrichment step, a concentration of 3 cfu/10 mL *E. coli* in surface water was detected by the electrochemical immunoassay (Zhang et al. 2009a).

7.7.3 FLUOROIMMUNOASSAYS AND QUANTUM DOT-BASED DETECTION OF PATHOGENS AND THEIR TOXINS

FIAs make use of the organic dyes such as rhodamine, sulforhodamine, fluorescin, and others. Instead of these dyes, QDs maybe suitable to obtain better specificity and sensitivity.

1. Cholera toxin B (CTB) is frequently used as an indicator of the presence of pathogenic *V. cholerae* and binds to the GM1 ganglioside on the surface of epithelial cells. To study *V. cholerae* virulence (CTB expression) in the presence of human epithelia, Edwards et al. (2008) devised an inexpensive, simple, and rapid method for quantifying CTB bound on epithelial surfaces in microtiter plates.

   GM_1 ganglioside was incorporated into the lipid bilayer of liposomes both encapsulating the fluorescent dye sulforhodamine B (SRB) and with SRB tagged to lipids in the bilayer encapsulated gangliosides (BEGs). In addition, GM_1-embedded liposomes encapsulating SRB only encapsulated gangliosides (EGs) and with SRB in their bilayers only bilayer gangliosides (BGs) were synthesized. The three types of liposomes were compared with respect to their efficacy for both visualizing and quantifying CTB attached to the surface of Caco-2 cells.

   The BEGs were the most effective overall, providing both visualization under a fluorescence microscope and quantification after lysis in a microtiter plate reader. A limit of detection corresponding to 0.28 μg/mL applied CTB was attained for the on-cell assay using the microtiter plate reader approach, whereas as low as 2 μg/mL applied CTB could be observed under the fluorescence microscope (Edwards et al. 2008).

   In the same way, instead of using SRB as a dye, fluorescin was used as a label for the detection of CTB by Edwards and March. A limit of detection for CTB using the liposomes was 340 pg/mL (Edwards and March 2007).

2. *E. coli* can be detected using carbohydrate-functionalized CdS QDs based on the recognition of their mannosides. Mannose-coated CdS QDs (Man-QDs) were prepared based on the self-assembly of thiolated mannose in the presence of CdS under reducing conditions, in a facile aqueous single step. The QDs induce luminescent aggregates of *E. coli* that can be used to detect bacteria in cell suspensions containing as few as 10^4 *E. coli*/mL. The luminescence emission is centered at 550 nm (Mukhopadhyay et al. 2009).

3. Cadmium-selenide-zinc sulfide (CdSe–ZnS) nanoparticles were employed together with molecular adaptor proteins that had been self-assembled on the QD surface. The molecular adaptor proteins constitute the link between QD and the antibody. SEB was estimated by a sandwich immunoassay using polyclonal sheep anti-SEB–QD conjugate and microtitre plates (MTPs) coated with monoclonal anti-SEB antibody. The lowest concentration of SEB to give useful signal over background was ~2 ng/mL. Signal saturation was reached at ~30 ng/mL (Goldman et al. 2002).

4. Warner et al. (2009) have developed a fluorescence sandwich immunoassay for the detection of botulinum neurotoxin serotype A (BoNT/A). In this assay QDs and high-affinity antibodies, AR4 and RAZ, were used as reporters, and a nontoxic recombinant fragment of the holotoxin (BoNT/A-H_c-fragment) was used as a structurally valid stimulant for the full toxin molecule. The antibodies will bind to the nonoverlapping epitopes present on the toxin and the recombinant fragment. Immunoassay was carried out in a 96-well plate using AR4 as the capture antibody and QD-coupled

RAZ1 as the reporter antibody. The reading was taken in a standard plate reader. Total incubation time of 3 h was required for detection to 31 pM. The immunoassay was repeated in microcentrifuge tubes by immobilizing the AR4 antibody on Sepharose beads. After an incubation period of 1 h the beads were captured and concentrated in a rotating rod "renewable surface" flow cell equipped with a fiber-optic system for fluorescence measurements. The concentrations of BoNT/A-H_c-fragment detected by the fluidic measurement approach were found to be as low as 5 pM (Warner et al. 2009).

7.8 MICROARRAYS

A microarray consists of a support onto which hundreds to thousands of different molecular reporter probes are attached or immobilized at fixed locations in either a two-dimensional or three-dimensional format. They provide powerful tools for biomedical and clinical applications because they can be configured to monitor the presence of molecular signatures (Situma et al. 2006). They can be configured to search for either nucleic acids (DNA microarrays) or protein (antibody-based microarray) or different types of cells. The probes may consist of antibodies to search for specific proteins or oligonucleotides/PCR products to recognize unique DNA sequences through Watson-Crick base pairing. However the DNA microarray is most commonly used.

7.8.1 DNA Microarrays

In a DNA microarray a mixture of labeled nucleic acids called *targets* are parallel hybridized with thousands of individual nucleic acid species called *probes*. The location of a specific probe on the array is termed a *spot* or *feature*. The probes are immobilized on a solid substrate, and the targets are applied as a solution onto the array for hybridization after fluorescent labeling (Brown and Botstein 1999).

Selectivity is imparted by targets and probes that are noncomplementary and thus do not form thermodynamically strong hybrids, which can be selectively tuned out through stringent washing with buffers or temperature regulation. DNA microarrays are most commonly used in gene expression analysis (Zhang et al. 1991), drug screening/therapeutic studies (Lebrun 2005), proteomics (MacBeath and Schreiber 2000; MacBeath 2002), cell analysis (McClain et al. 2003), and DNA sequencing/fragment analysis (Pease et al. 1994).

Three main types of DNA microarrays are widely used (Ehrenreich 2006):

1. Microarrays where the probes are synthesized *in situ* directly onto the surface of the chip
2. Double-stranded DNA microarrays
3. Oligonucleotide DNA microarrays

In the last two types, the probes are synthesized individually and printed on special glass slides. Oligonucleotide DNA arrays are used as an alternative to double-stranded DNA because they need much less logistics and are less error-prone because of their automatic manufacturing and well-documented delivery into the microtiter plates (Southern et al. 1999; Kane et al. 2000). These microarrays are advantageous because of their miniature size, high performance, ability to process samples in parallel, and ease of automation. Based on the probe set, a wide variety of food-borne pathogenic bacteria including *S. aureus, V. parahaemolyticus, Proteus* spp., *L. monocytogenes, Y. enterocolitica, Enterococcus faecalis, V. fluvialis, V. cholerae, Bacillus cereus, P. aeruginosa, Mycobacterium tuberculosis, Campylobacter jejuni, Clostridium tetani, Clostridium perfringens, A. hydrophila, P. cocovenenans* subsp. *farinofermentans, Clostridium botulinum,* and β- *Streptococcus haemolyticus* can be clearly detected and discriminated using the DNA microarray. Some of the detection systems, developed based on this technology, are as follows:

1. The insertion element (*Iel*) gene of *Salmonella enteritidis* can be rapidly detected with the help of a highly amplified biobarcode DNA assay. The biosensor transducer is composed of two nanoparticles: gold nanoparticles and magnetic nanoparticles. The gold nanoparticles are coated with the target-specific DNA probe, which can recognize the target gene and fluorescein-labeled barcode DNA in a 1:100 probe-to-barcode ratio. The magnetic nanoparticles are coated with the second target-specific DNA probe. After mixing the nanoparticles with the first target DNA, the sandwich structure (magnetic nanoparticle–second DNA probe/target DNA/first DNA probe–gold nanoparticle-barcode DNA) is formed. A magnetic field is applied to separate the sandwich from the unreacted materials. Then the biobarcode DNA is released from the gold nanoparticles.

Because the gold nanoparticles have a large number of barcode DNA per DNA probe binding event, there is substantial amplification. The released barcode DNA is measured by fluorescence. Using this technique, the detection limit of this biobarcode DNA assay is as low as $2.15 \times 10 - 16$ mol (Zhang et al. 2009b).

2. Multicolor colocalization of QD nanoprobes was used for the detection of a pathogenic strain of *Bacillus anthracis* by Ho et al. (2005). Two QD nanoprobes, each with discernible emission wavelengths, are designed to bind in juxtaposition to the same target DNA of interest to form a sandwiched nanoassembly.

Since the physical size of the nanoassembly (~50 nm) is smaller than the diffraction-limited resolution (~250 nm) of an optical imaging system, the nanoassembly is imaged as a combined color due to spatial colocalization of the two linked QD nanoprobes. As a result, target presence can be determined based on colorimetric measurements of the nanoassemblies, and multiplexed sequence identification can be achieved through the use of multiple combinations of different color QD nanoprobes (Ho et al. 2005).

To identify this pathogenic strain simultaneous detection of three genes, rpoB, pagA, and capC is required. RpoB is a chromosomal marker with species–species sequences that can be used to differentiate *B. anthracis* from other *Bacillus* species, while pagA and capC are the plasmid markers for determining the presence of anthrax toxin. Three synthetic oligonucleotides, each derived from conserved sequences from each of the anthrax-related

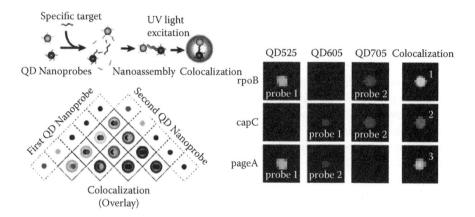

FIGURE 7.20 (a) QD nanoprobes prepared by surface-functionalizing QDs with target-specific oligonucleotide probes. Two target-specific QD nanoprobes with different emission wavelengths sandwich a target, forming a QD probe-target nanoassembly. The nanoassembly is detected as a blended color (orange) due to the colocalization of both QD nanoprobes. (b) The color combination scheme of multiplexed colocalization detection. (c) Color palette for the three pairs of target-specific QD nanoprobes and their resulting colocalized fluorescent images upon sandwich hybridization. (From Ho, Y.P., M.C. Kung, S. Yang, and T.H. Wang, 2005, *Nano Letters* 5:1693–1697. With permission.)

genes, were used as simulated targets for analysis with three pairs of target-specific probes conjugated to three QDs with distinct emission wavelengths: 525 QD, 605 QD, and 705 QD. The detection of the fluorescent spots of indigo, magenta, and orange signifies the presence of rpoB, pagA, or capC, respectively (Figure 7.20).

7.8.2 PROTEIN ARRAYS

Protein arrays are based on the biospecific binding such as antibody–antigen interactions and so forth. The analysis can be accomplished with a very less sample and reagent volume and also the speed of analysis is also faster than the DNA microarrays.

Karoonuthaisiri et al. (2009) used antibody-based protein array for the simultaneous detection of *E. coli* O157:H7 and *Salmonella* spp. Initially, the capture antibodies (goat anti–*E. coli* O157H:7 and goat anti-*Salmonella*) were used to fabricate a spot on a NC membrane and Poly-L-lysine slides. They were incubated overnight. Then they were treated with heat-killed bacteria. HRP-labeled goat anti–*E. coli* and HRP-labeled goat anti-*Salmonella* were then added to the spots. Luminol-based CL substrate was added to detect the signal. The sensitivity of the array is $10^5–10^6$ cfu/mL for *E. coli* and $10^6–10^7$ cfu/mL for *Salmonella* detections.

7.9 CONCLUSIONS

In brief, quality assurance and safety of food are the essential requirements in the food industry to market the food product with good manufacturing practice. Food and water contamination, especially by pesticides and foodborne pathogens and their toxins, are posing major health problems today all over the world. The increased use of pesticides and their tendency for bioaccumulation and bioconcentration are causing widespread ecotoxicity and adverse environmental effects. Moreover, they cause chronic and acute toxicity in human beings, not only when contaminated foods are consumed but also due to prolonged occupational exposure to treated commodities. Foodborne pathogens and their toxins are the leading cause of illness and diseases the world over. They are responsible for millions of cases of infectious gastrointestinal diseases each year. New foodborne pathogens and diseases are emerging every day because of pathogen evolution and changes in agricultural and food manufacturing practices.

The conventional methods of detection for the pesticide and foodborne pathogens are lacking in terms of sensitivity, specificity, reproducibility, and speed. The tests currently used for their detection require lots of time, expensive and complicated equipment, and skilled experts. Biosensors are proving to be favorable tools for screening of food contaminants like pesticide residue and foodborne pathogens and their toxins because of their high sensitivity, selectivity, simplicity, reliability, and applicability. Moreover, they can be used for online or on-field applications, results can be obtained in very short time, and the systems are fully automated and hence are less laborious.

Potentials of biosensors for monitoring of several analytes have increased to apply nanoparticles for *in vivo* and *in vitro* diagnoses. Gold nanoparticles are, so far, considered to be the most versatile and the most extensively studied nanoparticles, followed by semiconductor QDs. Gold nanoparticles are being widely used in bioassays because of their unique photophysical properties like SPR and enhanced light scattering. They can be easily synthesized in a wide range of sizes and shapes, and their surface can be tailored by ligand functionalization with a variety of chemical and biomolecular ligands. Modulation of their physicochemical properties and their robustness make them excellent for use as optical labels in bioassays.

Quantum dots are mainly popular as fluorescent labels, as they are ideal donors in FRET and have longer fluorescence lifetime and resistance to photo-bleaching. QDs are being used for biosensing in microarray systems. The spectral properties of QDs are being successfully applied for the multianalyte detection of DNA, proteins, pesticides, and so forth. Hence, gold nanoparticle and QD-based detection systems are bringing about a revolution in the field of food and environmental safety and analysis.

ACKNOWLEDGMENTS

The authors are thankful to the director, CFTRI, and CSIR (New Delhi) for providing necessary laboratory facilities.

REFERENCES

Adanyi, N., M. Varadi, N. Kim, and I. Szendro. 2006. Development of new immunosensors for determination of contaminants in food. *Current Applied Physics* 6:279–286.
Allen, M.J., S.C. Edberg, and D.J. Reasoner. 2004. Heterotrophic plate count bacteria—What is their significance in drinking water? *International Journal of Food Microbiology* 92:265–274.

Alvarez, M., A. Calle, J. Tamayo, L.M. Lechuga, A. Abad, and A. Montoya. 2003. Development of nanomechanical biosensors for detection of the pesticide DDT. *Biosensors and Bioelectronics* 18:649–653.

Anh, T.M., S.V. Dzyadevych, M.C. Van, N.J. Renault, C.N. Duc, and J.M. Chovelon. 2004. Conductometric tyrosinase biosensor for the detection of diuron, atrazine and its main metabolites. *Talanta*. 63:365–370.

Ayyagari. M.S., S. Kamitekar, R.A. Pande et al. 1995. Chemiluminescence-based inhibition kinetics of alkaline phosphatase in the development of a pesticide biosensor. *Biotechnology Progress* 11:699–703.

Bange, A., H.B. Halsall, and W.R. Heineman. 2005. Microfluidic immuno-sensor systems. *Biosensors and Bioelectronics* 20:2488–2503.

Baumner, A.J., and R.D. Schmid. 1998. Development of a new immunosensor for pesticide detection: A disposable system with liposome-enhancement and amperometric detection. *Biosensors and Bioelectronics* 13:519–529.

Bond, A.M. 1980. *Modern polarographic methods in analytical chemistry.* New York: Marcel Dekker Inc.

Brecht, A., J. Piehler, G. Lang, and G. Gauglitz. 1995. A direct optical immunosensor for atrazine detection. *Analytica Chimica Acta* 311:289–99.

Brown, P.O., and D. Botstein. 1999. Exploring the new world of the genome with DNA microarrays. *Nature Genetics* 21(Suppl 1):33–37.

Burtscher, C., and S. Wuertz. 2003. Evaluation of the use of PCR and reverse transcriptase *PCR* for detection of pathogenic bacteria in biosolids from anaerobic digestors and aerobic composters. *Applied and Environmental Microbiology* 69:4618–4627.

Chen, H., J. Jiang, Y. Huang et al. 2006. Electrochemical immune-biosensor for immunoglobulin G based bioelectrocatalytic reaction on micro-comb electrodes. *Sensors and Actuators B: Chemical* 117:211–218.

Chouhan, R.S., A.C. Vinayaka, and M.S. Thakur. 2010b. Thiol-stabilized luminescent CdTe quantum dot as biological fluorescent probe for sensitive detection of methyl parathion by a fluoroimmunochromatographic technique. *Analytical and Bioanalytical Chemistry* DOI: 10.1007/s00216–009–3433–1.

Chouhan, R.S., K.V. Babu, M.A. Kumar et al. 2006. Detection of methyl parathion using immuno-chemiluminescence based image analysis using charge-coupled device. *Biosensors and Bioelectronics* 21:1264–1271.

Chouhan, R.S., K.V.S. Rana, C.R. Suri, K.R. Thampi, and M.S. Thakur. 2010a. Trace-level detection of atrazine using immuno-chemiluminescence: Dipstick and automated flow injection analyses formats. *Journal of AOAC International* 93:28–35.

Ciumasu, I.M., P.M. Kramer, C.M. Weber et al. 2005. A new versatile field immunosensor for environmental pollutants: Development and proof of principle with TNT, diuron, and atrazine. *Biosensors and Bioelectronics* 21:354–364.

Darain, F., D.S. Park, J.S. Park, and Y.B. Shim. 2004. Development of an immunosensor for the detection of vitellogenin using impedance spectroscopy. *Biosensors and Bioelectronics* 19:1245–1252.

Dasary, S.S.R., U.S. Rai, H.T. Yu, Y. Anjaneyulu, M. Dubey, and P.C. Ray. 2008. Gold nanoparticle based surface enhanced fluorescence for detection of organophosphorus agents. *Chemical Physics Letters* 460:187–90.

Deng, A., and H. Yang. 2007. A multichannel electrochemical detector coupled with an ELISA microtiter plate for the immunoassay of 2,4-dichlorophenoxyacetic acid. *Sensors and Actuators B: Chemical* 124:202–208.

Dennison, M.J., and A.P.F. Turner. 1995. Biosensors for environmental monitoring. *Biotechnology Advances* 13:1–12.

Deo, R.P., J. Wang, I. Block et al. 2005. Determination of organophosphate pesticides at carbon nanotube/organophosphorus hydrolase electrochemical biosensor. *Analytica Chimica Acta* 530:185–189.

Dessy, R., L. Arney, L. Burgess, and E. Richmond. 1990. A comparison of three thermal sensors based on fiber optics and polymer films for biosensor applications. In: *Biosensor technology: Fundamentals and applications*, ed. R.P. Buck, W.F. Hatfield, M. Umana, and E.F. Bowden, 251–283. New York: Marcel Dekker, Inc.

Dungchai, W., W. Siangproh, W. Chaicumpa, P. Tongtawe, and O. Chailapakul. 2008. *Salmonella typhi* determination using voltametric amplification of nanoparticles: A highly sensitive strategy for metalloimmunoassay based on a copper-enhanced gold label. *Talanta* 77:727–732.

Dutra, R.F., and L.T. Kubot. 2007. An SPR immunosensor for human cardiac troponin T using specific binding avidin to biotin at carboxymethyldextran-modified gold chip. *Clinica Chimica Acta* 376:114–120.

Dzantiev, B.B., E.V. Yazynina, A.V. Zherdev et al. 2004. Determination of the herbicide chlorsulfuron by amperometric sensor based on separation-free bienzyme immunoassay. *Sensors and Actuators B: Chemical* 98:254–261.

Dzantiev, B.B., A.V. Zherdev, M.F. Yulaev, R.A. Sitdikov, N.M. Dmitrieva, and I.Y. Moreva. 1996. Electrochemical immunosensors for determination of the pesticides 2,4-dichlorophenoxyacetic and 2,4,5–tricholorophenoxyacetic acids. *Biosensors and Bioelectronics* 11:179–185.

Edwards, K.A., and J.C. March. 2007. GM_1-functionalized liposomes in a microtiter plate assay for cholera toxin in Vibrio cholerae culture samples. *Analytical Biochemistry* 368:39–48.

Edwards, K.A., F. Duan, A.J. Baeumner, and J.C. March. 2008. Fluorescent labeled liposomes for monitoring cholera toxin binding to epithelial cells. *Analytical Biochemistry* 380:59–67.

Ehrenreich, A. 2006. DNA microarray technology for the microbiologist: An overview. *Applied Microbiology and Biotechnology* 73:255–273.

Engstrom, H.A., P.O. Andersson, and S. Ohlson. 2006. A label-free continuous total-internal-reflection-fluorescence based immunosensor. *Analytical Biochemistry* 357:159–166.

Escamilla-Gomez V., S. Campuzano, M. Pedrero, and J.M. Pingarron. 2009. Gold screen-printed-based impedimetric immunobiosensors for direct and sensitive *Escherichia coli* quantisation. *Biosensors and Bioelectronics* 24:3365–371.

Farre, M., E. Martinez, J. Ramon et al. 2007. Part per trillion determination of atrazine in natural water samples by a surface plasmon resonance immunosensor. *Analytical and Bioanalytical Chemistry* 388:207–214.

Farre, M., J. Ramon, R. Galve, M.P. Marco, and D. Barcelo. 2006. Evaluation of a newly developed enzyme-linked immunosorbent assay for determination of linear alkyl benzenesulfonates in wastewater treatment plants. *Environmental Science and Technology* 40:5064–5070.

Gobi, K.V., H. Tanaka, Y. Shoyama, and N. Miura. 2005. Highly sensitive regenerable immunosensor for label-free detection of 2,4-dichlorophenoxyacetic acid at ppb levels by using surface plasmon resonance imaging. *Sensors and Actuators B: Chemical* 111:562–571.

Goldman, E.R., G.P. Anderson, P.T. Tran, H. Mattoussi, P.T. Charles, and J.M. Mauro. 2002. Conjugation of luminescent quantum dots with antibodies using an engineered adaptor protein to provide new reagents for fluoroimmunoassays. *Analytical Chemistry* 74(4):841–847.

Gonzalez-Martinez, M.A., R. Puchades, and A. Maquieira. 2006. Optical immunosensors for environmental monitoring: How far have we come? *Analytical and Bioanalytical Chemistry* 387:205–318.

Gonzalez-Martinez, M.A., S. Morais, R. Puchades, A. Maquieira. A. Abad, and A. Montova. 1997. Monoclonal antibody-based flow-through immunosensor for analysis of carbaryl. *Analytical Chemistry* 69:2812–3818.

Gosling, J.P. 1990. A decade of development in immunoassay methodology. *Clinical Chemistry* 36:1408–1427.

Gouda, M.D., M.S. Thakur, and N.G. Karanth. 1997. A dual enzyme amperometric biosensor for monitoring organophosphorous pesticides. *Biotechnology Techniques* 11:653–655.

Gouda, M.D., M.S. Thakur, and N.G. Karanth. 2001a. Optimization of the multienzyme system for sucrose biosensor by RSM. *World Journal of Microbiology and Biotechnology* 17:595–600.

Gouda, M.D., M.S. Thakur, and N.G. Karanth. 2001b. Stability studies on immobilized glucose oxidase using an amperometric biosensor- effect of protein based stabilizing agents. *Electroanalysis* 13:849–855.

Grennan, K., G. Strachan, A.J. Porter, A.J. Killard, and M.R. Smith. 2003. Atrazine analysis using an amperometric immunosensor based on single-chain antibody fragments and regeneration-free multi-calibrant measurement. *Analytica Chimica Acta* 500:287–298.

Grunert, K.G. 2005. Food quality and safety: Consumer perception and demand. *European Review of Agricultural Economics* 32:369–391.

Gulla, K.C., M.D. Gouda, M.S. Thakur, and N.G. Karanth. 2002. Reactivation of immobilized acetylcholinesterase in an amperometric biosensor for organophosphorus pesticide. *Biochimica et Biophysica Acta (BBA)—Protein Structure and Molecular Enzymology* 1597:133–139.

Guo, Y.R., S.Y. Liu, W.J. Gui, and G.N. Zhu. 2009. Gold immunochromatographic assay for simultaneous detection of carbofuran and triazophos in water samples. *Analytical Biochemistry* 38:932–939.

Halamek, J., M. Hepel, and P. Skladal. 2001. Investigation of highly sensitive piezoelectric immunosensors for 2,4-dichlorophenoxyacetic acid. *Biosensors and Bioelectronics* 16:253–260.

Hall, J.C., R.J.A. Deschamps, and K.K. Krieg. 1989. Immunoassays for the detection of 2,4-D and picloram in river water and urine. *Journal of Agricultural and Food Chemistry* 37:981–984.

Hermanson, G.T., A.K. Mallia, and P.K. Smith. 1992. *Immobilized affinity ligand techniques.* San Diego: Academic Press.

Hleli, S., C. Martelet, A. Abdelghani, N. Burais, and R.N. Jaffrezic. 2006. Atrazine analysis using an impedimetric immunosensor based on mixed biotinylated self-assembled monolayer. *Sensors and Actuators B: Chemical* 113:711–717.

Ho, Y.P., M.C. Kung, S. Yang, and T.H. Wang. 2005. Multiplexed hybridization detection with multicolor colocalization of quantum dot. *Nano Letters* 5:1693–97.

Hock, B. 1997. Antibodies for immunosensors: A review. *Analytica Chimica Acta* 347:177–86.

Homola, J. 2003. Present and future of surface plasmon resonance biosensors. *Analytical and Bioanalytical Chemistry* 377:528–539.

Horacek, J., and P. Skladal. 1997. Improved direct piezoelectric biosensors operating in liquid solution for the competitive label-free immunoassay of 2,4-dichlorophenoxyacetic acid. *Analytica Chimica Acta* 347:43–50.

Hu, J.T., L. Li, W. Yang, L. Manna, L. Wang, and A.P. Alivisatos. 2001. Linearly polarized emission from colloidal semiconductor quantum rods. *Science* 292:2060–2063.

Huang, S.H. 2006. Gold nanoparticle-based immunochromatographic test for identification of *Staphylococcus aureus* from clinical specimens. *Clinica Chimica Acta* 373:139–143.

Huang, S.H., H.C. Wei, and Y.C. Lee. 2007. One-step immunochromatographic assay for the detection of *Staphylococcus aureus*. *Food Control* 18:893–897.

Jiang, X., D. Li, X. Xua et al. 2008. Immunosensors for detection of pesticide residues. *Biosensors and Bioelectronics* 23:1577–1587.

Johansson, A., B. Mattiasson, and K. Mosbach. 1976. Immobilized enzymes in microcalorimetry. *Methods in Enzymology* 44:659–667.

Kandimalla, V.B., N.S. Neeta, N.G. Karanth et al. 2004. Regeneration of ethyl parathion antibodies for repeated use in immunosensor: A study on dissociation of antigens from antibodies. *Biosensors and Bioelectronics* 20:903–906.

Kane, M.D., T.A. Jatkoe, C.R. Stumpf, J. Lu, J.D. Thomas, and S.J. Madore. 2000. Assessment of the sensitivity and specificity of oligonucleotide (50mer) microarrays. *Nucleic Acids Research* 28(22):4552–4557.

Karamus, W., S. Asakevich, A. Indurkhva, J. Witten, and H. Kruse. 2002. Childhood growth and exposure to dichlorodiphenyl dichloroethene and polychlorinated biphenyls. *The Journal of Pediatrics* 140:33–39.

Karoonuthaisiri, N., R. Charlermroj, U. Uawisetwathana, P. Luxananil, K. Kirtikara, and O. Gajanandana. 2009. Development of antibody array for simultaneous detection of foodborne pathogens. *Biosensors and Bioelectronics* 24:1641–1648.

Kaur, J., K.V. Singh, R. Boro et al. 2007. Immunochromatographic dipstick assay format using gold nanoparticles labeled protein-hapten conjugate for the detection of atrazine. *Environmental Science and Technology* 41:5028–5036.

Kaur, J., K.V. Singh, A.H. Schmid, G.C. Varshney, C.R. Suri, and M. Raje. 2004. Atomic force spectroscopy-based study of antibody pesticide interactions for characterization of immunosensor surface. *Biosensors and Bioelectronics* 20:284–293.

Keay, R.W., and C.J. McNeil. 1998. Separation-free electrochemical immunosensor for rapid determination of atrazine. *Biosensors and Bioelectronics* 13:963–970.

Killard, A.J., L. Micheli, K. Grennan et al. 2001. Amperometric separation-free immunosensor for real-time environmental monitoring. *Analytica Chimica Acta* 427:173–180.

Killard, A.J., M.R. Smyth, K. Grennan, L. Micheli, and G. Palleschi. 2000. Rapid antibody biosensor assays for environmental analysis. *Biochemical Society Transactions* 28:81–84.

Kim, K.S., and J.K. Park. 2005. Magnetic force-based multiplexed immunoassay using superparamagnetic nanoparticles in microfluidic channel. *Lab on a Chip* 5:657–664.

Klotz, A., A. Brecht, C. Barzen et al. 1998. Immunofluorescence sensor for water analysis. *Sensors and Actuators B: Chemical* 51:181–187.

Knopp, D., P. Nuhn, and E. Mittag. 1986. Preparation of radioactively labeled tracers of 2,4-dichlorophenoxyacetic acid for radioimmunoassay. *Pharmazie* 41:143–152.

Ko, S., S. Gunasekaran, and S. Yu. 2010. Self-indicating nanobiosensor for detection of 2,4-dinitrophenol. *Food Control* 21:155–161.

Kolosova, A.Y., J.H. Park, S.A. Eremin, S.J. Kang, and D.H. Chung. 2003. Fluorescence polarization immunoassay based on a monoclonal antibody for the detection of the organophosphorus pesticide parathion-methyl. *Journal of Agricultural and Food Chemistry* 51:1107–1114.

Kumar, M.A., R.S. Chouhan, M.S. Thakur, B.E. Amita Rani, B. Mattiasson, and N.G. Karanth. 2006. Automated flow enzyme-linked immunosorbent assay (ELISA) system for analysis of methyl parathion. *Analytica Chimica Acta* 560:30–34.

Lange, K., B.E. Rapp, and M. Rapp. 2008. Surface acoustic wave biosensors: A review. *Analytical and Bioanalytical Chemistry* 391:1509–519.

Laschi, S., D. Ogonczyk, I. Palchetti, and M. Mascini. 2007. Evaluation of pesticide-induced acetylcholinesterase inhibition by means of disposable carbon-modified electrochemical biosensors. *Enzyme and Microbial Technology* 40:485–489.

Lazcka, O., F.J. Del Campob, and F.X. Munoz. 2002. Pathogen detection: A perspective of traditional methods and biosensors. *Biosensors and Bioelectronics* 22:1205–1217.

Lebrun, S.J. 2005. The use of microarrays for highly sensitive, multiplex immunoassays in biomarker discovery and drug efficacy studies. *Pharmace. Discov.* (Suppl. Vol.):53–56.

Lechuga, L.M., A.T.M. Lenferink, R.P.H. Kooyman, and J. Greve. 1995. Feasibility of eva-nescent wave interferometer immunosensors for pesticide detection: Chemical aspects. *Sensors and Actuators B: Chemical* 25:762–765.

Lisa, M., R.S. Chouhan, A.C. Vinayaka, H.K. Manonmani, and M.S. Thakur. 2009. Gold nanoparticles based dipstick immunoassay for the rapid detection of dichlorodiphe-nyltrichloroethane: An organochlorine pesticide. *Biosensors and Bioelectronics* 25:224–227.

Lisha, K.P., Anshup, and T. Pradeep. 2009. Enhanced visual detection of pesticides using gold nanoparticles. *Journal of Environmental Science and Health Part B—Pesticides Food Contaminants and Agricultural Wastes* 44:697–705.

Luppa, P.B., L.J. Sokoll, and D.W. Chan. 2001. Immunosensors—Principles and applications to clinical chemistry. *Clinica Chimica Acta* 314:1–26.

MacBeath, G. 2002. Protein microarrays and proteomics. *Nature Genetics* 32:526–32.

MacBeath, G., and S.L. Schreiber. 2000. Printing proteins as microarrays for high throughput function determination. *Science* 289:1760–1763.

Mallat, E., C. Barzen, R. Abuknesha, G. Gauglitz, and D. Barcelo. 2001a. Fast determination of paraquat residues in water by an optical immunosensor and validation using capillary electrophoresis-ultraviolet detection. *Analytica Chimica Acta* 427:165–171.

Mallat, E., C. Barzen, R. Abuknesha, G. Gauglitz, and D. Barcelo. 2001b. Part per trillion level determination of isoproturon in certified and estuarine water samples with a direct opti-cal immunosensor. *Analytica Chimica Acta* 426:209–216.

Mastichiadis, C., S.E. Kakabakos, I. Christofidis, M.A. Koupparis, C. Willetts, and K. Misiakos. 2002. Simultaneous determination of pesticides using a four-band disposable *optical* capillary immunosensor. *Analytical Chemistry* 74:6064–6072.

Mauriz, E., A. Calle, A. Abad et al. 2006b. Determination of carbaryl in natural water sam-ples by a surface plasmon resonance flow-through immunosensor. *Biosensors and Bioelectronics* 21:2129–36.

Mauriz, E., A. Calle, J.J. Manclús et al. 2006c. Single and multi-analyte surface plasmon reso-nance assays for simultaneous detection of cholinesterase inhibiting pesticides. *Sensors and Actuators B: Chemical* 118:399–407.

Mauriz, E., A. Calle, A. Montoya, and L.M. Lechuga. 2006d. Determination of environmental organic pollutants with a portable optical immunosensor. *Talanta* 69:359–364.

Mauriz, E., A. Calle, L.M. Lechuga, J. Quintana, A. Montoya, and J.J. Manclus. 2006a. Real-time detection of chlorpyrifos at part per trillion levels in ground, surface and drink-ing water samples by a portable surface plasmon resonance immunosensor. *Analytica Chimica Acta* 561:40–47.

McClain, M.A., C.T. Culbertson, S.C. Jacobson, N.L. Allbritton, C.E. Sims, and J.M. Ramsey. 2003. Microfluidic devices for the high-throughput chemical analysis of cells. *Analytical Chemistry* 75(21):5646–55.

Medintz, I.L., H.T. Uyeda, E.R. Goldman, and H. Mattoussi. 2005. Quantum dot bioconju-gates for imaging, labelling and sensing. *Nature Materials* 4:435–446.

Milner, K.R., A.P. Brown, D.W.E. Allsopp, and W.B. Betts. 1998. Dielectrophoretic clas-sification of bacteria using differential impedance measurements. *Electronics Letters* 34:66–68.

Moulin, A.M., S.J. O'Shea, and M.E. Welland. 2000. Microcantilever based biosensors. *Ultramicroscopy* 82:23–31.

Mukhopadhyay, B., M.B. Martins, R. Karamanska, D.A. Russell, and R.A. Field. 2009. Bacterial detection using carbohydrate-functionalised CdS quantum dots: A model study exploiting *E. coli* recognition of mannosides. *Tetrahedron Letters* 50:886–889.

Mulchandani, A., W. Chen, P. Mulchandani, J. Wang, and K.R. Rogers. 2001. Biosensors for direct determination of organophosphate pesticides. 16:225–230.

Mullett, W.M., E.P.C. Lai, and J.M. Yeung. 2000. Surface plasmon resonance-based immuno-assays. *Methods* 22: 77–91.

Namsoo, K., P. In-Seon, and K. Woo-Yeon. 2007. Salmonella detection with a direct-binding optical grating coupler immunosensor. *Sensors and Actuators B: Chemical* 121:606–615.

Navartilova, I., and P. Skladal. 2004. The immunosensors for measurement of 2,4-dichlorophe-noxyacetic acid based on electrochemical impedance spectroscopy. *Bioelectrochemistry* 62:11–18.

Nichkova, M., D. Dosev, A.E. Davies, S.J. Gee, I.M. Kennedy, and B.D. Hammock. 2007. Quantum dots as reporters in multiplexed immunoassays for biomarkers of exposure to agrochemicals. *Analytical Letters* 40:1423–1433.

Park, J.W., S. Kurosawa, H. Aizawa, S. Wakida, S. Yamada, and K. Ishihara. 2003. Comparison of stabilizing effect of stabilizers for immobilized antibodies on QCM immunosensors. *Sensors and Actuators B: Chemical* 91:158–162.

Patel, P.D. 2002. Biosensors for measurement of analytes implicated in food safety: A review. *TrAC Trends in Analytical Chemistry* 21:96–115.

Patra, M.K., K. Manzoor, M. Manoth, S.C. Negi, S.R. Vadera, and N. Kumar. 2008. Nano-technology applications for chemical and biological sensors. *Defence Science Journal* 58:636–649.

Pease, A.C., D. Solas, E.J. Sullivan, M.T. Cronin, C.P. Holmes, and S.P. Fodor. 1994. Light-generated oligonucleotide arrays for rapid DNA sequence analysis. *Proceedings of the National Academy of Sciences of the United States of America* 91(11): 5022–5026.

Pribyl, J., M. Hepel, J. Halamek, and P. Skladal. 2003. Development of piezoelectric immu-nosensors for competitive and direct determination of atrazine. *Sensors and Actuators B: Chemical* 91:333–41.

Prieto, F., B. Sepulveda, A. Calle, A. Llobera, C. Dominguez, and L.M. Lechuga. 2003. Integrated Mach–Zehnder interferometer based on ARROW structures for biosensor applications. *Sensors and Actuators B: Chemical* 92:151–158.

Proll, G., M. Kumpf, M. Mehlmann et al. 2004. Monitoring an antibody affinity chromatog-raphy with a label-free optical biosensor technique. *Journal of Immunological Methods* 292:35–42.

Raiteri, R., M. Grattarola, H.J. Butt, and P. Skladal. 2001. Micromechanical cantilever-based biosensors. *Sensors and Actuators B: Chemical* 79:115–126.

Regnault, B., S. Martin-Delautre, M. Lejay-Collin, M. Lefevre, P.A. Grimont. 2000. Oligonucleotide probe for the visualization of *Escherichia coli/Escherichia ferguso-nii* cells by *in situ* hybridization: Specificity and potential applications. *Research in Microbiology* 151:521–533.

Rekha, K., M.S. Thakur, and N.G. Karanth. 2000a. Biosensors for organophosphorus pesticide monitoring. *CRC Critical Reviews in Biotechnology* 20:213–235.

Rekha, K., M.D. Gouda, M.S. Thakur, and N.G. Karanth. 2000b. Ascorbate oxidase based amperometric biosensor for organophosphorous pesticide monitoring. *Biosensors and Bioelectronics* 15:499–502.

Rodriguez-Mozaz, S., S. Reder, M.J. Lopez de Alda, G. Gauglitz, and D. Barceló. 2004. Simultaneous multi-analyte determination of estrone, isoproturon and atrazine in natu-ral waters by the RIver ANAlyser (RIANA), an optical immunosensor. *Biosensors and Bioelectronics* 19:633–640.

Rong-Hwa, S., T. Shiao-Shek, C. Der-Jiang, and H. Yao-Wen. 2010. Gold nanoparticle-based lateral flow assay for detection of staphylococcal enterotoxin B. *Food Chemistry* 118:462–466.

Sadik, O.A., and J.M. Van Emon. 1996. Applications of electrochemical immunosensors to environmental monitoring. *Biosensors and Bioelectronics* 11:i–x.

Sadik, O.A., H. Xu, E. Gheorghiu et al. 2002. Differential impedance spectroscopy for monitoring protein immobilization and antibody-antigen reactions. *Analytical Chemistry* 74:3142–3150.

Salmaina, M., N. Fischer-Durand, and C.M. Pradier. 2008. Infrared optical immunosensor: Application to the measurement of the herbicide atrazine. *Analytical Biochemistry* 373:61–70.

Samsonova, J.V., M.Y. Rubtsova, A.V. Kiseleva, A.A. Ezhov, and A.M. Egorov. 1999. Chemiluminescent multiassay of pesticides with horseradish peroxidase as a label. *Biosensors and Bioelectronics* 14:273–81.

Santra, S., J. Xu, K. Wang, and W. Tan. 2004. Luminescent nanoparticle probes for bioimaging. *Journal of Nanoscience and Nanotechnology* 4:590–599.

Schmid, A.H., S.A. Stance, M.S. Thakur, K.R. Thampi, and C.R. Suri. 2006. Site-directed antibody immobilization on gold substrate for surface plasmon resonance sensors. *Sensors and Actuators B: Chemical* 113:297–303.

Schobel, U., C. Barzen, and G. Gauglitz. 2000. Immunoanalytical techniques for pesticide monitoring based on fluorescence detection. *Fresenius' Journal of Analytical Chemistry* 366:646–658.

Shim, W.B., J.G. Choi, J.Y. Kim et al. 2007. Production of monoclonal antibody against Listeria monocytogenes and its application to immunochromatography strip test. *Journal of Microbiology and Biotechnology* 17:1152–1161.

Situma, C., M. Hashimoto, and S.A. Soper. 2006. Merging microfluidics with microarray-based bioassays. *Biomolecular Engineering* 23:213–231.

Skládal, P., and T. Kaláb. 1995. A multichannel immunochemical sensor for determination of 2,4-dichlorophenoxyacetic acid. *Analytica Chimica Acta* 316:73–78.

Skottrup, P.D., M. Nicolaisen, and A.M. Justesen. 2008. Towards on-site pathogen detection using antibody-based sensors. *Biosensors and Bioelectronics* 24:339–348.

Soler, C., B. Hamilton, A. Furey, K.J. James, J. Mañes, and Y. Picó. 2007. Liquid chromatography quadrupole time-of-flight mass spectrometry analysis of carbosulfan, carbofuran, 3-hydroxycarbofuran, and other metabolites in food. *Analytical Chemistry* 79:1492–1501.

Southern, E., K. Mir, and M. Shchepinov. 1999. Molecular interactions on microarrays. *Nature Genetics Supplement* 21(Suppl. 1):5–9.

Starodub, N.F., B.B. Dzantiev, V.M. Starodub, and A.V. Zherdev. 2000. Immunosensor for the determination of herbicide. *Analytica Chimica Acta* 424:37–43.

Suri, C.R., M. Raje, and G.C. Varshney. 2002. Immunosensors for pesticide analysis: Antibody production and sensor development. *Critical Reviews in Biotechnology* 22:15–32.

Székács, A., N. Trummer, N. Adányi, M. Váradi, and I. Szendrő. 2003. Development of a non-labeled immunosensor for the herbicide trifluralin *via* optical waveguide lightmode spectroscopic detection. *Analytica Chimica Acta* 487:31–42.

Thakur, M.S., and N.G. Karanth. 2003. Research and development of biosensors for food analysis in India. In *Advances in biosensors,* ed. B.D. Malhotra, and A.P.F. Turner, 5:131–160. Amsterdam: Elsevier Science.

Turner, A.P.F. 1992. *Advances in biosensors,* Vol. 2. London: JAI Press.

Valera, E., J. Ramón-Azcón, F.J. Sanchez, M.P. Marco, and A. Rodríguez. 2008. Conductimetric immunosensor for atrazine detection based on antibodies labelled with gold nanoparticles. *Sensors and Actuators B: Chemical* 134:95–103.

Van Dyck, E., M. Ieven, S. Pattyn, L. Van Damme, and M. Laga. 2001. Detection of chlamydia trachomatis and neisseria gonorrhoeae by enzyme immunoassay, culture, and three nucleic acid amplification tests. *Journal of Clinical Microbiology* 39:1751–1756.

Vinayaka, A.C., and M.S. Thakur. 2010. Focus on quantum dots as potential fluorescent probes for monitoring food toxicants and foodborne pathogens. *Analytical and Bioanalytical Chemistry* DOI: 10.1007/s00216-010-3683-y.

Vinayaka, A.C., S. Basheer, and M.S. Thakur. 2009. Bioconjugation of CdTe quantum dot for the detection of 2,4-dichlorophenoxyacetic acid by competitive fluoroimmunoassay based biosensor. *Biosensors and Bioelectronics* 24:1615–1620.

Warner, M.G., J.W. Grate, A. Tyler et al. 2009. Quantum dot immunoassays in renewable surface column and 96-well plate formats for the fluorescence detection of botulinum neurotoxin using high-affinity antibodies. *Biosensors and Bioelectronics* 25:179–184.

Yang, H., H. Li, and X. Jiang. 2008. Detection of foodborne pathogens using bioconjugated nanomaterials. *Microfluidics and Nanofluidics* 5:571–583.

Yang, L., and Y. Li. 2005. AFM and impedance spectroscopy characterization of the immobilization of antibodies on indium–tin oxide electrode through self-assembled monolayer of epoxysilane and their capture of *Escherichia coli* O157:H7. *Biosensors and Bioelectronics* 20:1407–1416.

Yang, M., Y. Kostov, H.A. Bruck, and A. Rasooly. 2009. Gold nanoparticle-based enhanced chemiluminescence immunosensor for detection of Staphylococcal Enterotoxin B (SEB) in food. *International Journal of Food Microbiology* 133:265–271.

Yulaev, M.F., R.A. Sitdikov, N.M. Dmitrieva, E.V. Yazynina, A.V. Zherdev, and B.B. Dzantiev. 2001. Development of a potentiometric immunosensor for herbicide simazine and its application for food testing. *Sensors and Actuators B: Chemical* 75:129–135.

Zhang, D., D.J. Carr, and E.C. Alocilja. 2009b. Fluorescent bio-barcode DNA assay for the detection of *Salmonella enterica* serovar Enteritidis. *Biosensors and Bioelectronics* 24:1377–1381.

Zhang, X., P. Geng, H. Liu et al. 2009a. Development of an electrochemical immunoassay for rapid detection of *E. coli* using anodic stripping voltammetry based on Cu@Au nanoparticles as antibody labels. *Biosensors and Bioelectronics* 24:2155–2159.

Zhang, Y., M.Y. Coyne, S.G. Will, C.H. Levenson, and E.S. Kawasaki. 1991. Single-base mutational analysis of cancer and genetic diseases using membrane bound modified oligonucleotides. *Nucleic Acids Research* 19(14):3929–3933.

Zhao, X., L.R. Hilliard, S.J. Mechery et al. 2004. A rapid bioassay for single bacterial cell quantitation using bioconjugated nanoparticles. *Proceedings of the National Academy of Sciences of the United States of America* 101:15027–15032.

Zhelev, Z., R. Bakalova, H. Ohba, and Y. Baba. 2006. Quantum dot-based nanohybrids for fluorescent detection of molecular and cellular biological targets. In *Nanomaterials for Biosensors*, ed. C.S.S.R. Kumar, 175–197. Weinheim: Wiley-VCH.

Zhou, P., Y. Lu, J. Zhu et al. 2004. Nanocolloidal gold-based immunoassay for the detection of the n-methylcarbamate pesticide carbofuran. *Journal of Agricultural and Food Chemistry* 52:4355–4359.

8 Impedance Biosensors/ Biochips for Detection of Foodborne Pathogens

Liju Yang

CONTENTS

8.1　INTRODUCTION

8.1.1　FOODBORNE PATHOGENS AND THEIR DETECTION

Foodborne diseases caused by foodborne pathogens have been a serious threat to public health and food safety for decades and remain one of the major concerns of our society. Today, more than 250 different foodborne diseases have been described. The Centers for Disease Control and Prevention (CDC) estimate that 76 million foodborne illness cases occur in the United States every year, accounting for 325,000 hospitalizations and more than 5000 deaths (Mead et al. 1999). Pathogenic bacteria are the most common foodborne pathogens, accounting for 91% of the total outbreaks of foodborne illness in the United States (Beran et al. 1991; Potter et al. 2007). The U.S. Department of Agriculture (USDA) Economic Research Service (ERS) estimates that the annual cost of five foodborne illnesses caused by *Camploybacter*, *Salmonella*, *Escherichia coli* O157:H7, *Listeria monocytogenes*, and *Toxoplasma gondii* is about $6.9 billion in medical costs, lost productivity, and premature deaths (USDA/ERS 2002).

Effective detection and inspection methods are necessary to control pathogens in food products and to prevent foodborne diseases. Conventional microbiological methods have been the gold standard for identification and detection of pathogens in food for nearly a century and continue to be a reliable standard for ensuring food safety. However, these methods are labor intensive and time-consuming, generally requiring 5 to 7 days to get a confirmed result for a particular pathogenic organism (Swaminathan and Feng 1994; Vasavada 1997). They are therefore unsuited for modern food quality assurance, in which timely responses to possible risks is critical. As a result, during the past 30 years, scientists have made great efforts to develop a variety of rapid methods to reduce assay times. These methods include the miniaturized biochemical tests, physicochemical methods that measure bacterial metabolites, highly specific nucleic acid–based tests, antibody-based methods, and fully automated instrumental diagnostic systems (Silley 1994; Swaminathan and Feng 1994; Van der Zee and Huis In't Veld 1997; Vasavada 1997; Ivnitski et al. 1999; Hall 2002). Two well-studied methods, enzyme linked immunosorbent assay (ELISA) and polymerase chain reaction (PCR), have been widely used in the food industry. They have reduced the assay time to 10~24 h and 4~6 h, respectively, and have achieved detection limits varying from 10^1 to 10^6 cfu/mL (cfu = colony forming units). More recently, biosensor-based methods have gained great attention in this field due to their rapid detection speed, high sensitivity, and reliability. These biosensor methods have reduced assay times to 10 min to 2 h with detection limits varying from 10^2 to 10^7 cfu/mL (Ivnitski et al. 1999, 2000; Rand et al. 2002; Su and Li 2004).

Despite the significant improvements in the assay times and detection limits, important concerns still need to be addressed to meet the needs for pathogen detection. These concerns include the throughput of a method (the number of samples per operation), differentiation of live and dead cells, automation, cost, simplicity, training, and accuracy. A sensitive method is expected to be able to detect a very low number of pathogens since the presence of even a single pathogenic organism in food may be an infectious dose. For example, the infectious dosage of coliform standard for *E. coli* in water is 4 cells/100mL (Federal Register 1990, 1991; Greenberg et al. 1992). In many food samples, the demand for detection limits is less than 1 cell per 25 g of food. However, to date, an ideal rapid method that meets all the needs mentioned above does not yet exist. Thus, there is still a great need for scientists and engineers to work together to develop new methods or to promote existing methods for the rapid detection of foodborne pathogens.

8.1.2 IMPEDANCE TECHNIQUE FOR FOODBORNE PATHOGEN DETECTION

The impedance technique is one of the earliest physicochemical methods that has been applied to the field of microbiology as a means to detect and/or quantify bacteria. The first impedance method for detection of bacteria was described one century ago (Stewart 1899) and was based on the measurement of the change in electrical impedance of a medium resulting from bacterial growth. In the mid-1970s to 1980s, an increasing number of papers were published on this method, including notable work by Ur and Brown (1974, 1975), Cady (1975, 1978), Cady et al. (1978), and the important work of Eden and Torry Research Station (Richards et al. 1978; Eden and Eden 1984), to promote it as a rapid method that can detect bacteria within 24 h. This classic impedance method, called *Impedance Microbiology*, was approved by the Association of Official Analytical Chemists International (AOAC) in 1992 as a first-action method for screening *Salmonella* in food samples (Gibson et al. 1992; AOAC 1996).

In recent years, the impedance technique has been expanding rapidly and becoming a more valuable technique for the detection of bacteria. Recent advances in microfabrication technology and especially biosensor technology offer great opportunities for integrating the impedance technique with these technologies, opening new avenues for the development of impedance-based methods for rapid detection of bacteria. A number of factors contribute to the popularity of the impedance technique for biological detection: (1) the distinct electrical properties of bacterial cells and their electrophysiology well suit the impedance technique for detection of bacteria based on these properties; (2) the impedance technique allows label-free, real-time, and noninvasive detection; and (3) the impedance technique, as an electrical method, can be easily integrated into miniaturized electronic devices to meet the growing need of smaller portable systems that replace reliance on large laboratory-based instruments.

8.1.3 IMPEDANCE PROPERTIES OF BACTERIAL CELLS

Bacterial cells consist of adjacent structures that have very different electrical properties. Here we consider two major structural components of a bacterial cell: the cell membrane and the interior of the cell. The cell membrane consists of a lipid bilayer in which lipid molecules are oriented with their polar groups facing outward into the aqueous environment, and their hydrophobic hydrocarbon chains pointing inward to form the membrane interior. It contains numerous proteins that are primarily responsible for transport of ions, nutrients, and waste across the membrane. This membrane is general highly insulated with its conductivity around 10^{-7} S/m (Pethig and Markx 1997). The inside of a cell contains DNA, nutrient storage granules, ribosomes, and many dissolved charged molecules and is therefore much more conductive than the cell membrane. Its conductivity can be as high as 1 S/m (Pethig and Markx 1997).

Besides these electrical properties of different structures of a bacterial cell, metabolic activity is another property that can be linked to impedance measurements. As is well known, the medium consists generally of uncharged or weakly charged substances that are metabolized into small and highly charged molecules during bacterial growth. This process increases both the conductance and capacitance of the medium, leading to a decrease in impedance.

8.1.4 MECHANISMS FOR IMPEDANCE DETECTION OF BACTERIA

Based on these electrical-related properties of bacterial cells, three mechanisms have been reported for developing impedance-based methods for detection/quantification of bacterial cells (Yang 2008):

1. Impedance detection based on bacterial metabolism. This approach is represented by impedance microbiology, a technique based on the measurements of changes in electric impedance in a medium or a reactant solution resulting from bacterial metabolism (Silley and Forsythe 1996; Wawerla et al. 1999). The impedance changes are mainly caused by the release of ionic metabolites from live cells into the medium, including both the main energy metabolism (catabolism) and minor ion exchange through ion channels on the cell membrane (such as K^+, Na^+ ion channels) (Owicki and Parce 1992).

2. Impedance detection based on the insulating properties of the cell membrane. Because of the highly insulated cell membrane, bacterial cells attached to an electrode surface can effectively reduce the electrode area that the current reaches and thus increase the interface impedance. Most impedance biosensors for bacterial detection are based on this principle, in which specific antibodies are immobilized on an electrode surface to

facilitate the attachment of bacterial cells to the electrode surface and to provide selectivity to the sensor. The impedance response of the sensor is related to the number of the attached cells (Ruan et al. 2002; Yang et al. 2004a; Radke and Alocilja 2005).
3. Impedance detection based on the release of ionic cytoplasm substances. Since the inside of a cell contains many dissolved charged molecules and is highly conductive, cell lysis or release of intracellular ions to a low-conductance buffer can change the impedance of the buffer. This change in impedance can be used to detect bacterial cells (Yang 2008).

In this chapter, we describe the fundamentals of impedance technique, present the microelectrode/microchip-based impedance methods that we have recently developed using each of these mechanisms for detection/quantification of foodborne pathogenic bacteria, summarize representative experimental data, and discuss the applications of these methods.

8.2 BASICS OF IMPEDANCE TECHNIQUE

8.2.1 DEFINITION OF IMPEDANCE

Impedance is a measure of the total opposition to current flow in an alternating current (AC) circuit, symbolized as Z. It is a complex variable that also accounts for phase shifts. It is made up of two components, ohmic resistance, R, and reactance, X, and can be represented in complex notation as Equation (8.1) (Yang and Guiseppi 2008):

$$Z = R + jX \qquad (8.1)$$

The resistance term, R, describes the part of the circuit that behaves as a resistor, while the reactance, X, describes the part of the circuit that behaves as a capacitor or inductor. The magnitude of the impedance is determined by the square root of the sum of the squares for the resistance and reactance (Equation 8.2) and the phase angle (θ) is determined by Equation (8.3):

$$|Z| = \sqrt{R^2 + X^2} \qquad (8.2)$$

$$\theta = \arctan X/R \qquad (8.3)$$

Impedance is usually measured by applying an AC potential to a system and measuring the resulting current. The measurement can be performed using either a lock-in amplifier and function generator, a frequency response analyzer with a two-electrode system, or an additional front-end potentiostat with a three-electrode system. Alternatively, a commercial impedance analyzer can be used with both two-electrode and three-electrode systems. Figure 8.1a shows the simplified setup for impedance measurement using a three-electrode system. In general, a two-electrode system includes the working electrode (WE) and counter electrode (CE), which may be of equal or different areas. A three-electrode system includes an additional electrode, the reference electrode (RE), which contributes a half cell potential, to which

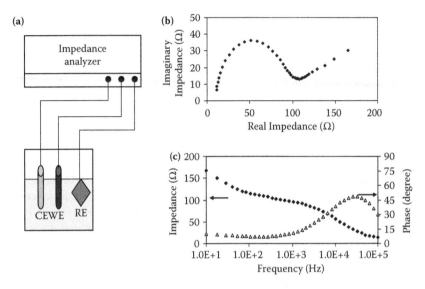

FIGURE 8.1 (a) The simplified experimental setup for the measurement of electrical/electrochemical impedance; (b) a representative Nyquist plot (Z_{im} vs. Z_{re}); and (c) Bode plot (magnitude of impedance and phase vs. frequency) from electrochemical impedance spectroscopic measurement in phosphate buffered solution (PBS) containing 0.1 mM $[Fe(CN)_6]^{-3/-4}$.

the potentiostat references its interrogating voltage. In all cases the instrumentation delivers a sinusoidally potential (V) of specific frequency to the CE and the ensuing AC current (I) is measured at the WE. The applied excitation voltage is sinusoidal and given by

$$V = V_0 \sin \omega t \tag{8.4}$$

in which ω is the radial frequency, V_0 is the amplitude of the voltage signal, V is voltage at any given time. The AC response (I) is characterized by both its amplitude (I_0) and its phase shift (θ) with respect to the applied AC voltage, as shown below.

$$I = I_0 \sin(\omega t + \theta) \tag{8.5}$$

The impedance of the system is determined by the ratio of the amplitudes of the applied and the response signal (V/I) and is expressed in terms of the magnitude, Z_0, and the phase shift between these signals (θ):

$$Z = \frac{V}{I} = \frac{V_0 \sin \omega t}{I_0 \sin(\omega t + \theta)} = Z_0 \frac{\sin \omega t}{\sin(\omega t + \theta)} \tag{8.6}$$

8.2.2 Electrical/Electrochemical Impedance Spectroscopy

When the applied potential is scanned in a frequency range, the technique is called electrical/electrochemical impedance spectroscopy (EIS). EIS can be performed either in the presence or absence of a redox probe, which is referred to as faradaic or nonfaradaic impedance measurement, wherein one or the other type dominates the impedance signal (Bard and Faulkner 2001).

The impedance spectrum, measured as a function of the interrogating frequency, can be presented by two popular plots: Nyquist (Cole-Cole) plot (Figure 8.1b) and Bode ($|Z|$ and θ) plot (Figure 8.1c). The Nyquist plot shows the relationship between the imaginary component of impedance, Zim (on Y axis), and the real component of the impedance, Zre (on X axis), at each frequency. The Nyquist plot usually contains one or more semicircles, but often only a portion of one or more of the semicircles can be seen. Each semicircle represents the characteristics of a single *time constant*. The Nyquist plot sometimes also contains a diagonal line with a slope of 45°, which represents the Warburg impedance and is characteristic of diffusion. The Bode plot presents the magnitude of the impedance, $|Z|$, and the phase shift, θ, in relation to the frequency. On the Bode plot, specific regions of the spectrum can be linked to characteristic frequency ranges.

8.2.3 EQUIVALENT CIRCUIT

It is common to analyze EIS data using an appropriate equivalent circuit model that best fits the acquired frequency-dependent impedance spectrum. An equivalent circuit usually consists of a specific arrangement of resistors, capacitors, and inductors either in series, in parallel, or in a combination of both. To be useful, the electrical elements in an equivalent circuit should always have physicochemical significance in the physical electrochemistry of the tested system. For example, in many impedance measurement systems, the solution resistance is represented by a resistor in the equivalent circuit, and the double-layer capacitance of an electrode is represented by a capacitor because it closely resembles a pure capacitance.

Using a typical three-electrode impedance measurement system as an example, when it contains electrochemical redox species, the system can be represented by an electrical equivalent circuit (Randles model) that includes ohmic resistance of the electrolyte solution (R_s), faradaic impedance (Z_f), and double-layer capacitance (C_{dl}) (Figure 8.2a) (Bard and Faulkner 2001). Since all the current must pass through the resistance of the solution and the electrode surface, the R_s is connected with the electrode part (i_f and i_c) in series. Since the total current through the electrode surface is the sum of faradaic current (i_f) and the double-layer capacitance current (i_c), the i_f and i_c are connected in parallel. The faradaic impedance includes the Warburg impedance (Z_w) and the electron-transfer resistance (R_{et}). Among these components, R_s and Z_w represent the properties of the bulk solution and the diffusion of the redox probe, respectively. C_{dl} and R_{et} are related to the dielectric and insulting properties at the electrode/electrolyte interface. If the impedance measurement system does not contain electrochemically active species, the equivalent circuit can be simplified as a serial combination of the solution resistance (R_s) and the double-layer capacitance (C_{dl}) (Figure 8.2b), because there is no electrochemical reaction on the electrode surface, the faradaic path is inactive, and only nonfaradaic impedance is operative (Yang et al. 2003).

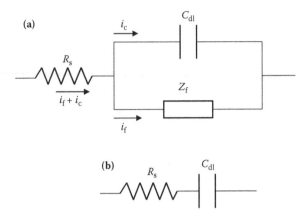

FIGURE 8.2 (a) An equivalent circuit of a general electrochemical cell containing redox species; and (b) an equivalent circuit of the electrochemical cell without the presence of redox species. R_s, Z_f, and C_{dl} stand for the resistance of the electrolyte solution, faradaic impedance, and double-layer capacitance, respectively. i_c and i_f are nonfaradaic current and faradaic current, respectively. (From Yang, L., C. Ruan, and Y. Li, 2003, *Biosensors and Bioelectronics* 19:495–502. With permission.)

8.3 MICROFABRICATED INTERDIGITATED MICROELECTRODES FOR IMPEDANCE MEASUREMENTS

In recent years, the microlithographically fabricated, coplanar, interdigitated micro-electrodes (IMEs) have been a suitable format for the two-electrode arrangement used in impedance measurements. An IME consists of a pair of opposing microband array electrodes with each array having tens to hundreds of finger electrodes. Figure 8.3a shows the schematic of an IME. The finger electrodes are of width and space that may be equal or unequal in the range of 100 nm to 25 microns with electrode length of 1 mm to 5 mm. The IMEs can be platinum, gold, or indium tin oxide (ITO) electrodes fabricated on silicon or glass substrates. Figure 8.3b shows a picture of a gold IME from ABtech Scientific Inc. (Richmond, Virginia). It has 50 pairs of equidistantly spaced finger electrodes of 15 μm width. The image in the circle is the microscopic graph showing the finger electrodes.

IMEs, as a type of microelectrode, possess advantages over conventional electrodes for analytical measurements, such as low ohmic drop, high signal-to-noise ratio, rapid attained steady state, and higher sensitivity (Stulik et al. 2000). With regard to impedance measurement, IMEs have unique features that favor impedance measurement, when compared with conventional two-electrode systems used in classic impedance microbiology. These features include (1) multiple electrode pairs, rather than the single pair of electrodes in conventional electrode systems, (2) progressively shorter distances between the finger electrodes (in μm) than those in the two electrodes used in classic impedance microbiology (usually in mm or cm), and (3) a relatively large electrode surface.

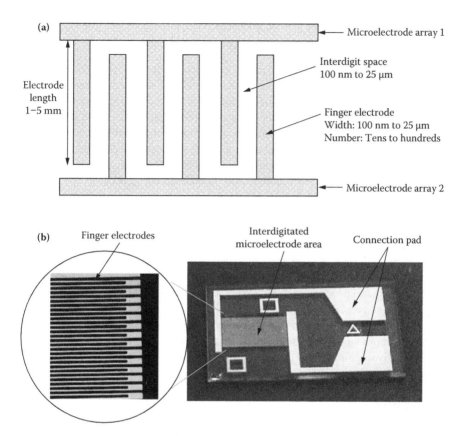

FIGURE 8.3 (a) The schematic structure of an interdigitated microelectrode (IME); and (b) the microscopic pictures of a gold IME on glass substrate and its finger electrodes.

Such IMEs enable the sensitive detection of the electrical property changes in the immediate neighborhood of their surfaces. Van Gerwan et al. (1998) reported a calculated current distribution of a nano-sized IME, that is, curves under which a certain amount of current is flowing. Their result showed that 95% of the current flowing between the finger electrodes flows within a region just above the electrode surface in a distance that equals the sum of the electrode width and space. IMEs have received great attention in the areas of impedimetric immunosensing and biosensing (Van Gerwen et al. 1998; Laureyne et al. 1999, 2000) and impedimetric studies of biological cell behaviors (Ehret et al. 1997, 1998).

We have explored the unique properties of IMEs for their promising applications in impedance detection of bacterial pathogens compared with the classic impedance microbiology method (Yang et al. 2004b; Yang and Li 2006) and other microchip platforms for bacterial detection. We present here three IME-based microchips that we have developed in the past few years for impedance detection of foodborne bacteria based on the three different mechanisms described in Section 8.1.4.

8.4 MICROCHIP IMPEDANCE DETECTION OF *SALMONELLA* BASED ON BACTERIAL METABOLISM

Salmonella is considered one of the major foodborne pathogenic bacteria. The CDC estimates that 1.4 million people are infected with salmonellosis each year in the United States, causing about 1000 deaths. The salmonellosis infections result in approximately $1 billion in direct medical costs and products annually (Mead et al. 1999). *Salmonella typhimurium* is recognized as the second most common serotype of *Salmonella* found in humans (Foodborne Disease, National Institute of Allergy and Infectious Diseases [NIAID]). In this section, we describe a microchip-based impedance method that can be used for detection of *Salmonella* based on their metabolic activity.

8.4.1 The Principle of Metabolism-Based Impedance Detection

The metabolism-based impedance detection technique is based on the measurement of impedance changes in a medium or a reactant solution due to bacterial growth (Silley and Forsythe 1996). During growth, bacteria metabolize oxygen and sugars in the medium and produce carbon dioxide and organic acids. For example, a non-ionized glucose can be converted to two molecules of lactic acid. Further metabolism will convert the lactic acid and three oxygen molecules to carbonic acid. The smaller and more mobile bicarbonate ion is a more effective ionic conductor than the lactate ion. Hydrogen ions are nearly seven times more effective as an ionic conductor than sodium ions (Eden and Eden 1984). Such metabolic processes cause changes in the ionic composition of the medium and consequent changes in the impedance of the medium.

In the metabolism-based impedance method, the impedance change is usually measured over time using a pair of electrodes submerged in the growth medium or the reactant solution. The measured electrical signals are then graphically plotted on the ordinate against the incubation times on the abscissa, producing impedance growth curves. Figure 8.4a illustrates a typical impedance growth curve together

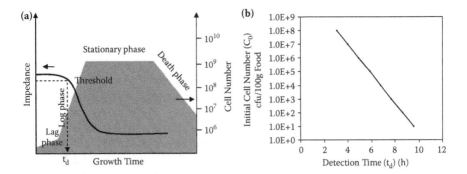

FIGURE 8.4 (a) A generalized impedance growth curve with the threshold and the detection time, along with the typical bacterial growth phases; and (b) the relationship between the initial cell concentration and the detection time. (Modified from Yang, L., and R. Bashir, 2008, *Biotechnology Advances* 26:135–150. With permission.)

with typical bacterial growth phases. The shape of the impedance growth curve well reflects the bacterial growth phases. As we can see, the impedance is stable in the initial period, which corresponds to the lag phase where bacteria metabolize but do not multiply; the impedance decreases sharply in the log or exponential growth phase, when the bacteria multiply exponentially; and in the stationary phase, when the bacterial cell number remains relatively constant, the impedance becomes relatively stable again.

The time corresponding to a point at which the decrease in impedance exceeds a threshold is defined as the *detection time*, t_d (Figure 8.4a). Generally, the detection time does not appear until the bacteria number reaches approximately 10^6–10^7 cfu/mL (as determined by the plating method). The detection time (t_d) is related to the initial cell concentration C_0. Their relationship can be expressed by Equation (8.7),

$$\log (C_0) = -\alpha\, t_d + \beta \qquad (8.7)$$

where, α ($\alpha > 0$) and β are constants that associate with the certain microorganism, the medium, growth conditions, and so forth. Using the values of $\alpha=1.08$ and $\beta=11.33$ for a set of food samples tested for *E. coli* and C_0 as cfu/100g food, reported by Dupont et al. (1996), Figure 8.4b presents a typical calibration curve representing the relation between the initial cell number (C_0) and the detection time (t_d) expressed in Equation (8.7). The detection time in classic impedance microbiology ranges from about 1 h to 8 h for initial cell concentrations of 10^7 to 10^1 cfu/mL.

8.4.2 The Microchip and Methods

An IME microchip from ABtech Scientific Inc. (Richmond, Virginia) was used for impedance measurement. The IME chip consisted of 50 pairs of ITO finger electrodes with 15-µm wide electrodes and spaces, and the finger electrode length of approximately 5 mm. The entire chip was 1.0 cm × 0.5 cm.

Figure 8.5a shows the experimental setup for the IME microchip impedance measurement of *Salmonella typhimurium* (ATCC 14028) growth. Selenite-cystine (SC) broth supplemented with mannitol (M, 5.0 g/L) and trimethylamine oxide (TMAO, 5.0 g/L) (Sigma-Aldrich, St. Louis, Missouri) was used as the growth medium. For growth, 4.5 mL of SC/M/TMAO broth inoculated with 0.5 mL *Salmonella* culture was placed in a 15 mL water-jacketed glass vial (Bioanalytical System, West Lafayette, Indiana). The temperature of the broth was maintained at 37°C by circulating water through the water jacket using a thermostatic water bath. The IME chip was immersed into the growth broth. Impedance was monitored immediately after the inoculation.

Impedance measurements were performed using an IM-6 impedance analyzer (Bioanalytical System, West Lafayette, Indiana) with IM-6/THALES-software. The EIS program was used to run the impedance spectroscopy. A sine-modulated AC potential was scanned in the frequency range from 0.2 Hz to 100 kHz or 0.2 Hz to 5 MHz with amplitude of ±5 mV. The bode plots (impedance and phase vs. frequency) were recorded. The impedance growth curves (impedance vs. growth time) were recorded at fixed frequencies.

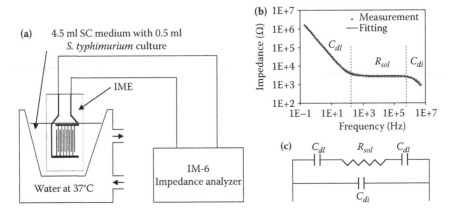

FIGURE 8.5 (a) Experimental setup of the microchip impedance measurement of *Salmonella* growth; (b) a representative impedance spectrum obtained during *Salmonella* growth using microchip, together with its fitting curve; (c) the equivalent circuit of the microchip impedance measurement system. C_{dl} is the double-layer capacitance at each set of the finger electrodes; R_{sol} is the resistance of the medium; and C_{di} is the dielectric capacitance of the medium. (From Yang, L., Y. Li, C.L. Griffis, and M.G. Johnson, 2004b, *Biosensors and Bioelectronics* 19:1139–1147. With permission.)

8.4.3 IMPEDANCE SPECTRUM OF THE IME SYSTEM AND ITS EQUIVALENT CIRCUIT

Figure 8.5b shows the Bode impedance spectrum of the IME microchip system with *Salmonella* in SC/M/TMAO medium. The impedance spectrum has three regions: the low-frequency region from 0.2 Hz to approximately 50 Hz, where the impedance decreases with increasing frequency; the middle-frequency region from 1 kHz to 1 MHz, where impedance does not change with frequency; and the high-frequency region (>1 MHz), where impedance decreases again with increasing frequency.

The impedance spectroscopic behavior of the IME system can be interpreted by the equivalent circuit shown in Figure 8.5c (Van Gerwen et al. 1998; Laureyn et al. 2000; Yang et al. 2004b). It includes two identical double-layer capacitances (C_{dl}) and the medium resistance (R_{sol}) connected in series, and the dielectric capacitance of the medium (C_{di}) connected in parallel with these series elements. This equivalent circuit is the combination of two partial branches: one is $C_{dl}+R_{sol}+C_{dl}$, and the other is C_{di}. Contributions from each partial circuit to the impedance and the total impedance could be expressed with the following equations (Yang et al. 2004b):

$$|Z_1| = \sqrt{R_{sol}^2 + \frac{1}{(\pi f C_{dl})^2}} \quad \text{(For branch } C_{dl} + R_{sol} + C_{dl}) \tag{8.8a}$$

$$|Z_2| = \sqrt{\frac{1}{(2\pi f C_{di})}} \quad \text{(For branch } C_{di}) \tag{8.8b}$$

$$\frac{1}{Z_{tol}} = \frac{1}{Z_1} + \frac{1}{Z_2} \quad \text{(For two branches in parallel)} \tag{8.8c}$$

The three regions in the impedance spectrum correspond to the responses from the three electrical elements in the equivalent circuit. At frequencies lower than 1 MHz, only one of the two partial branches, $C_{di}+R_{sol}+C_{dl}$, is active. The total impedance was determined by Equation (8.8a). Both the solution resistance and the double-layer capacitances are included, but they dominate at different frequencies. At frequencies lower than 50 Hz, the double-layer capacitances offer a major contribution to the impedance, and the impedance from the solution resistance can be neglected. Therefore, the impedance value increases with decreased frequencies (double-layer region). Up to frequencies higher than 1 kHz, double-layer capacitances offer almost no impedance, and the medium resistance is the only contribution to the total impedance. Thus, the total impedance was frequency-independent (resistive region). At frequencies higher than 1 MHz (dielectric region), the impedance from the medium resistance and the double-layer capacitance can be ignored; and the partial branch C_{di} is active, the dielectric capacitance is the major contribution to the impedance. Therefore, the impedance value was inversely proportional to the frequency (Equation 8.8b).

8.4.4 IMPEDANCE CHANGE DUE TO THE GROWTH OF *S. TYPHIMURIUM*

Figure 8.6 shows the impedance spectra measured before and after 16 h growth of *Salmonella* in SC/M/T medium. The difference in the impedance before and after bacterial growth was observed below 100 Hz but was negligible at frequencies higher than 1 kHz. By simulating the measured two spectra using the equivalent circuit above, we can obtain the values of R_{sol} and C_{dl}. (Table 8.1). As shown in Table 8.1,

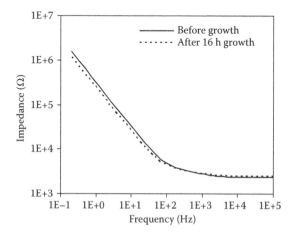

FIGURE 8.6 Impedance spectra recorded before and after *Salmonella* growth in SC/M/T medium using the IME microchip. Initial cell concentration: 1.76×10^2 cfu/mL. (From Yang, L., Y. Li, C.L. Griffis, and M.G. Johnson, 2004b, *Biosensors and Bioelectronics* 19:1139–1147. With permission.)

TABLE 8.1

**Simulated Values of the Double-Layer Capacitance
and Medium Resistance before and after *Salmonella*
Growth, Obtained by Fitting the Measured
Impedance Spectral Data into the Equivalent Circuit**

	$C_{dl}(\mu F)$	$R_{sol}(k\Omega)$
Before bacterial growth	1.06 ± 0.047	3.47 ± 0.52
After 16 h growth	1.38 ± 0.11	3.45 ± 0.48
Change in percentage (%)	30.19	−0.58

Source: Yang, L., C. Ruan, and Y. Li, 2003, *Biosensors and Bioelectronics* 19:495–502. (With permission.)

upon the growth of *Salmonella*, C_{dl} increased significantly from 1.06 µF to 1.38 µF by 30.37%; whereas R_{sol} changed little.

In theory, the value of double-layer capacitance can be expressed as follows:

$$C_{dl} = (\varepsilon_{dl} A)/d \tag{8.9}$$

Where ε_{dl} is the dielectrical permittivity of the double-charged layer, $\varepsilon_{dl} = \varepsilon_0 \varepsilon_p$, ε_0 is the permittivity of the free space and ε_p is the effective dielectric constant of the layer separating the ionic charges and the electrode; A is the electrode area, and d is the thickness of the layer. In the case of *Salmonella* growth in SC/T/M medium, TMAO was reduced to trimethylamine cations. This process increased the number of polar molecules and small molecules in the double layer, resulting in an increase in the dielectric permittivity and a decrease in the thickness of the double layer at the same time. These changes in combination led to an increase in the double-layer capacitance, which is the main contribution to the total impedance at frequencies lower than 50 Hz. Therefore, the observed impedance decreases at frequencies lower than 100 Hz came from the increase in the double-layer capacitance due to the bacterial growth.

Though changes in the ionic composition of the medium should make the medium more conductive and decrease R_{sol}, the IME detection system did not show a significant change in R_{sol}. This was because bacterial cells were attached to the IME surface during the growth, and the membrane resistance of these attached bacterial cells compensated for the decrease in medium resistance. But the double-layer capacitance was not influenced by these attached cells because the attached cells were actually separated by an aqueous gap of 10–20 nm from the electrode surface, thus preventing direct influence of the cell membrane capacitance on the electrode impedance (Yang et al. 2004b; Yang and Bashir 2008).

8.4.5 Impedance Detection of *S. typhimurium*

To quantify the bacterial concentration, a calibration curve between the impedance and the cell concentration needed to be established. To do so, the impedance growth

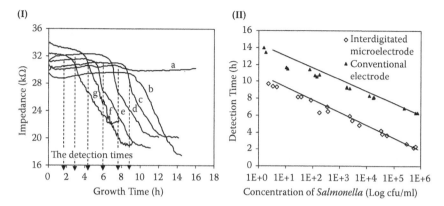

FIGURE 8.7 (I) A group of impedance growth curves recorded at 10 Hz using the IME, Initial *Salmonella* cell concentration: (a) control, (b) 4.8×10^0, (c) 2.84×10^1, (d) 1.75×10^2, (e) 5.4×10^3, (f) 5.3×10^4, (g) 5.4×10^5 cfu/mL. (From Yang, L., Y. Li, C.L. Griffi's, and M.G. Johnson, 2004b, *Biosensors and Bioelectronics* 19:1139–1147. With permission.) (II) The linear relationship between the logarithmic values of the initial cell concentration and the detection time obtained using the microchip system, in comparison with that obtained using a conventional three-electrode system. (From Yang, L., and R. Bashir, 2008, *Biotechnology Advances* 26:135–150. With permission.)

curves of a series of samples containing different concentrations of *S. typhimurium* were tested at a fixed frequency of 10 Hz . Figure 8.7a shows a representative group of impedance growth curves (impedance vs. growth time) for cell concentrations from 10^0 to 10^5 cfu/mL. On each growth curve, the detection time for the particular sample was determined. Figure 8.7b (blank dots) shows the plot of the detection time as a function of the logarithmic values of initial *Salmonella* cell concentration. The linear relationship between the detection time (t_d, h) and the logarithmic value of the initial cell concentration (*N*, cells/mL) was found to be $t_d = -1.38 \log N + 10.18$ with $R^2 = 0.99$.

Figure 8.7b (filled triangle) also shows the calibration curve obtained using a conventional three-electrode detection system. The comparison of the two linear curves indicated that the detection time for the same initial concentration of bacteria could be reduced by 3–4 h by using the IME system. The detection time of the IME system was close to that of an electrochemical cyclic voltammetry method (Ruan et al. 2002) and more rapid than the classic impedance microbiology (16 to 20 h) (Van Der Zee and Huis In't Veld 1997). Another advantage of the IME system is reduction of the sample volume: the test volume can be reduced from 10–15 mL to 1–2 mL with the IME system.

8.5   MICROCHIP IMPEDANCE DETECTION OF *SALMONELLA* BASED ON ION RELEASE

It was reported that most bacterial cell walls are negatively charged at neutral pH. These charges are compensated by counterions that penetrate into the porous cell wall and to a minor extent by coions that are expelled from it, thereby conferring electrostatic charge to the cell periphery (Wilson et al. 2001; Van Der Wal et al. 1997; Mozes et al. 1990; Carstensen et al. 1968). The charge density of a bacterial cell wall can be as high as 0.5–1.0 C/m^2 (Van Der Wal et al. 1997). When bacterial cells are suspended in a proper solution, these charged ions on bacterial cell surfaces can be released into the solution. This process will alter the impedance of the solution. Another possible source for bacterial cells to release ions or ionic substances into the solution is from the highly conductive interior of the cell. When bacterial cells are suspended in a solution that is not physiologically favorable, they may experience an osmotic shock. In response to fluctuations in environmental osmolarity, bacterial cells adjust their intracellular solute concentrations to maintain constant turgor pressure and to ensure continuation of cellular activity. Other properties of cells such as cell size and buoyant density might also be altered in response to the osmotic shock (Baldwin et al. 1988). Impedance changes of the solution may also be caused by these responses. The resulting impedance change in the solution caused by the two possible sources of ion or ionic substance release from cells can be used to quantify the concentration of bacterial cells in the suspensions. It has been reported that when bacterial cells experience an osmotic shock, processes such as leakage of ions through the cytoplasmic membrane, negative adsorption of electrolyte, or ion uptake into the cytoplasm, and specific adsorption of ions can occur. These processes can influence the conductivity of the low-conductivity solution (Van Der Wal et al. 1997).

The solution in which bacterial cells are suspended plays an important role in this method. Two characteristics of the solution are required to observe the changes in impedance when cells are suspended in the solution. First, the solution must be a low-conductivity solution; second, it should be able to induce ion release from the bacterial cells.

We used deionized (DI) water as such a solution to demonstrate the principle of this method. The conductivity of DI water is in a range from about 1–2 μS/cm to 10–15 μS/cm. When bacterial cells suspended in DI water reach a sufficient concentration, the charges on the cell walls, ion release, and the other combined responses to the osmotic shock can result in a substantial impedance change in DI water, which can be used to measure the concentration of bacterial cells in suspension. In this section, *Salmonella typhimurium*, a Gram-negative foodborne bacterial pathogen, was used as an example to demonstrate a microchip device for impedance detection of bacterial cells in suspensions based on this ion release mechanism.

8.5.1 The Microchip and Methods

The microchip device used in this experiment consists of a silica or glass chip patterned with an IME and a microchamber (~25 μl capacity) right above the electrode area formed by silicone rubber. The IMEs have a total of 50 pairs of finger electrodes each of 15 μm electrode width and space. For impedance measurements, 20 μL of bacterial cell sample was placed into the microchamber and covered with a glass cover. Two micromanipulators were used to connect the microchip to the IM-6 impedance analyzer. Figure 8.8a shows the side view of the microchip device and the schematic experimental setup for impedance detection.

Impedance spectral measurements were carried out in the frequency range from 1 Hz to 100 kHz. Impedance at a fixed frequency of 1 kHz was measured using the capacitance-potential (C/E) program with an amplitude of ±50 mV.

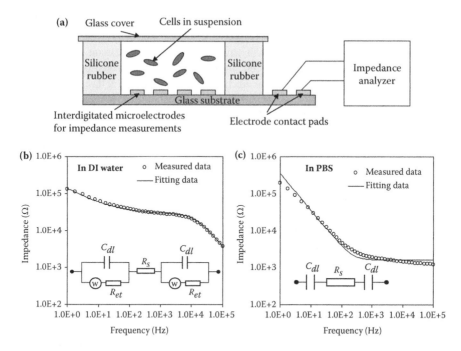

FIGURE 8.8 (a) The schematic experimental setup of the microchip system for impedance measurement of bacterial cells in suspension; (b) impedance spectra of *Salmonella* cell suspensions in DI water and (c) in PBS together with their fitting spectra and the equivalent circuits. Frequency range: 1 Hz–100 kHz. Amplitude: ± 50 mV. *Salmonella* concentration: 1.93 × 106 cfu/mL. (From Yang, L., 2009, *Talanta* 80:551–558. With permission.)

The impedance responses of *Salmonella* cell suspensions in DI water and in a phosphate buffered saline (PBS) solution were studied over a wide range of frequencies. The bacterial samples were prepared using the following procedure. *Salmonella typhimurium* cells were grown in brain–heart infusion (BHI) broth at 37°C for 16–18 h and were centrifuged at 6000 × g for 2 min. After removal of the supernatant, the cell pellet was resuspended in sterilized DI water or PBS. The cells were washed three times with DI water or PBS in order to get rid of residues from the growth medium. Then they were serially (1:10) diluted with DI water or PBS to the desired concentrations for further experimental use.

8.5.2 Impedance Spectra of *Salmonella* Cell Suspensions in Deionized (DI) Water and in a Phosphate Buffered Solution (PBS)

Figure 8.8b,c shows the Bode impedance spectra of *Salmonella* suspensions in DI water (b) and PBS (c) with a cell concentration of 1.93 × 10^6 cfu/mL (blank dots lines), along with their respective equivalent circuits. It can be seen that the two impedance spectra are different in shape. The impedance spectrum of the cell suspension in DI water presents a typical Bode plot of a system that combines kinetic and diffusion processes (Figure 8.8b). The impedance spectrum of cell suspension in PBS only shows two major regions: a double-layer region (from 1 Hz to ~500 Hz) and a resistive region (from 500 Hz to 100 kHz). The different impedance responses of cell suspensions in DI water and in PBS can be modeled by their equivalent circuits. The equivalent circuit of the microchip measuring the impedance of the cell suspension in DI water consists of a solution resistance between the two sets of finger electrodes (R_s) and the double-layer capacitance (C_{dl}), Warburg impedance (Z_w), and electron-transfer resistance (R_{et}) around each set of the finger electrodes. The equivalent circuit of the microchip measuring the impedance of cell suspension in PBS only includes the solution resistance between the two sets of finger electrodes (R_s) and the double-layer capacitance (C_{dl}) around each set of the finger electrodes. When the data from the measured impedance spectra were input to these equivalent circuits, fitting spectra could be obtained (solid lines in Figure 8.8b,c), and they matched well with their respective measured spectra. This indicates that the equivalent circuit could provide a feasible model to represent the behavior of the IME microchip in DI water suspension and in PBS suspension. The difference between the two circuit models is the presence of electron-transfer resistance (R_{et}) and Warburg impedance (Z_w) in the equivalent circuit for cell suspensions in DI water, but not in PBS suspensions. These two elements indicate the presence of electrochemical reactions and the diffusion of the electrochemical active species in the DI water cell suspension. This implies that cells suspended in DI water may release some electrochemical active composites to DI water. However, in the cell suspension in PBS, the impedance spectrum does not show any characteristics related to electrochemical active parameters, which implies cells do not release active electrochemical species into PBS.

8.5.3 IMPEDANCE RESPONSE TO ION RELEASE FROM BACTERIAL CELLS IN SUSPENSIONS

The impedance response measured at a fixed frequency (1 kHz) as a function of time provides clear evidence of ion release from cells that decreases the impedance. Figure 8.9 shows the plots of impedance at 1 kHz with time for the samples of *Salmonella* suspensions in DI water and PBS, using water and PBS as controls. The cell concentrations in the suspensions were at 1.9×10^9 cfu/mL. For pure DI water, a slight decrease trend in impedance with time was observed, probably due to the slight temperature change at the locality of the electrodes. For *Salmonella* suspension in DI water, impedance decreased significantly by 15% in the first 10 min, and decreased by 22.5% at 25 min. It became relatively stable after 1 h. The results indicate that DI water can induce the release of ions or other ionic intracellular and/or extracellular substances from bacterial cells into DI water, which decreases the impedance of the suspension. However, the impedances of both PBS solution and *Salmonella* cells in PBS suspension were constant with time, indicating that bacterial cells suspended in PBS did not cause impedance change in PBS.

Release of ion or other ionic intracellular and/or extracellular substances in certain conditions has been reported in various types of cells. For example, yeast cells released glycerol upon hypo-osmotic shock, and possibly arabitol and erythritol as well, depending on cell type (Kayingo et al. 2001). Other studies reported that osmotic shock could cause the release of extracellular enzymes including invertase, a

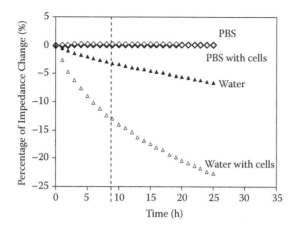

FIGURE 8.9   The percentage of impedance change with time for bacterial cells in DI water and PBS suspensions. *Salmonella* concentration: 2.1×10^9 cfu/mL. Impedance was measured at 1 kHz with an amplitude of 50 mV. (From Yang, L., 2009, *Talanta* 80:551–558. With permission.)

Co^{2+} activated 5'-nucleotidase, acid phosphatase, and alkaline pyrophosphatase from yeast cells (Schwencke et al. 1971), and decreases in the Na^+, and Cl^- levels in isolated axons of *Carcinus maenas* (Kevers et al. 1979). The responses to osmotic shock are rapid (Schwencke et al. 1971; Kayingo et al. 2001), which is consistent with the observation of rapid impedance decrease in the first 10 min.

8.5.4   IMPEDANCE DETECTION OF *SALMONELLA* CELLS IN SUSPENSIONS

Figure 8.10a,b presents the Bode impedance spectra of *Salmonella* suspensions in DI water and in PBS with different cell concentrations from 10^4 to 10^9 cfu/mL. For cell suspensions in DI water (Figure 8.10a), observed impedance values in the frequency range between 100 Hz to 10 kHz decreased with the increasing cell concentrations, whereas for cell suspensions in PBS (Figure 8.10b), no significant difference in impedance was observed over the entire frequency range for the samples with different cell concentrations. This again confirms that ion release from bacterial cells to DI water can cause impedance change, and the change in impedance is related to the cell concentration in the suspension. The results also suggest that the measurement of impedance value at a fixed frequency within the frequency range from 100 Hz to 10 kHz would allow us to estimate the cell concentration in DI water suspensions. We identified 1 kHz as the best representative test frequency to investigate the relationship between impedance value and cell concentration in DI water suspensions.

Figure 8.10c shows the plot of the impedance values measured at 1 kHz as a function of the bacterial concentrations. Impedance decreases with increasing cell concentration. There is a linear relationship between the impedance and the logarithmic value of the cell concentration in the range from 10^4 to 10^8 cells/20 µl (10^6 to 10^{10} cfu/mL). The linear regression equation is $Z\ (k\Omega) = -2.06 \log C$ (cells/20µl) $+ 5.23$ with $R^2 = 0.98$. The detection limit was calculated to be 6.9×10^4 cells/20µl (3.45×10^6 cfu/mL).

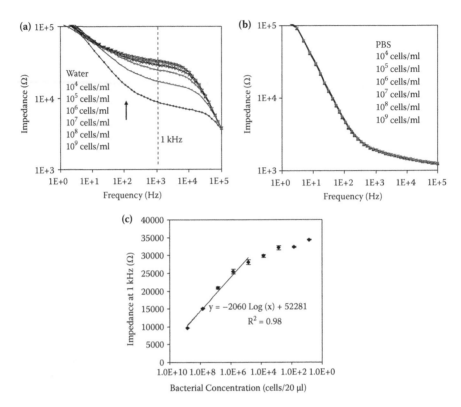

FIGURE 8.10 (a) Impedance spectra of *Salmonella* suspension in DI water and (b) in PBS with cell concentrations ranging from 10^4 to 10^9 cfu/mL; and (c) the linear relationship between the logarithmic value of the concentration of *Salmonella* cells and the impedance measured at 1 kHz. Error bars are standard deviations of 3–5 measurements. (From Yang, L., 2009, *Talanta* 80:551–558. With permission.)

These results indicated that the cell concentrations in DI water suspensions can be determined from the impedance of the bacterial suspensions. This can be a better alternative impedance approach to quantify bacterial cells in suspensions than impedance microbiology. This method does not require any label or amplification steps. The detection limit of this method is comparable with many label-free immunosensors for detection of pathogenic bacteria using different transducer techniques.

8.6 INTERDIGITATED MICROELECTRODE (IME)-BASED IMPEDANCE IMMUNOSENSORS FOR DETECTION OF *ESCHERICHIA COLI* O157:H7

E. coli O157:H7 is another most concerning foodborne pathogenic bacteria. Infection by *E. coli* O157:H7 may cause life-threatening complications—hemolytic uremic syndrome and hemorrhagic colitis in humans. The CDC estimates that each year

73,000 cases of infection and 61 deaths associated with *E. coli* O157:H7 occur in the United States (Mead et al. 1999). The detection of *E. coli* O157:H7 in food samples is important for preventing *E. coli* infections. In recent years, the integration of impedance technique with biosensor technology has led to rapidly expanding development of impedance biosensors for detection of bacteria (Ruan et al. 2002; Yang et al. 2004a; Radke and Alocilja 2005). Advances in microfabrication technologies have launched the use of microfabricated microarray electrodes, which have made the impedance biosensors enter into a microchip platform era. In this section, we describe an IME-based label-free electrochemical impedance immunosensor for the detection of *E. coli* O157:H7.

8.6.1 THE IME MICROCHIP AND ANTIBODY IMMOBILIZATION

The IME used in the experiments consists of 25 pairs of opposing ITO finger electrodes. The finger electrodes are of 15 μm width and 15 μm interdigit space with digit length of 2985 μm.

The first step in constructing the immunosensor is to immobilize specific antibodies for the target bacterial cells on the IME electrode surface. The immobilized antibodies are used to selectively capture the target cells to the electrode surface. In our experiment, the IMEs were first cleaned with acetone, alcohol, and 20% ethanolamine solution in water, and with DI water. Then, they were treated with a solution of 1:1:5 (v/v) H_2O_2:NH_4OH:H_2O for 30 min, rinsed thoroughly with DI water, and dried with nitrogen. This cleaning treatment step generated abundant hydroxyl groups on the ITO surface (~12–13 OH groups/nm^2). Antibodies can be linked to the ITO surface through the formation of M^+COO^- covalent linkages (where M = indium or tin) between their free carboxyl groups and the reactive hydroxyl groups on the ITO surface. The OH groups on the ITO surface also help to stabilize the immobilization of the antibodies through noncovalent hydrogen bonding with the antibody carboxyl or amine groups (Fang et al. 2000, 2001; Ng et al. 2002). In this immobilization method, anti-*E. coli* antibody solution (5 μl, 4–5 mg/mL) was directly spread on the surface of the cleaned IME surface. The electrode was then kept at 4°C overnight. After this, excess and unbound antibodies were removed from the electrode by slowly dipping it into a PBS buffer (pH 7.4) for 30 s; then it was extensively rinsed with PBS containing 1% BSA and deionized water and dried with nitrogen, and was ready for use.

8.6.2 PRINCIPLE OF THE IMPEDANCE IMMUNOSENSOR

The immunosensor probes the attachment of bacterial cells by measuring the change in electrical properties of the sensor due to the insulating properties of the cell membrane. The presence of intact cell membranes on the electrodes effectively reduces the electrode area that the current reaches and hence increases the interface impedance, and thus determines the resulting sensor signals. The measurement of the impedance can be performed in the presence or absence of a redox probe, such as $[Fe(CN)_6]^{-3/-4}$, which is referred to as a faradaic or nonfaradaic impedance biosensor. Without a redox probe, the measured impedance signal

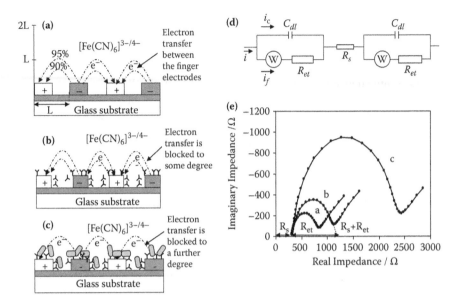

FIGURE 8.11 (a)–(c) The principle of the label-free impedance immunosensor constructed using an IME: (a) bare electrode, (b) with antibody immobilization, (c) with *E. coli* cell binding; (d) the equivalent circuit of the immunosensor impedance detection system; (e) the Nyquist plots of stepwise impedance responses of the immunosensor: (a) bare electrode, (b) after antibody immobilization, and (c) after *E. coli* cell binding. Electrolyte 10 mM $[Fe(CN)_6]^{-3/-4}$ (1:1) in 0.01M PBS, *E. coli* O157:H7, 2.6×10^7 cfu/mL. Frequency range: 1Hz to 100 kHz, Amplitude: 5mV. (From Yang, L., Y. Li, and G. Erf, 2004a, *Analytical Chemistry* 76:1107–1113. With permission.)

results directly from the intact bacteria cells that are adherently growing on or physically attached to the electrode surface, mainly due to the insulating effects of the cell membranes (Ehret et al. 1997, 1998). The impedance is influenced by the changes in the number, the growth, and the morphological behavior of adherent cells. In the presence of a redox probe, the sensor probes the biological events occurring on its surface by measuring the induced changes in faradaic impedance, such as electron-transfer resistance.

Figure 8.11 shows the principle of the electrochemical impedance imunosensor, which the electrochemical impedance was measured by in the presence of $[Fe(CN)_6]^{3-/4-}$ as a redox probe. As shown in the Figure 8.11a, when a bare IME was immersed into an electrolyte solution containing the redox couple, the faradaic process of oxidation and reduction of the redox couple occurred, and then electrons were transferred between the two sets of microelectrodes through the redox couple. When antibodies were immobilized onto the electrode surface (Figure 8.11b), they formed a layer that would inhibit the electron transfer between the electrodes to some degree. An increase in the electron transfer resistance would be expected. If bacterial cells were attached to the electrode surface (Figure 8.11c), the cell membrane could create a further barrier for electrons to transfer between the electrodes, thereby increasing the electron-transfer resistance.

8.6.3 The Equivalent Circuit and the Measurement of Electron-Transfer Resistance

Figure 8.11d shows the equivalent circuit of the impedance immunosensor detection system. It consists of the resistance (R_s) of the electrolyte between two electrodes and the double-layer capacitance (C_{dl}), electron-transfer resistance (R_{et}), and the Warburg impedance (Z_w) around each set of electrodes. The two elements R_s and Z_w represent the properties of the bulk solution and the diffusion of the redox probe; thus they are not affected by the reaction occurring at the electrode surface. The other two elements, C_{dl} and R_{et}, depend on the dielectric and insulating features at the electrode/electrolyte interface, and they are affected by the cell attachment at the electrode surface.

The electron transfer resistance, R_{et}, is the parameter measured in the immunosensor. Attachment of bacterial cells to the sensor surface increases the electron-transfer resistance. The total electron-transfer resistance after cell attachment can be expressed as

$$R_{et} = R_e + R_{cell} \tag{8.10}$$

where R_e and R_{cell} are the electron-transfer resistance of the antibody immobilized electrode and the variable electron-transfer resistance introduced by the attached bacterial cells.

The Nyquist plot $(Z_{im}$ vs. $Z_{re})$ is the best way to visualize and determine the electron transfer resistance, R_{et}. In a typical Nyquist plot, the semicircle portion observed at the higher frequencies corresponds to the electron-transfer-limited process. The intercept of the semicircle with the Z_{re} axis at high frequency is equal to R_s. Extrapolation of the semicircle to lower frequency yields another intercept with the Z_{re} axis equaling R_s+R_{et}. Therefore, the diameter of the semicircle is equal to the electron-transfer resistance, R_{et} (Figure 8.11e, spectrum b for example). Figure 8.11e shows the representative Nyquist plot of the bare IME (curve a), after antibody immobilization (curve b), and after *E. coli* O157:H7 (2.6×10^7 cells) cell binding (curve c). The electron transfer resistances (the diameters of the semicircles) increase significantly after each step. The R_{et} of the bare electrode after antibody immobilization and cell binding was 563 Ω, 850 Ω, and 2120 Ω, respectively. This result demonstrated that the electron-transfer resistance is a feasible parameter to measure the change on the electrode surface due to the immobilization of antibodies and the binding of *E. coli* O157:H7 cells.

8.6.4 Detection of *Escherichia coli* O157:H7 Cells

Figure 8.12a presents a set of the impedance spectroscopic responses of the IME-based immunosensor to different cell numbers of *E. coli* O157:H7 antibodies (without cells) (a), (b) 4.36×10^5 cfu/mL, (c) 4.36×10^6 cfu/mL, (d) 4.36×10^7 cfu/mL, and (e) 4.36×10^8 cfu/mL of *E. coli* cells on their surfaces. Significant increases in the diameters of the semicircles can be seen as cell concentration increases, indicating that electron-transfer resistance increases with increasing numbers of bacterial cells bound to the immunosensor surface. Figure 8.12b shows the linear relationship

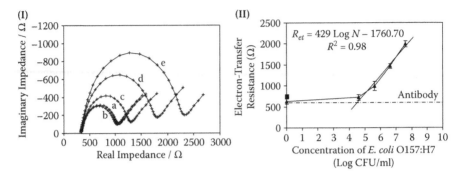

FIGURE 8.12 (I) Nyquist plot of impedance responses of the immunosensor to different concentrations of *E. coli* O157:H7: (a) antibodies, (b) 4.36×10^5, (c) 4.36×10^6, (d) 4.36×10^7, and (e) 4.36×10^8 cfu/mL. (II) The linear relationship between the logarithmic value of the concentration of *E. coli* O157:H7 and the electron transfer resistance. (From Yang, L., Y. Li, and G. Erf, 2004a, *Analytical Chemistry* 76:1107–1113. With permission.)

between the electron-transfer resistance derived from Figure 8.12a and logarithmic value of *E. coli* concentrations. This is the basis for enumeration of bacterial cell number using the immunosensor. Using the electron-transfer resistance of the antibody-immobilized IME microelectrode as the threshold, the detection limit of this immunosensor was 10^6 cfu/mL, which is comparable to other label-free immunosensors for detection of pathogenic bacteria using different transducer techniques. For example, Park et al. (1998, 2000) reported two quartz crystal microbalance immunosensors for detection of *Salmonella*, achieving detection limits of 3.2×10^6 cfu/mL and 9.9×10^5 cfu/mL; Koubova et al. (2001) developed surface plasmon resonance (SPR) immonusensors for detection of *Salmonella enteritidis* and *Listeria monocytogens*, which had a detection limit of 10^6 cfu/mL; Fratamico et al. (1998) reported an SPR sensor for detection of *E. coli* O157:H7 with a detection limit of 10^7 cfu/mL.

8.7 ENHANCED IMMUNOCAPTURE OF BACTERIAL CELLS ON INTERDIGITATED MICROELECTRODES BY DIELECTROPHORESIS

When using biosensors and biochips for bacterial cell detection, the immuno-capture of target cells by the immobilized antibodies on the sensor surface is the critical step. This step determines the number of target cells bound to the sensor surface, and thus the sensitivity of the sensor. Unfortunately, the immunocapture efficiency of the immobilized antibodies on the solid sensor/chip surfaces to bacterial cells is very low, partly because bacterial cells do not diffuse and it is difficult for them to access and interact with immobilized antibodies on the sensor surfaces. The reported immunocapture efficiencies of *Salmonella* and other foodborne bacteria cells ranged from 0.01% to less than 20% (Brewster et al. 1996; Ruan et al. 2002). Further improvements in the immunocapture efficiencies in biosensor and biochip settings require the integration of available technologies to address this challenge.

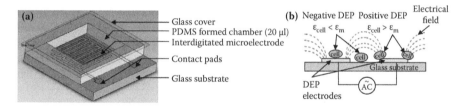

FIGURE 8.13   (a) The microchip for DEP-enhanced immunocapture of *Salmonella* cells, which consists of a chamber formed by PDMS and a set of interdigitated microelectrodes (IMEs) on a glass substrate; (b) schematic of the nonuniform electrical field generated by the IME, and the positive and negative DEP for bacterial cells in this electrical field. (From Yang, L., 2009, *Talanta* 80:551–558. With permission.)

Dielectrophoresis (DEP) is the electrokinetic motion of dielectrically polarized particles in nonuniform electric fields (Pohl 1979). Since most biological cells behave as dielectrically polarized particles in an external electric field, DEP has been recognized as a powerful tool to manipulate biological cells. Gomez et al. (2005) demonstrated DEP-concentrating *Listeria* cells in a lab-on-a-chip impedance microbiology, achieving a concentration factor of 10^4 to 10^5. This DEP concentration step eliminated the time-consuming growth-based enrichment steps used in traditional microbial methods. In our previous study (Yang et al. 2006), we used DEP in a flow through microchip to concentrate *Listeria monocytogenes* cells from the flow and used immobilized antibody to capture the cells in the microfluidic channel. This microfluidic chip achieved an improved immunocapture efficiency of ~27% to *Listeria* cells. The latest study reported that 5-min DEP treatment at the beginning of the 1-h incubation step improved the capture efficiency of *L. monocytogenes* by heat shock protein 60 (Hsp60) immobilized on silicon dioxide by 60% (Koo et al. 2009). These studies have demonstrated that DEP can concentrate bacterial cells from the solution and enhance the capture efficiency to bacterial cells in microfluidic chips.

Here, we present an IME biochip integrated with DEP to enhance the immunocapture of *Salmonella typhimurium* (Yang 2009). Two significant functions of DEP in the biochip platform to improve the detection of *Salmonella* are demonstrated: (1) DEP can concentrate bacterial cells from the suspension to different locations on the chip surface, making it useful for manipulation of bacterial cells in biosensors and biochips; (2) DEP can make bacterial cells in close contact with the immobilized antibodies on the chip surface, which can effectively improve the immunocapture efficiency.

8.7.1   THE MICROCHIP DEVICE

Figure 8.13a shows the assembled biochip, which include two parts: a gold IME on a glass substrate and a micro-chamber right above the electrode area formed by polydimethylsiloxane (PDMS). The IME contained 50 pairs of opposing gold finger electrodes with 15 μm of electrode width and space, and 5 mm of electrode length. The PDMS layer (approximately 2 mm thick) was made by mixing the elastomer monomer and the curing agent at the ratio of 15:1 (w/w) and was cured at 70°C

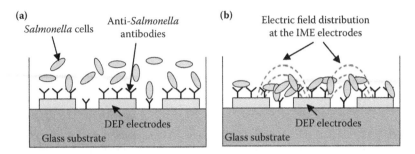

FIGURE 8.14 The mechanisms of immunocapture of *Salmonella* cells on the microchip (a) without DEP and (b) with DEP enhancement. (From Yang, L., 2009, *Talanta* 80:551–558. With permission.)

overnight in a petri dish. The PDMS layer was then cut into 1 cm × 1 cm pieces. A rectangle-shaped chamber (~3 mm × 5 mm, equivalent to the IME electrode area) was made in the center of each PDMS piece. The PDMS piece was then aligned with the IME chip to form a chamber right above the electrode area.

An anti-*Salmonella* antibody was immobilized onto the chip surface using a sandwich structure consisting of biotinylated BSA—streptavidin—biotinylated anti-*Salmonella* antibody.

8.7.2 PRINCIPLE OF DIELECTROPHORESIS-ENHANCED IMMUNOCAPTURE OF BACTERIAL CELLS ON THE CHIP

DEP uses nonuniform electrical fields to manipulate biological cells. When a particle is placed in a nonuniform electrical field, the time-averaged DEP force acting on it can be approximated as Equation (8.11) (Gascoyne and Vykoukal 2002),

$$F_{DEP} = 2\pi r^3 \varepsilon_m Re[f_{CM}(\omega)]\nabla E_{rms}^2 \qquad (8.11)$$

where r is the particle radius; ε_m, the permittivity of the suspending medium; E_{rms}, the root mean square value of the electric field; and $Re[f_{CM}(\omega)]$, the real part of the Clausius-Mossotti factor (f_{CM}). Depending on the relative polarizability of the particle with respect to its surrounding medium, the particle can either experience positive DEP $(Re[f_{CM}] > 0)$, which will move the particle toward the region where the electrical field is the strongest, or negative DEP $(Re[f_{CM}] < 0)$, which will move the particle toward the region where the electrical field is the weakest.

Figure 8.13b shows the nonuniform electrical field generated by the IME, which has its strongest electrical field zones at the edges of the finger electrodes and its weakest electrical field zones at the centers of the finger electrodes and the gaps between the finger electrodes. *Salmonella* cells can either experience negative DEP, which moves them toward the centers of the finger electrodes or the gaps between the finger electrodes, or experience positive DEP, which moves them to the edges of the finger electrodes.

Figure 8.14 shows the mechanism of DEP-enhanced immunocapture of *Salmonella* cells on the chip in comparison with non-DEP assisted immunocapture. In a typical diffuse solution of *Salmonella* cells placed in the chamber without DEP

(Figure 8.14a), many of the bacteria never reach the surface of the chip and thus do not interact with the immobilized antibodies. The resulting immobilized antibody capture efficiency is low, especially when using a small sensor in a large sample volume. When a DEP voltage is applied, however, *Salmonella* cells are pulled down to the chip surface. Many more cells make contact with the immobilized antibodies on the chip surface (Figure 8.14b). Therefore, the immunocapture efficiency of the immobilized antibody can be increased.

8.7.3 Dielectrophoresis-Enhanced Immunocapture of *Salmonella* Cells on the Interdigitated Microelectrode

Figure 8.15 presents the representative fluorescence images of immunocaptured *Salmonella* cells on the chip with positive DEP at 100 kHz and 3 Vpp for 5 min and without DEP. Images on the left side show the immunocaptured cells on the chips with DEP enahcement, while images on the right side show the immunocaptured

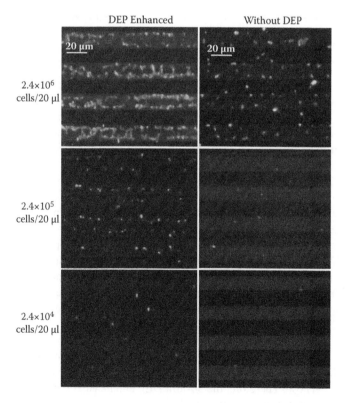

FIGURE 8.15 The representative fluorescence images of *Salmonella* cells captured by immobilized antibodies on the chip with DEP treatment at 100 kHz and 3 Vpp for 5 min in comparison with those without DEP. Images on the left side are immunocaptured cells with DEP assistance, and the images on the right side are immunocaptured cells without DEP assistance. The three samples contained different cell concentrations at 2.4×10^6, 2.4×10^5, and 2.4×10^4 cells/20 µl. (From Yang, L., 2009, *Talanta* 80:551–558. With permission.)

TABLE 8.2

The Immunocapture Efficiency of the Immobilized Antibody on the Nonflow-Through Chip to *Salmonella* Cells with 15-Minute and 30-Minute DEP Assistance in Comparison with Those without DEP Assistance

	Cell Concentration in Suspension			
Reaction Time (min)	Original Sample (cfu/mL)	After Immunoreaction with DEP (cfu/mL)	After Immunoreaction Capture without DEP (cfu/mL)	Immunocapture Efficiency (%)
15 min	$(1.93\pm0.21)\times10^6$	$(8.5\pm0.48)\times10^5$	—	~56.0%
	$(1.93\pm0.21)\times10^6$	—	$(1.73\pm0.15)\times10^6$	~10.4%
30 min	$(4.03\pm0.43)\times10^6$	$(1.45\pm0.13)\times10^6$	—	~64.0%
	$(1.42\pm0.46)\times10^6$	—	$(1.17\pm0.33)\times10^6$	~17.6%

Note: DEP: 100 kHz and 3 Vpp.

cells without DEP. It can be seen that almost all the cells being captured with positive DEP are at the edges of the finger electrodes, while without DEP treatment the cells are distributed randomly on the chips. At each concentration, more cells were captured by the immobilized antibodies with DEP enhancement than without DEP.

Table 8.2 shows the immunocapture efficiency of the immobilized antibodies to *Salmonella* cells with DEP enhancement in comparison with those without DEP. As an example, when 20 μl of a sample containing ~1.93×10^6 cfu/mL was placed into the chamber, after a 15-min DEP treatment at 100 kHz and 3 Vpp, the cell concentration in the suspension was reduced to ~8.50×10^5 cfu/mL. The capture efficiency was ~56.0%. However, a 15-min immunoreaction with no DEP treatment only reduced the cell concentration to ~1.73×10^6 cfu/mL, the calculated capture efficiency was ~10.4%. The results demonstrated that the 15 min DEP enhancement increased the immunocapture efficiency approximately fivefold. A similar experiment using a 30-min DEP treatment (100 kHz and 3 Vpp) decreased the cell concentration from 4.03×10^6 cfu/mL to 1.45×10^6 cfu/mL, achieving an immunocapture efficiency of ~64.0%; while 30-min immunoreaction with no DEP treatment reduced the cell concentration from 1.42×10^6 cfu/mL to 1.17×10^6 cfu/mL, with an immunocapture efficiency of ~17.6%. Therefore, the 30-min DEP enhancement increased the immunocapture efficiency more than threefold.

Typically, the immunocapture efficiency of immobilized antibodies immobilized to bacterial cells is very low. For example, Brewster et al. (1996) reported that the binding efficiency of anti-*Salmonella* antibodies immobilized on a roughened glassy carbon electrode surface was less than 0.01% after immersing the electrode into a 1 mL solution containing 10^7 cells, but it increased to ~ 0.5% by placing a droplet (containing ~10^5 cells) directly on the electrode surface. Ruan et al. (2002) achieved a capture efficiency of ~16% of anti-*E. coli* antibody immobilized on an ITO electrode surface to *E. coli* cells, when 50 μl of 6.0×10^6 cfu/mL sample was dropped on the sensor surface twice and incubated for 1 h at 37°C. We achieved capture efficiencies of 18% to 27% for *Listeria* cells with concentrations ranging from 10^1 to 10^3 cells in a flow-through microchip with DEP at 1 MHz, 20 Vpp (Yang et al. 2006). The results of this study demonstrated that DEP can effectively improve the immunocapture efficiency to 56–64% with a considerably short reaction time (15–30 min) in the designed IME microchip device.

8.8 CONCLUSION

The impedance technique as a principle of transduction has become a fertile area for developing rapid and effective methods for the detection of bacteria. Advances in microfabrication technology have paved the way for miniaturization of impedance-based methods using microelectrode and microchip platforms. These microchips have proved to be successful in maximizing the impedance signal, minimizing the volume of testing sample, increasing sensitivity, and saving assay time for detection of foodborne pathogenic bacteria.

To date, impedance biosensors for bacterial detection are comparable to current well-studied rapid methods such as ELISA and PCR, which achieve detection limits varying from 10^1 to 10^6 cfu/mL, and other various biosensors, which achieve detection limits of 10^2 to 10^7 cfu/mL, with the assay times around 2 h under ideal conditions.

The IME microchip-based method allows the integration of other electrical methods, such as DEP, to enhance immunocapture of bacterial cells, which has the potential to advance the state-of-the-art in chip-based immunoassay methods for microbial detections. Impedance biosensors/biochips also offer the promise of development of biosensor arrays for multiplex analysis of different bacteria simultaneously, providing label-free, online, and high-throughout devices for bacterial detection. By integrating micro- and nanofluidics with biosensors, many of the unit operations associated with sample preparation, such as separation, mixing, incubation, concentration, and so forth, may be performed directly in a lab-on-a-chip format.

ACKNOWLEDGMENTS

Professor Yang acknowledges the funding from Golden LEAF Foundation and North Carolina BIOIMPACT Initiative State through the Biomanufacturing Research Institute and Technology Enterprise (BRITE) Center for Excellence at North Carolina Central University.

REFERENCES

Association of Official Analytical Chemists (AOAC). 1996. *Salmonella* in food, automated conductance methods: AOAC official method 991.38. In *Official methods of analysis of AOAC International*, 16th ed. Gaithersburg, MD: AOAC International.

Baldwin, W.W., M.J.T. Sheu, P.W. Bankston, and C.L. Woldringh. 1988. Changes in buoy-ant density and cell size of *Escherichia coli* in response to osmotic shocks. *Journal of Bacteriology* 170:452–455.

Bard, A.J. and L.R. Faulkner. 2001. *Electrochemical methods: Fundamentals and applications.* New York: John Wiley & Sons.

Beran, G.W., H.P. Shoeman, and K.F. Anderson. 1991. Food safety: An overview of problems. *Dairy, Food and Environmental Sanitation* 11:189–194.

Brewster, J.D., A.G. Gehring, R.S. Mazenko, L.J. Van Houten, and C.J. Crawford. 1996. Immunoelectrochemical assays for bacteria: Use of epifluorescence microscopy and rap-id-scan electrochemical techniques in development of assay for *Salmonella*. *Analytical Chemistry* 68:4153–159.

Cady, P. 1975. Rapid automated bacterial identification by impedance measurement. In *New approaches to the identification of microorganisms*, ed. C.G. Heden, 73–99. New York: John Wiley.

Cady, P. 1978. Progress in impedance measurements in microbiology. In *Mechanizing microbiology*, ed. A.N. Sharp and D.S. Clark, 199–239. Springfield: Charles C Thomas.

Cady, P., D. Hardy, S. Martins, S.W. Dufour, and S.J. Kraeger. 1978. Electrical impedance measurements: Rapid method for detecting and monitoring microorganisms. *Journal of Clinical Microbiology* 7:265–272.

Carstensen, E.L., and R.E. Marquis. 1968. Passive electrical properties of microorganisms: III. Conductivity of isolated bacterial cell walls. *Biophysical Journal* 8:536–548.

Dupont, J., D. Menard, C. Herve, F. Chevalier, B. Beliaeff, and B. Minier. 1996. Rapid estimation of *Escherichia coli* in live marine bivalve shellfish using automated conductance measurement. *Journal of Applied Microbiology* 80:81–90.

Eden, R., and G. Eden. 1984. *Impedance microbiology.* Herts, UK: Research Studies Press Ltd.

Ehret, R., W. Baumann, M. Brischwein, A. Schwinde, and B. Wolf. 1998. On-line control of cellular adhesion with impedance measurements using interdigitated electrode structures. *Medical and Biological Engineering and Computing* 36:365–370.

Ehret, R., W. Baumann, M. Brischwein, A. Schwinde, K. Stegbauer, and B. Wolf. 1997. Monitoring of cellular behavior by impedance measurements on interdigitated electrode structures. *Biosensors and Bioelectronics* 12:29–41.

Fang, A., H.T. Ng, and S.F.Y. Li. 2001. Anchoring of self-assembled hemoglobin molecules on bare indium–tin oxide surfaces. *Langmuir* 17:4360–4366.

Fang, A, H.T. Ng, X. Su, and S.F.Y. Li. 2000. Soft-lithography-mediated submicrometer patterning of self-assembled monolayer of hemoglobin on ITO surfaces. *Langmuir* 16:5221–5226.

Federal Register. 1990. Assessing defined-substrate technology for meeting monitoring requirements of the total coliform rule. *Journal American Water Works Association* 82:83–87.

Federal Register. 1991. Drinking water: National primary drinking water regulations; total coliform proposed rule. *Federal Register* 54:27544–27567.

Foodborne Disease, NIAID (National Institute of Allergy and Infectious Diseases), http://www.niaid.nih.gov/topics/salmonellosis/pages/cause.aspx (accessed September 9, 2010).

Fratamico, P.M., T.P. Strobaugh, M.B. Medina, and A.G. Gehring. 1998. Detection of *E. coli* O157:H7 using a surface-plasmon resonance biosensor. *Biotechnology Techniques* 12:571–576.

Gascoyne, P.R.C., and J. Vykoukal. 2002. Particle separation by dielectrophoresis. *Electrophoresis* 23:1973–1983.

Gibson, D.M., P. Coombs, and D.W. Pimbley. 1992. Automated conductance method for the detection of *Salmonella* in foods: Collaborative study. *Journal of AOAC International* 75:293–2302.

Gomez, R., D.T. Morisette, and R. Bashir. 2005. Impedance microbiology-on-a-chip: Microfluidic bioprocessor for rapid detection of bacterial metabolism. *Journal of Microelectromechanical Sytems* 14:829–838.

Greenberg, A.E., R.R. Trussel, L.S. Clesceri, and M.A.H. Franson. 1992. *Standard methods for the examination of water and waste.* Washington, DC: American Public Health Association.

Hall, R.H. 2002. Biosensor technologies for detecting microbiological foodborne hazards. *Microbes and Infection* 4:425–432.

Ivnitski, D., I. Abdel-Hamid, P. Atanasov, and E. Wilkins. 1999. Biosensors for detection of pathogenic bacteria. *Biosensors and Bioelectronics* 14:599–624.

Ivnitski, D., I. Abdel-Hamid, P. Atanasov, E. Wilkins, and S. Stricker. 2000. Application of electrochemical biosensors for detection of food pathogenic bacteria. *Electroanalysis* 12:317–325.

Kayingo, G., S.G. Kilian, and B.A. Prior. 2001. Conservation and release of osmolytes by yeast during hypo-osmotic stress. *Archives of Microbiology* 177:29–35.

Kevers, C., A. Péqueux, and R. Gilles. 1979. Effects of a hypo-osmotic shock on Na^+, K^+ and Cl^- levels in isolated axons of *Carcins maenas. Journal of Comparative Physiology B: Biochemical, Systemic, and Environmental Physiology* 129:365–371.

Koo, O.K., Y. Liu, S. Shuaib et al. 2009. Targeted capture of pathogenic bacteria using a mammalian cell receptor coupled with dielectrophoresis on a biochip. *Analytical Chemistry* 81:3094–3101.

Koubova, V., E. Brynda, L. Karasova et al. 2001. Detection of foodborne pathogens using surface plasmon resonance biosensors. *Sensors and Actuators B: Chemical* 74:100–105.

Laureyn, W., D. Nelis, P. Van Gerwen et al. 2000. Nanoscaled interdigitated titanium electrodes for impedimetric biosensing. *Sensors and Actuators B: Chemical* 68:360–370.

Laureyn, W., F. Frederix, P. Van Gerwen, and G. Maes. 1999. Nanoscaled interdigitated gold electrodes for impedimetric immunosensing. *Transducers '99, Digest of Technical Papers*, Sendai, Japan, 1884–1885.

Mead, P.S., L. Slutsker, V. Dietz, et al. 1999. Food-related illness and death in the United States. *Emerging Infectious Diseases* 5:607–625.

Mozes, N., and P.G. Rouxhet. 1990. Microbial hydrophobicity and fermentation technology. In *Microbial cell surface hydrophobicity*, ed. R.J. Doyle, and M. Rosenberg, 75–105. Washington, DC: American Society for Microbiology.

Ng, H.T., A. Fang, L. Huang, and S.F.Y. Li. 2002. Protein microarrays on ITO surfaces by a direct covalent attachment scheme. *Langmuir* 18:6324–6329.

Owicki, J., and J. Parce. 1992. Biosensors based on the energy metabolism of living cells: The physical chemistry and cell biology of extracellular acidification. *Biosensors and Bioelectronics* 7:257–272.

Park, I., and N. Kim. 1998. Thiolated *Salmonella* antibody immobilization onto the gold surface of piezoelectric quartz crystal. *Biosensors and Bioelectronics* 13:1091–1097.

Park, I., W. Kim, and N. Kim. 2000. Operational characteristics of an antibody-immobilized QCM system detecting *Salmonella* spp. *Biosensors and Bioelectronics* 15:167–172.

Pethig, R., and G.H. Markx. 1997. Application of dielectrophresis in biotechnology. *Trends in Biotechnology* 15:426–432.

Pohl, H.A. 1979. *Dielectrophoresis.* Cambridge: Cambridge University Press.

Potter, M.E., S. Gonzalez-Ayala, and N. Silarug. 2007. Epidemiology of foodborne diseases. In: *Food microbiology: Fundamentals and frontiers*, ed. M.P. Doyle, L.R. Beuchat, and T.J. Montville, 376–390. Washington, DC: ASM Press.

Radke, S.M., and E.C. Alocilja. 2005. A high density microelectrode array biosensor for detection of *E. coli* O157:H7. *Biosensors and Bioelectronics* 20:1662–1667.

Rand, A.G., J. Ye, C.W. Brown, and S.V. Letcher. 2002. Optical biosensors for food pathogen detection. *Food Technology* 56:32–39.

Richards, J.C.S., A.C. Jason, G. Hobbs, D.M. Gibson, and R.H. Christie. 1978. Electronic measurement of bacteria growth. *Journal of Physics E: Scientific Instruments* 11:560–568.

Ruan C., L. Yang, and Y. Li. 2002. Immunobiosensor chips for detection of *Escherichia coli* O157:H7 using electrochemical impedance spectroscopy. *Analytical Chemistry* 74:4814–20.

Schwencke, J., G. Farís, and M. Rojas. 1971. The release of extracellular enzymes from yeast by "osmotic shock." *European Journal of Biochemistry* 21:137–143.

Silley, P. 1994. Rapid microbiology—Is there a future? *Biosensors and Bioelectronics* 9: xv–xxi.

Silley, P., and S. Forsythe. 1996. Impedance microbiology—A rapid change for microbiologists. *Journal of Applied Bacteriology* 80:233–243.

Stewart, G.N. 1899. The changes produced by the growth of bacteria in the molecular concentration and electrical conductivity of the culture media. *Journal of Experimental Medicine* 4:235–243.

Stulik, K., C. Amatore, K. Holub, V. Marecek, and W. Kutner. 2000. Microelectrodes. Definitions, characterization, and applications. *Pure and Applied Chemistry* 72:1483–1492.

Su, X.L., and Y. Li. 2004. A self-assembled monolayer-based piezoelectric immunosensor for rapid detection of *Escherichia coli* O157:H7. *Biosensors and Bioelectronics* 19:563–574.

Swaminathan, B., and P. Feng. 1994. Rapid detection of foodborne pathogenic bacteria. *Annual Reviews in Microbiology* 48:401–426.

Ur, A., and D.F.J. Brown. 1974. Rapid detection of bacterial activity using impedance measurements. *Biomedical Engineering* 8:18–20.

Ur, A., and D.F.J. Brown. 1975. Impedance monitoring of bacterial activity. *Journal of Medical Microbiology* 8:19–28.

USDA/ERS (Economic Research Service). Bacterial foodborne disease: Medical costs and productivity loss, http://www.ers.usda.gov/publications/aer741/AER741fm.PDF (accessed August 31, 2010).

Van Der Wal, A., M. Minor, W. Norde, A. Zehnder, and J. Lyklema. 1997. Conductivity and dielectrical dispersion of Gram-positive bacterial cells. *Journal of Colloid and Interface Science* 186:71–79.

Van Der Zee, H., and J.H.J. Huis In't Veld. 1997. Rapid and alternative screening methods for microbiological analysis. *Journal of AOAC International* 4:934–940.

Van Gerwen, P., W. Laureyn, W. Laureys et al. 1998. Nanoscaled interdigitated electrode arrays for biochemical sensors. *Sensors and Actuators B: Chemical* 49:73–80.

Vasavada, P.C. 1997. Advances in pathogen detection. *Food Testing and Analysis* 47:18–23,47–48.

Wawerla, M., A. Stolle, B. Schalch, and H. Eisgruber. 1999. Impedance microbiology: Applications in food hygiene. *Journal of Food Protection* 62:1488–1496.

Wilson, W.W., M.M. Wade, S.C. Holman, and F.R. Champlin. 2001. Status of methods for assessing bacterial cell surface charge properties based on zeta potential measurements. *Journal of Microbiological Methods* 43:153–164.

Yang, L. 2008. Electrical impedance spectroscopy for detection of bacterial cells in suspensions using interdigitated microelectrodes. *Talanta* 74:1621–1629.

Yang, L. 2009. Dielectrophoresis assisted immuno-capture and detection of foodborne pathogenic bacteria in biochips. *Talanta* 80:551–558.

Yang, L., and R. Bashir. 2008. Electrical/electrochemical impedance for rapid detection of foodborne pathogenic bacteria. *Biotechnology Advances* 26:135–50.

Yang, L., and A. Guiseppi. 2008. Impedimetric biosensors for nano and microfluidics. In *Encyclopedia of microfluidics and nanofluidics*, ed. D. Li, vol 2, 811–823. Verlag GmbH Berlin Heidelberg: Springer.

Yang, L., and Y. Li. 2006. Detection of viable *Salmonella* using microelectrode-based capacitance measurement coupled with immunomagnetic separation. *Journal of Microbiological Methods* 64:9–16.

Yang, L., Y. Li, and G. Erf. 2004a. An interdigitated array microelectrode-based electrochemical impedance immunosensor for rapid detection of *Escherichia coli* O157:H7. *Analytical Chemistry* 76:1107–1113.

Yang, L., C. Ruan, and Y. Li. 2003. Detection of viable *Salmonella typhimurium* by impedance measurement of electrode capacitance and medium resistance. *Biosensors and Bioelectronics* 19:495–502.

Yang, L., Y. Li, C.L. Griffis, and M.G. Johnson. 2004b. Interdigitated microelectrode (IME) impedance sensor for the detection of viable *Salmonella* typhimurium. *Biosensors and Bioelectronics* 19:1139–1147.

Yang, L., P. Banada, R. Chatni, A. Bhunia, and R. Bashir. 2006. A multifunctional microfluidic system for dielectrophoretic concentration coupling with immuno-capture of biological cells. *Lab-on-a-Chip* 6:896–905.

9 Application of Biosensors for the Quality Assurance of Dairy Products

P. Narender Raju and K. Hanumantha Rao

CONTENTS

9.1 INTRODUCTION

With the growing awareness of food quality and safety among consumers, food quality control and assurance have become essentials in the food processing industry. Consumers expect an adequate quality of food product at a reasonable price with longer shelf-life and high product safety, while food inspectors look for and emphasize safe manufacturing practices, adequate product labeling, and compliance with the statutory regulations. *Quality* is defined as the degree to which a set of inherent characteristics fulfills a need or expectation that is stated, generally implied, or obligatory (Hoyle 2001). As per the ISO 9000:2000 standards, quality control is a part of the quality management system focused on providing confidence that quality requirements will be fulfilled. In other words, *quality control* means the control of "goodness" or excellence. Food quality comprises mainly three aspects: (1) food safety, (2) shelf life, and (3) consistency. Critical factors that affect the product quality and safety are plant conditions, manufacturing practices, housekeeping, sanitary standards, personal hygiene, work habits of employees, and visitors. Food producers are increasingly demanding efficient control methods, particularly through online or at-line quality sensors to satisfy and ensure the quality and safety to consumers and regulatory authorities also. This would help to improve the feasibility of automated food processing and quality of sorting, and also to reduce the production time (increase throughput) and final product cost (Takhistov 2006).

In the dairy industry, quality control has the mandatory objective of ensuring the quality and safety of milk and milk products offered to the consumer. For a dairy plant to be successful in manufacturing safe and wholesome milk and dairy products, it is imperative to maximize customer satisfaction, minimize product losses, and comply with food laws. The quality of the dairy product equates survival and growth of the dairy business. Milk contains fat, protein, lactose, and minerals as principal constituents. In addition to these, it contains several minor compounds, many of which, for example, vitamins, metal ions, and flavor compounds, have a major impact on the nutritional, technological, and sensory properties of milk and dairy products. Apart from these constituents, the preparation of milk products requires use of certain food additives like emulsifiers, stabilizers, and food colors. With the rapid developments in food and nutrition science, a large number of dairy-based foods with functional ingredients such as probiotics, prebiotics, phytochemicals, and others are developed. More recently, new-generation functional dairy foods have been introduced on the market, in which ingredients or components are added to a product to combat diseases beyond contributing to basic human nutrition, or from which possibly adverse components have been removed, thus introducing the need of monitoring an increasing number of various parameters. This has resulted in complexity of analytical methods, and regulatory mechanisms have become more and more complex.

In the present World Trade Organization (WTO) regime, as our food supply stems from diverse sources across the world, there is a need for safe and secure management system(s) along the entire food supply chain. This is especially true and becomes more critical for perishable food products such as milk and most dairy products. The quality of milk and milk products is evaluated through periodic chemical and microbiological analysis that involves usage of conventional techniques such as titration, chromatography, spectrophotometry, electrophoresis, and others. These methods are slow, expensive, need well-trained operators (in some cases), and not easy to monitor continuously; and some require steps of extraction or sample pretreatment, which increases the time of analysis (Mello and Kubota 2002). Further, these methods are less practical for quick food monitoring or diagnosis. Hence, there is a need for robust food analytical instruments such as biosensors. Biosensors are a subgroup of chemical sensors, which are analytical devices composed of a biological recognition element (such as an enzyme, antibody, receptor, or microorganism) coupled with a chemical or physical transducer (electrochemical, mass, optical, and thermal). These devices represent a promising tool for food analysis due to the possibility to fulfill some demand that the classic methods of analysis do not attain (Mello and Kubota 2002). The measurements performed with a biosensor are sensitive and selective due to the extraordinary nature of the biological interaction (Carlo et al. 2006). In this chapter, the application of biosensors in quality control and/or quality assurance of milk and milk products is briefly reviewed.

9.2 GENERAL PRINCIPLE FOR BIOSENSORS USED IN THE DAIRY INDUSTRY

Biosensors are small analytical devices that combine a transducer with a biologically active substance (sensing component). The aim of a biosensor is to produce a signal that is related to a single analyte or a group of analytes. The biologically active elements or components include enzymes, multienzyme systems, antibodies or antigens, receptors, whole bacteria, yeast or mold cells, and so forth. The choice of the biosensing element depends on the target analyte or substrate of interest. A schematic representation of the principle of a biosensor is depicted in Figure 9.1. The transducer, which is in intimate contact with the biologically sensitive material, transforms the variations occurring in the biosensing element as the result of a positive detection event into an electrical signal, which is then amplified and used. The biodetection principle can be schematically described as follows. A chemical, physical, or biological sensor produces a signal, namely, voltage, absorbance rate, heat, or current, in response to a detectable event such as binding between molecules. In

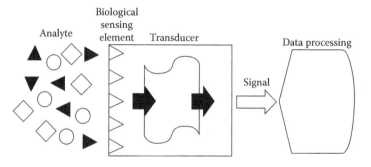

FIGURE 9.1   Schematic presentation of the biosensor principle.

case of a biological or chemical sensor, this event typically involves a receptor (e.g., enzyme or antibody) binding to a specific target molecule in a sample (Takhistov 2006). Physical sensors, on the contrary, measure inherent physical parameters of a sample, such as current or temperature, which can change due to reactions occurring in it. In any case, the signal is then transduced by passing it to an electronic circuit, where it is digitized, stored, displayed on a monitor, or made accessible for a later use. Substances such as sugars, amino acids, alcohols, lipids, nucleotides, and so forth can specifically be identified and their concentration measured by these sensors.

The transducers can be electrochemical (amperometric, potentiometric, conductometric/impedimetric), optical, piezoelectric, or calorimetric. Most biosensors use electrochemical or optical transducers. The characteristics and applications of different transducers are given by Mello and Kubota (2002). Among the various transducers, amperometric, optical, and potentiometric transducers are commonly used in biosensors (Table 9.1). Amperometric biosensors measure the produced current during the oxidation or reduction of a product or reactant, usually at a constant applied potential. They monitor either the consumed oxygen or the produced hydrogen peroxide. The current generated by the reaction will be proportional to the concentration of the measured substrate (Adanyi 2010). Potentiometric biosensors are based on monitoring the potential of an electrochemical cell at a working electrode with respect to an accurate reference electrode under conditions of essentially zero current flow. The electrochemical potential is usually proportional to the logarithm of the substrate concentration in the sample (Adanyi 2010). Three basic types of potentiometric electrodes are used in measuring electrodes: ion-selective electrodes (ISEs), gas-sensing electrodes (GSEs), and field-effect transistors (FETs). Conductometric biosensors measure the change in the conductance of the biological component between a pair of metal electrodes. Optical biosensors are based on the measurement of absorbed or emitted light as a consequence of a biochemical reaction. Optical sensors can employ different techniques to detect the presence of a substrate using the well-understood phenomenon of which surface plasmon resonance (SPR) is widely used. Piezoelectric quartz crystal microbalance (QCM) can be affected by a change of mass on the surface of the crystal, when the crystal is placed in an alternating field; any change in mass is detected by the change in oscillating frequency of the crystal (Adanyi 2010).

9.3 APPLICATIONS OF BIOSENSORS IN THE DAIRY INDUSTRY

9.3.1 BIOSENSORS FOR MILK COMPONENT ANALYSIS

9.3.1.1 Lactose and Other Milk Carbohydrates

Lactose is the principal carbohydrate in the milks of most mammals. Milk contains only trace amounts of other sugars, including glucose (50 mg L^{-1}), fructose, glucosamine, galactosamine, neuraminic acid, and neutral and acidic oligosaccharides (Fox and McSweeney 1998). The lactose content of cow and buffalo milks is about 4.8% and 5.1%, respectively (Pandya and Khan 2006). As lactose is the main carbohydrate in commercial milk, its determination is a basic indicator for quality control and detection of abnormal milk, such as the lower concentrations of lactose in mastitic

TABLE 9.1
Characteristics of Different Types of Transducers and Their Applications

Transducer	Advantages	Disadvantages	Applications
Amperometric	Simple, extensive variety of redox reaction for construction of the biosensors, facility to miniaturize	Low sensitivity, multiple membranes or enzymes can be necessary for selectivity and adequate sensitivity	Glucose, galactose, lactate, sucrose, aspartame, acetic acid, glycerides, biological oxygen demand, cadaverine, histamine, etc.
Ion-selective electrode (ISE)	Simple, reliable, easy to transport	Sluggish response, requires a stable reference electrode, susceptible to electronic noise	Amino acids, carbohydrates, alcohols and inorganic ions
Field effect transistors (FETs)	Low cost, mass production, stable output, requires very small amount of biological material, monitors several analytes simultaneously	Temperature sensitive, fabrication of different layer on the gate has not been perfected	Carbohydrates, carboxylic acids, alcohols and herbicide
Optical	Remote sensing, low-cost, miniaturizable, multiple modes, absorbance, reflectance, fluorescence, extensive electromagnetic range can be used	Interference from ambient light, requires high-energy sources, only applicable to a narrow concentration range, miniaturization can affect the magnitude of the signal	Carbohydrates, alcohols, pesticides, bacteria, and others
Thermal	Versatility, free from optical interferences such as color and turbidity	No selectivity with the exception of when used in arrangement	Carbohydrates, sucrose, alcohols, lipids, amines
Piezoelectric	Fast response, simple, stable output, low cost of readout device, no special sample handling, good for gas analysis, possible use in sensor arrays	Low sensitivity in liquid, interference due to non specific binding	Carbohydrates, vitamins, pathogenic microorganisms (e.g., *E. coli*, *Salmonella*, *Listeria*, *Enterobacter*), contaminants (e.g., antibiotics, fungicides, pesticides), toxic recognition as bacterial toxins

Source: Adapted from Mello, L.D., and L.T. Kubota, 2002, *Food Chemistry* 77:237–256. (With permission.)

milk. During heat treatment of milk, lactose is partially transformed into lactulose that may in turn be degraded to form galactose, tagatose, and saccharinic acids. As lactulose and galactose content increases with the severity of the heating process, their determination could be helpful in differentiating different types of processed milks (Olano et al. 1989). The amount and composition of milk oligosaccharides differ not only among the mammalian species but also during the course of lactation. Bovine colostrum has a relatively high content of oligosaccharides when compared with mature cows' milk that contains only trace amounts. Thus, colostrum is considered as a suitable raw material for the large-scale preparation of milk oligosaccharides to be used as the functional ingredients in the food industry.

During the production of yogurt, *dahi* (Indian yogurt), and other fermented functional dairy products, oligosaccharides are produced from lactose in variable amounts mainly depending on the strains used (Mahoney 1998; Yadav et al. 2007). Some cheeses are almost lactose-free, because during the manufacturing process, most of the lactose is removed with the whey, and the remaining lactose in the cheese is gradually consumed by the starter microorganisms during the fermentation process. However, in fresh cheeses and those made by coagulating fresh milk or cream with rennet or an acid, or even in a heat-and-acid coagulated dairy product such as *paneer* (Indian fresh cheese), lactose remains unchanged; therefore, these products may contain appreciable amounts of lactose. In cheese and most fermented dairy products, as most of the strains ferment only the glucose moiety of lactose, galactose remains in the curd. Both galactose and lactose may influence the characteristics of the cheese, and hence, its contents need to be monitored and controlled. Determination of lactose may also be useful for evaluating grated cheese adulterated with whey solids (Corzo et al. 2010). In functional foods development for lactose-intolerant people, the aim is to produce a lactose-free product, and hence measurement of lactose is critical. The determination of lactose and its derivatives in dairy products is important, and a number of methods have been available for many years, such as gravimetric, polarimetric, and colorimetric methods. Most of the conventional methods are time-consuming, as the quantitative determination of lactose can be performed only in the soluble serum samples obtained after prior removal of fat and protein fractions.

Some of the applications of biosensors in milk and dairy products analysis are given in Table 9.2. Most lactose biosensors are enzyme sensors that work by immobilization of an enzyme system (β-galactosidase and glucose oxidase) onto a transducer. Glucose oxidase is the most important model enzyme used in glucose biosensors. However, it is not used intensively in the dairy industry. Ivekovic et al. (2004) investigated an amperometric glucose biosensor for the analysis of yogurt drinks with glucose oxidase immobilized into palladium hexacyanoferrate hydrogel. Galactose oxidase enzyme is used to develop biosensors for the determination of galactose or lactose. Lundback and Olsson (1985) reported on amperometric determination of galactose and lactose using galactose oxidase in a flow injection analysis system with immobilized enzyme reactors. An enzymatic amperometric biosensor was prepared by Malhotra et al. (2005) to measure galactose in milk and milk products by immobilizing the enzymes in Langmuir-Blodgett films of poly-3-hexylthiophene mixed with stearic acid and depositing the film onto indium tin oxide–coated glass plates.

Amarita et al. (1997) developed a reusable hybrid biosensor based on carbon dioxide (CO_2) electrode for lactose estimation. The immobilized lactases (lactozym and β-galactosidase of *Enterococcus agglomerans*) were used for hydrolyzing lactose, while *Saccharomyces cerevisiae* aided in fermenting glucose and galactose to produce CO_2. The released CO_2 was measured by potentiometric detection. Marrakchi et al. (2008)

TABLE 9.2

Application of Biosensors in Milk and Dairy Product Analyses

Analyte	Application	Biocomponent	Transducer	Detection Range	Reference
Antibiotic (β-Lactams)	Milk	Antibody	Surface plasmon resonance	1.2 µg/kg (Penicillin G)	Gustavsson et al. (2004)
Antibiotic (β-Lactams)	Milk	Antibody	Surface plasmon resonance	2.6 µg/kg (Penicillin G)	Gustavsson et al. (2002)
Antibiotic (Sulfamethazine)	Raw Milk	Antibody	Surface plasmon resonance	< 1 part per billion	Strenesjo et al. (1995)
Antibiotic (Chloramphenicol)	Milk	Antibody	Surface plasmon resonance	0.1 µg L^{-1}	Gaudin and Maris (2001)
Antibiotic (Tetracycline)	Milk	Single strain DNA	Surface plasmon resonance	15 µg L^{-1}	Moeller et al. (2007)
Antibiotic (Penicillin G)	Milk	Antibody	Impedimetric	3×10^{-15} M	Thavarungkul et al. (2007)
Antibiotic (Levamisole)	Milk	Antibody	Surface plasmon resonance	—	Crooks et al. (2003)
Antibiotic (Chloramphenicol)	Milk	Chloramphenicol antibody	Surface plasmon resonance	—	Ferguson et al. (2005)
Antibiotic (Streptomycin)	Milk	Antibody	Surface plasmon resonance	20 ng mL^{-1}	Haasnoot et al. (2002)
Antibiotics	Milk	Antibodies	Surface plasmon resonance	—	Baxter et al. (2001)
Antibiotics	Milk	Antibodies	Surface plasmon resonance	—	Bergstrom et al. (1999)
Antibiotics	Milk	Antibodies	Surface plasmon resonance	—	Gaudin and Maris (2001)
Antibiotics (Neomycin and gentamycin)	Milk	Antibody	Surface plasmon resonance	8 ng mL^{-1} (gentamycin) 21 ng mL^{-1} (neomycin)	Raz et al. (2008)

(continued)

TABLE 9.2 (CONTINUED)

Application of Biosensors in Milk and Dairy Product Analyses

Analyte	Application	Biocomponent	Transducer	Detection Range	Reference
Antibiotics (Enrofloxacin and Ciprofloxacin)	Milk	Antibodies	Surface plasmon resonance	—	Mellgren and Sternesjo (1998)
Biotin and folate	Infant formula and milk	Antibiotin antibody and antifolic acid antibody	Surface plasmon resonance	2–70 ng/mL	Indyk et al. (2000)
Cobalamin	Infant formula	Vitamin-specific binding protein (antibody)	Surface plasmon resonance	—	Gao et al. (2008)
Folate-binding proteins	Milk	Antibody	Surface plasmon resonance	—	Nygren et al. (2003)
Folic acid	Milk powder and infant formula	Antifolic acid antibody	Surface plasmon resonance	—	Caselunghe and Linderberg (2000)
Folic acid	Infant formula	Vitamin-specific binding protein (antibody)	Surface plasmon resonance	—	Gao et al. (2008)
Fructose	Milk	D-fructose dehydrogenase	Amperometric	$50 \times 10^{-6} - 10 \times 10^{-3}$ mol/L	Stredansky et al. (1999)
Fructose and lactulose	Milk	D-fructose dehydrogenase and β-galactosidase	Amperometric	$1 \times 10^{-6} - 5 \times 10^{-3}$ mol/L (fructose) $1 \times 10^{-5} - 5 \times 10^{-3}$ mol/L (lactulose)	Moscone et al. (1999)
Glucose	Milk	Glucose oxidase	Amperometric	50–500 mM	Centonze et al. (1997)
Glucose	Milk	Glucose oxidase	Amperometric	100–1000 ppm	Mannino et al. (1997)
Glucose and galactose	Yogurt and milk	Glucose oxidase, galactose oxidase, and peroxidase	Amperometric	250–4000 mg/L	Mannino et al. (1999)
Glucose and lactose	Milk	Glucose oxidase, β-galactosidase, and mutarotase	Amperometric	4.44g/ 100g (lactose)	Liu et al. (1998)

Analyte	Sample	Biological element	Detection	Detection range/limit	Reference
Glucose and lactose	Milk	β-galactosidase and glucose oxidase	Amperometric	—	Pilloton and Mascini (1990)
Glucose and lactose	Whey	β-galactosidase and glucose oxidase	Amperometric	1×10^{-7} (glucose) 1×10^{-5} (lactose)	Lukacheva et al. (2007)
Glucose, galactose, and lactose	Milk	Glucose oxidase, galactose oxidase, and mixture of β-galactosidase and glucose oxidase	Amperometric	0.05–10 mM (glucose) 0.1–20 mM (galactose) 0.2–20 mM (lactose)	Rajendran and Irudayaraj (2002)
Heavy metals (Fe^{3+}, Pb^{2+}, Cd^{2+})	Milk	Single-stranded DNA	Amperometric	1×10^{-10} (Pb^{2+}) 1×10^{-9} (Cd^{2+}) 1×10^{-7} (Fe^{3+})	Babkina and Ulakhovich (2004)
Lactic acid	Cow milk, goat milk, and WPC* enriched goat milk yogurt	Horseradish peroxidise, lactate oxidase	Amperometric	—	Herrero et al. (2004)
Lactic acid	Yogurt and kefir	Dried cells of *Saccharomyces cerevisiae* (as source of flavocytochrome b_2 enzyme)	Amperometric	0.1–1 mM	Garjonyte et al. (2009)
Lactic acid	—	Lactate oxidase and horseradish peroxidase	—	—	—
Lactose	Milk	β-galactosidase, lactozym, and *Saccharomyces cerevisiae*	Potentiometric	—	Amarita et al. (1997)

* WPC: Whey protein concentrate.

(continued)

TABLE 9.2 (CONTINUED)

Application of Biosensors in Milk and Dairy Product Analyses

Analyte	Application	Biocomponent	Transducer	Detection Range	Reference
Lactose	Milk and instant dessert powder	β-galactosidase and glucose oxidase	Amperometric	—	Adanyi et al. (1999)
Lactose	Raw milk	β-galactosidase, glucose oxidase, and horseradish peroxidase	Amperometric	—	Eshkenazi et al. (2000)
Lactose	Milk	Glucose oxidase and β-galactosidase	Amperometric	$1 \times 10^{-4} - 3.5 \times 10^{-3}$	Goktug et al. (2005)
Lactulose	Milk	D-fructose dehydrogenase and β-galactosidase	Amperometric	1–30 μM	Sekine and Hall (1998)
Lactulose	UHT milk and pasteurized milk	β-galactosidase and D-fructose dehydrogenase	Amperometric	—	Mayer et al. (1996)
L-amino acids	Milk	D-Amino acid oxidase	Amperometric	0.47–2.5 mM (L-leucine) 0.20–2.0 mM (L-glycine)	Sarkar et al. (1999)
L-Lysine	Milk	L-Lysine-α-oxidase	Amperometric	$10^{-5} - 10^{-3}$ mol/L	Curulli et al. (1998)
L-Lysine	Milk	Lyase oxidase	Amperometric	2–25 μM	Kelly et al. (2000)
Nisin	Milk	Bacteria	Bioluminescence	3.0 pg/mL	Immonen and Karp (2007)
Pantothenic acid	Infant formula	Vitamin-specific binding protein (antibody)	Surface plasmon resonance	—	Gao et al. (2008)
Pesticide	Milk	Cholinesterase	Amperometric	$1 \times 10^{-11} - 5 \times 10^{-7}$ M	Medyantseva et al. (1998)
Riboflavin	Infant formula	Vitamin-specific binding protein (antibody)	Surface plasmon resonance	—	Gao et al. (2008)
Urea	Milk	Urease	Potentiometric	—	Verma and Singh (2003)
Urea	Milk	Urease	Potentiometric	2.5×10^{-5} mol/L	Trivedi et al. (2009)

reported a biosensor of β-galactosidase and glucose oxidase combined with conductometric detection. Adanyi et al. (1999) developed a biosensor by immobilizing three enzymes, namely, β-galactosidase, galactose oxidase, and glucose oxidase, on the surface of a measuring electrode. The lactose was decomposed and oxidized by the immobilized enzymes, and the hydrogen peroxide (H_2O_2) generated during the enzymatic reactions was determined by amperometric detection. Rajendran and Irudayaraj (2002) developed a biosensor based on enzyme-catalyzed reaction in combination with techniques like microdialysis sampling, flow injection analysis, and amperometric detection. An electrochemical biosensor developed by Eshkenazi et al. (2000) determines the lactose concentration in fresh raw milk based on serial reactions of three enzymes (β-galactosidase, galactose oxidase, and horseradish peroxidase) immobilized on a glassy carbon electrode. The sequential enzymatic reactions increased the selectivity and sensitivity of the biosensor. Goktug et al. (2005) prepared a biosensor by immobilizing β-galactosidase and glucose oxidase onto a glassy electrode coated with mercury thin film. It was reported that the biosensor response to lactose was in the linear range of 1×10^{-4} and 3.5×10^{-3} M. Lukacheva et al. (2007) used a biosensor for lactose determination based on Berlin blue (as a signal transducer) by immobilizing β-galactosidase and glucose oxidase on a perfluorosulfonated polymer and reported that such biosensor exhibited high sensitivity and fast response. Sharma et al. (2004) presented an amperometric biosensor by immobilizing lactase and galactose oxidase in Langmuir-Blodgett films of poly(3-hexylthiophene)/stearic acid and used the enzyme-immobilized film as a working electrode and platinum as a reference electrode. The enzyme electrodes showed linearity of 1–6 g lactose/dL and have a shelf life of more than 120 days. A manometric sensor was developed by Jenkins and Delwiche (2003) to measure glucose and lactose through enzymatic oxidation. The change in pressure in an enclosed cavity was correlated to the depletion of oxygen resulting from the enzymatic oxidation of glucose or lactose.

Lactulose is used as an indicator of heat treatment of milk and to distinguish between pasteurized and ultra-heat-treated milk and sterilized milk. Mayer et al. (1996) developed a rapid enzymatic procedure for the determination of lactulose by hydrolyzing it to fructose and galactose by soluble β-galactosidase from *Aspergillus oryzae*. The amount of fructose was determined by using immobilized fructose dehydrogenase (FDH) and potassium ferrocyanide as mediator. The reduced mediator was reoxidized at a screen-printed Pt-electrode at a potential of 385 mV versus a screen-printed Pt-pseudoreference electrode. A dual enzyme ring electrode was explored by Sekine and Hall (1998) onto which tetrathiafulvalen-tetracyanoquinodimetane salt was physically packed and FDH and β-galactosidase were immobilized with a dialysis membrane. The hydrolyzed D-fructose was oxidized by FDH, which was simultaneously reduced. The detection limit of the lactulose sensor was 1.0 mM. Moscone et al. (1999) reported a rapid flow system for detection of lactulose in milk samples based on the hydrolysis of lactulose to galactose and fructose by the enzyme β-galactosidase immobilized in a reactor. The amount of fructose produced was measured with a biosensor based on FDH, $K_3[Fe(CN)_6]$ as mediator, and a platinum-based electrochemical transducer. The fructose was determined to be in the range of 1×10^{-6} to 5×10^{-3} mol L^{-1}.

9.3.1.2 Milk Proteins

Normal bovine milk contains about 3.5% protein. The main constituents of the milk protein fraction are caseins, and α-lactalbumin and β-lactoglobin as minor components. The concentration changes significantly during lactation; and especially during the first few days postpartum, the greatest change occurs in the whey protein fraction. The properties of many dairy products, in fact their very existence, depend on the properties of milk proteins, although other constituents exert significant modifying influences (Fox and McSweeney 1998). Casein products are almost exclusively milk protein, while the production of most cheese varieties is initiated through specific modification by proteolytic enzymes or isoelectric precipitation. The high-heat treatments to which many milk products are subjected are possible only because of the exceptionally high heat stability of the principal milk proteins, the caseins. Changes in protein characteristics, for example, insolubility as a result of heat denaturation in milk powders or the increasing solubility of cheese proteins during ripening, are industrially important features of these products. Also some of the proteins cause allergy, and hence, quality control is an important task.

Haasnoot et al. (2004) developed inhibition biosensor immunoassays for the determination of bovine κ-casein for the detection of cow's milk, as well as in the milk of ewes and goats, on the basis of the SPR technique by immobilizing the proper antibodies raised against bovine κ-casein. Different forms of casein were determined separately with immunoassays, quantifying the intact form of αS_1-casein in milk based on the anti-αS_1-casein antibodies directed against each extremity of the molecule (Muller-Renaud et al. 2003). The proteolysis of β-casein during ripening of cheese was followed using a two-step sandwich strategy, with two anti-β-casein antibodies directed against each extremity of the casein (Muller-Renaud et al. 2004). Dupont and Muller-Renaud (2006) reported a method for the simultaneous quantification of the three major intact caseins with a two-step sandwich strategy, with two monoclonal antibodies directed against the N- and C-terminal extremities of each casein. The denaturation of proteins was widely used on the base of the SPR biosensor assay to control the heating systems. The denaturation index of α-lactalbumin was determined by quantifying separately the native and "heat-denatured" forms of α-lactalbumin with specific monoclonal antibodies (Dupont et al. 2004). Some of the proteins cause allergy, and therefore their concentration must be controlled in the commercial products. An optical biosensor has been developed both for direct and sandwich immunoassays using polyclonal antibodies that rise against the proteins immobilized on the biosensor chip. Proteins from the samples that bound to the antibodies on the surface were detected by a shift in resonance angle. By adding a second antibody in a sandwich assay, matrix effects could be overcome and the sensitivity and selectivity enhanced down to 1–12.5 $\mu g/g$ (Yman et al. 2006). An optical biosensor chip for detection of β-lactoglobulin, a major whey protein and also a major allergen in milk, based on resonance-enhanced absorption (REA) effect was reported by Hohensinner et al. (2007). REA effects are observed when light-absorbing noble metal clusters are positioned at a nanodistance from a highly reflective mirror.

Indyk (2009) developed a biosensor-based immunoassay for detection of α-lactalbumin in bovine milk using SPR detection. α-Lactalbumin content was estimated from the specific interaction with an antibovine α-lactalbumin antibody immobilized on the sensor surface. The detection limit was reported to be 0.12 mg mL^{-1} in milk. The developed biosensor was also applied to colostrum, whey protein concentrate, and infant formula. Rapid and sensitive biosensor-based immunoassays for determination of immunoglobulin G, lactoferrin, lactoperoxidase, and folate-binding protein (FBP) in bovine milk and colostrum were reported by Indyk et al. (2006). Lactoferrin was determined using goat and rabbit antibovine lactoferrin antibody, separately immobilized on a sensor chip, by SPR optical detection in bovine milk, colostrums, and infant formulas (Indyk and Filonzi 2005). Also, the optical biosensor-based immunoassay was applied to monitor the heat-induced denaturation of native bovine lactoferrin (Indyk et al. 2007) and immunoglobin G (IgG) (Indyk et al. 2008). The function of minor proteins such as FBPs is to trap folate from blood to secure an adequate supply via milk to neonates (Parodi 1997). Biotin- and folate-supplemented infant formulas and milk powders were analyzed by SPR biosensor-based inhibition immunoassay using monoclonal antibodies. The active concentration of FBP was estimated by SPR detection of the specific interaction with a pteroyl-L-glutamic (folic) acid (PGA) derivative immobilized on the sensor surface. The detection limit was 0.13 μg/mL in fluid milk (Indyk and Filonzi 2004). Nygren et al. (2003) observed that the biosensor method could also be exploited to specifically determine the amount of free FBP in different biological matrices.

9.3.1.3 Lipids

The alarming rise in the rate of clinical disorders such as heart disease, hypertension, arteriosclerosis, coronary artery disease, and so forth due to abnormal levels of cholesterol in blood have stimulated public concern about the determination of the cholesterol level in food products. Cholesterol is the most important lipid constituent in milk and dairy products. The recent advances in materials and techniques for cholesterol biosensor design and construction was reviewed by Arya et al. (2008). Ram et al. (2001) reported a biosensor in which cholesterol oxidase and cholesterol esterase enzymes were bound to collagen membrane or immobilized on conducting polymer matrix, and amperometric detection was carried out in a water phase by platinized electrodes versus Ag/AgCl (Vidal et al. 2000). Pena et al. (2001) investigated an amperometric composite biosensor for the determination of free and total cholesterol. Cholesterol oxidase and horseradish peroxidise together with potassium ferrocyanide as a mediator were incorporated into a graphite–70% Teflon matrix. Free and total cholesterol content was determined in butter samples with a thin-layer cell containing cholesterol oxidase and cholesterol esterase connected into a stopped-flow injection system, with amperometric detector using organic solvents (Adanyi and Varadi 2003). Schmidt et al. (1996) developed a biosensor for the estimation of free fatty acids in milk as butyric acid equivalents by immobilizing *Arthrobacter nicotianae* microorganism in calcium-aliginate gel directly on the electrode surface, while the respiratory activity was monitored by oxygen consumption at −600 mVvs Ag/AgCl reference electrode.

9.3.1.4 Milk Enzymes and Hormones

Like all other foods of plant or animal origin, milk contains several indigenous enzymes that are constituents of the milk as secreted. In addition, milk and most dairy products contain viable microorganisms that secrete extracellular enzymes or release intracellular enzymes after the cells have died and lysed. Some of these enzymes may cause undesirable changes and spoilage, for example, hydrolytic rancidity of milk and dairy products, bitterness and/or age gelation of ultrahigh temperature (UHT)-treated milks, off-flavors (due to lipolysis or proteolysis). However, if properly controlled, these changes may be desirable, for example, in ripening of cheese. Alkaline phosphatase is an important enzyme in quality control of milk, as it is an indicator for effective pasteurization of milk. An amperometric graphite-Teflon composite tyrosinase biosensor was built for the rapid monitoring of alkaline phosphatase as a substrate. The produced phenol was monitored at the tyrosinase composite electrode through the electrochemical reduction of the o-quinone to catechol. A linear calibration plot was obtained for alkaline phosphatase between 2.0×10^{-13} and 2.5×10^{-11} M, with a detection limit of 6.7×10^{-14} M (Serra et al. 2005).

Reproductive performance is a major factor affecting the production and economic efficiency of dairy herds. In general, for herds using artificial insemination (AI), failure to detect heat in cows is one of the major causes of a prolonged calving interval (Boettcher and Perera 2007). In India, it is observed that heat detection is the major constraint for reproductive performance of buffalo herds (Ingawale and Dhoble 2004; Mohan et al. 2010). *Estrus* is defined as the period of time in the reproductive cycle prior to ovulation. Measurement of hormone concentrations in the blood or milk is an accurate indicator of estrus. Elevated milk progesterone concentrations indicate luteal dominance, while low amounts of progesterone are associated with estrus. Claycomb and Delwiche (1998) developed a biosensor that uses enzyme immunoassay method for the online measurement of progesterone in bovine milk and detection of estrus. A rapid automated immunoassay to measure progesterone in bovine milk using an SPR biosensor was developed by Gillis et al. (2002). The limit of detection was reported to be 3.56 ng mL^{-1}. Kappel et al. (2007) developed a total internal reflectance fluorescence-based biosensor with antiprogesterone antibodies as a biological recognition element for the detection of progesterone in milk.

9.3.1.5 Vitamins

Milk is the only source of nutrients including vitamins for the neonates during the early stage of life until other foods are added to their diet. Milk is a good source of most vitamins except vitamin C. Milk is normally processed to a lesser or greater extent before consumption, and hence processing influences the vitamin status of milk and dairy products. The current international methods for estimation of especially water-soluble vitamins are mainly based on microbiological assays. SPR biosensor technology using a ligand-binding protein interaction is a promising method for analysis of several vitamins. The vitamin assay is based on the principle of inhibition in which a derivative of the analyte (vitamin) is covalently bound to the sensor surface. When a sample or standard containing the analyte is mixed with an excess of a vitamin-specific binding protein, free vitamin-specific binding protein that has not bound to the analyte binds to the surface and gives rise to a response. The response is measured as the difference in absolute response obtained immediately before and immediately after the injection of the sample (Blake 2007). Caelen et al. (2004) developed a biosensor-based method for determination of riboflavin (vitamin B_2) in milk-based products using the principle of SPR. Quantification was done indirectly by measuring the excess of riboflavin-binding protein that remains free after complex formation with riboflavin molecules originally present in the sample. Development of immunoaffinity-based optical biosensors was reported for the determination of folic acid in milk (Indyk et al. 2000) and milk powder (Caselunghe and Lindeberg 2000) and for riboflavin, pantothenic acid (Gao et al. 2008), cobalamin (Indyk et al. 2002; Gao et al. 2008), biotin, and folic acid (Indyk et al. 2000; Gao et al. 2008) in infant formula.

9.3.1.6 Minerals

The salts of milk are mainly the phosphates, citrates, chlorides, sulphates, carbonates, and bicarbonates of sodium, potassium, calcium, and magnesium. Approximately 20 other elements are found in milk in trace amounts including copper, iron, silicon, zinc, and iodine. The major elements are of importance in nutrition, in the preparation, processing, and storage of milk products, due to their marked influence on the conformation and stability of milk proteins, especially caseins. The mineral content of food is usually determined from the ash prepared by heating a sample at 500–600°C in a muffle furnace for about 4 h to oxidize organic matter. Selective minerals are estimated using the atomic absorption spectrophotometer. Milk and most dairy products are good sources of calcium. Akyilmaz and Kozgus (2009) developed a biosensor based on the activation of catalase enzyme by calcium ion for the determination of calcium in milk. Catalase converts hydrogen peroxide into hydrogen dioxide and carbon dioxide in the presence of oxygen. The principle of the measurement of biosensor was based on the determination of the changes in the dissolved oxygen concentration in the enzymatic reaction.

9.3.1.7 Lactate/Lactic Acid

Milk, by its very nature, is an excellent growth medium for microorganisms. The sources of microbial contamination of raw milk are lactating animals, the production environment, the milk handling equipment, and the environmental sources such as air, the milk handler, and the water supply. Microbial contamination of milk results in an increased lactate content as a consequence of homo- or heterolactic fermentation. Lactate determination can therefore be useful for the early detection of microbial contamination, especially for milk thermally treated for long-term storage: Typically, this contamination is only detected by measuring the pH (Palmisano et al. 2001). Also, lactic acid is the principal product of milk fermentation by starter cultures or lactic acid bacteria. It is responsible for the typical sour taste of yogurt, *dahi*, and other fermented milk products and therefore an important indicator of quality. The various conventional analytical procedures for determining lactic acid in fermented dairy products are slow and laborious, and some do not differentiate between D- and L-lactic acid isomers. Electrodes modified with a layer of baker's and wine yeast *Saccharomyces cerevisiae* and dried (at room temperature) were used as amperometric biosensors in the presence of potassium ferricyanide as mediator for determination of lactic acid in yogurt and kefir. It was reported that such biosensor was suitable for determination of lactic acid in the range of 0.1 to 1 mM (Garjonyte et al. 2009). Herrero et al. (2004) developed a rapid method for the determination of L-lactic acid in different milk and milk products (cows' milk, goats' milk, and whey protein concentrate–enriched goat milk yogurt) using an amperometric graphite-Teflon biosensor in which horseradish peroxidise, L-lactate oxidase, and the mediator ferrocene were immobilized. It was reported that the correlation between the L-lactic acid results obtained using the bienzyme biosensor method and a standard colorimetric enzymatic method was 0.95. Amperometric sensors based on lactate oxidase and platinized carbon were constructed entirely by screen-printing and were used to estimate the lactate content in yogurt and buttermilk (Janssen and Hart 1996). Palmisano et al. (2001) investigated a disposable biosensor, where the biosensing layer of lactate oxidase was cast on an underlying electropolymerized layer of overoxidized polypyrole. The introduction of a microdialysis membrane-based sampler increased the sensitivity to 7.9 ± 0.2 mM. Choi (2005) developed a biosensor using L-lactate oxidase enzyme-immobilized eggshell membrane and an oxygen electrode for L-lactate determination in milk and yogurt. The detection was based on the depletion of dissolved oxygen content, as monitored by oxygen electrode, upon exposure to L-lactate solution. The shelf-life of the biosensor was reported to be at least a year. Zaydan et al. (2004) studied the fermentation capability of milk samples by using a bioreactor employing bacteria (*Streptococcus thermophilus*) encapsulated in calcium-alginate beads coupled with an L-lactate biosensor containing horseradish peroxidise/ferrocene-modified electrode (–100 mV vs. Ag/AgCl). A bienzyme sensor for the determination of D-lactate in milk, yogurt, and cheese was developed by Montagne and Marty (1995). The D-lactate dehydrogenase, NADH oxidase, and NAD+ coupled to dextran was fixed and covered by a dialysis membrane. The measurement was conducted using hexacyanoferrate (III) as an electron relay.

9.3.2 BIOSENSORS FOR MILK ADULTERANT AND PRESERVATIVE ANALYSES

9.3.2.1 Urea

The recent reports on synthetic milk indicate an emerging quality concern of the dairy industry, especially in developing countries. It was observed that bovine milk is adulterated with synthetic milk. Synthetic milk is a product closely resembling milk but having none of the nutrients found in natural milk. Synthetic milk is prepared by mixing appropriate amounts of vegetable oil, urea, detergent, powdered sugar and salt, and skim milk powder to water, followed by thorough blending in a mixer. The liquid formed has the appearance of thick, rich, creamy milk and visually it is identical to natural milk (Paradkar et al. 2000). Most of the ingredients of synthetic milk are the same as used in adulteration of milk. The only component that appears to be exclusive to synthetic milk is detergent. Urea, on the other hand, is a natural component whose concentration varies from 20 to 70 mg/100 mL in milk. However, this concentration is quite low compared to that in synthetic milk in which the concentration is nearly 20 times more (Bansal and Bansal 1997). The urease enzyme was immobilized on three different support materials, namely, glass wool, glass beads, and arylamine-controlled pore glass beads for use in the development of a fiber-optic urea biosensor for determination of urea in milk (NDRI, 2008). It was reported that the immobilized enzyme had better thermal and storage stability compared with soluble urease. Jenkins et al. (2002) presented a biosensor for online measurement of urea in milk during the milking process. Verma and Singh (2003) developed a biosensor using immobilized urease enzyme coupled to the ammonium ion selective electrode of a potentiometric transducer for monitoring the presence of urea in milk. It was reported that the response time of the biosensor was as low as 2 min. A monometric sensor was constructed for the carbon dioxide generated by the enzymatic hydrolysis with urease. Recently, Trivedi et al. (2009) developed a urea biosensor by immobilizing the urease enzyme onto the NH_4^+ ion selective membrane using a polymer matrix of polycarbamoylsulphonate and polyethyleneimine for determination of urea in milk. The detection limit was reported to be 2.5×10^{-5} mol/L.

9.3.2.2 Nisin

Nisin is a bacteriocin produced by a strain of the dairy starter culture *Lactococcus lactis* spp. *lactis*. It is a peptide composed of 34 amino acid moieties with a molecular size of 3.5 kDa. Stability of nisin in a food system during storage is dependent upon three factors: incubation temperature, length of storage, and pH. Greater nisin retention occurs at lower temperatures. Nisin is approved as a food preservative in more than 50 countries and is used in processed cheese and canned dairy products. The importance of nisin as an effective preservative lies in its wide spectrum of activity, which includes almost all Gram-positive bacteria and their spores (Delves-Broughton 2005). As labeling has become mandatory for food trade both in domestic and international markets, the quantification of added food additives is essential to address food quality and safety concerns. Several methods are available for nisin detection and quantification such as agar diffusion assays, turbidometric bioassays, flow cytometry, capillary zonal electrophoresis, and immunoassays. Immonen and Karp (2007) developed a method for determining ultralow amounts of nisin in milk samples based on luminescent biosensor bacteria. Modified bacterial luciferase operon luxABCDE was placed under control of the nisin-inducible nisA promoter in plasmid pNZ8048 and transformed into *Lactococcus lactis* strains. The resulting luminescence could be directly measured in living bacteria without the addition of exogenous substrates. The sensitivity of the nisin bioassay was reported to be 3.0 pg/mL in milk.

9.3.3 Biosensors for Milk Contaminant Analysis

9.3.3.1 Antibiotics and Veterinary Drug Residues

Veterinary and chemical drugs having anabolic effects are used for therapeutic and prophylactic purposes as well as for improved breeding efficiency. Most of them are banned in many parts of the world and can only be administered in specific circumstances (therapeutic purposes) under strict control. It was observed that these substances are illegally added to act as growth promoters, improving feed conversion efficiency and increasing the lean-to-fat ratio in muscle foods (Toldra and Reig 2006). In principle, all preparations administered to production animals may lead to residues in animal products such as milk, eggs, and edible tissues. The most common antibiotics used for the treatment of farm animals are the β-lactam antibiotics including penicillin. Passage of antibiotics into milk from medicated animals causes major problems in the quality of raw milk and also public health concerns. Antibiotic residues can provoke allergic reactions in sensitized individuals, and their continuous exposure may lead to an increase in the number of individuals resistant to antibiotics (Dewdney et al. 1991). In the dairy industry, antibiotics inhibit the activity of starter cultures of fermented dairy products and negatively influence fermentation. The antibiotic residues in raw milk are initially screened by microbial or enzymatic methods. The positive samples are analyzed for detecting the presence of antibiotics with conventional chemical methods like high-performance liquid chromatography, coupled with UV detector or gas chromatography, coupled with mass-spectrophotometer. The whole multistep examination takes several days (Rinken and Riik 2006), and hence biosensors have been developed as an alternative to commonly used immunoassays.

Eshkenazi et al. (2000) developed a disposable amperometric flow injection biosensor composed of screen-printed electrodes including a working electrode, a reference electrode, and a homemade microflow cell (cell volume 30μL) for the detection of organophosphates. Gustavsson and Sternesjo (2004) developed SPR biosensor assays for the detection of β-lactams in milk. A microbial receptor protein with carboxypeptidase activity was used as detection molecule in an SPR-based biosensor in the detection of penicillin G. The limit of detection of the assay was reported to be 2.6 μg kg^{-1} for antibiotic-free milk (Gustavsson et al. 2002). A label-free impedimetric flow injection immunosensor for the direct detection of penicillin G was developed by Thavarungkul et al. (2007) by immobilizing antipenicillin G on a gold electrode modified with a self-assembled nanolayer of thioctic acid. The real-time monitoring of impedance was carried out at the optimum frequency of 160 Hz. The detection limit of penicillin G in milk was reported to be 3.0×10^{-15} M. An assay was developed by Cacciatore et al. (2004) for the detection of penicillins and cephalosporins in milk using an SPR biosensor. The assay was based on the inhibition of the binding of digoxigenin-labeled ampicillin to a soluble penicillin-binding protein 2x derivative (2x*) of *Streptococcus pneumonia*. Haasnoot et al. (2003) developed an optical biosensor (Biacore 3000) with four flow channels for the simultaneous detection of five relevant aminoglycosides, namely, gentamycin, kanamycin, streptomycin, dihydrostreptomycin, and neomycin, in reconstituted skimmed milk in combination with a mixture of four specific antibodies. Gaudin et al. (2001) constructed an SPR biosensor using commercial antibody against ampicillin, which had much higher affinity for open beta-lactam than for ring. It was reported that between the enzymatic and chemical pretreatments tested to open the ring, enzymatic pretreatment resulted in improved detection limit. An immune-based screening method was developed to detect levamisole residues in milk using an optical SPR biosensor by Crooks et al. (2003). Levamisole was derivatized to aminolevamisole, conjugated to a carrier protein, and the immunogen generated was used to raise polyclonal antibodies. Fodey et al. (2007) characterized polyclonal antibodies against chloramphenicol raised in three species (camel, donkey, and goat) using a chloramphenicol biosensor. Gaudin and Maris (2001) developed a fast biosensor immunoassay for direct screening of chloramphenicol residues in milk based on the inhibition of the binding of polyclonal antibodies against chloramphenicol. The detection limit was reported to be 0.1 μg L^{-1}. Ferguson et al. (2005) developed an immunochemical screening assay using SPR for chloramphenicol and chloramphenicol glucuronide residues in cows' milk using a sensor coated with a chloramphenicol derivative and an antibody. The detection capability was reported to be 0.05 μg kg^{-1}.

A new strategy was developed for establishing an indirect tetracycline assay using SPR based on the resistance mechanism of tetracycline in Gram-negative bacteria (Moeller et al. 2007). The estimated limit of detection of tetracycline in raw milk was 15μg L^{-1}. To determine the sulfamethazine residues in milk, an SPR sensor was built by covalently immobilizing sulfamethazine onto a carboxymethyldextran-modified gold film on the surface. Polyclonal sulfamethazine antibodies were added to samples, and free antibodies were detected. No cross-reactivity of antibodies with other antibiotics was found (Sternesjo et al. 1995). An immunosensor was developed for the detection of sulphonamide antibiotics in milk (Adrain et al. 2009). The immunossensor combines generic immunoreagents with a wave-guide-interrogated immune sensor label-free optical biosensor. It was reported that the developed system allowed discrimination between milk contaminated with sulphonamide antibiotics (sulfapyridine) at or above the maximum residue level (MRL) (100 μg L^{-1}). Recently, a microarray biosensor was developed using an imaging SPR system for detecting neomycin and gentamycin (Raz et al. 2008). A biosensor for determination of enrofloxacin was developed using SPR technology and employed a gold-coated glass surface on which denatured DNA was immobilized in alternate layers with a cationic polymer. The detection limit for the antibiotic in milk samples was reported to be 3 μg/mL. In another application an SPR biosensor was developed where heat-denatured DNA was immobilized on the gold-coated glass surface layer by layer with a cationic polymer (Cao et al. 2007).

9.3.3.2 Pesticide Residues and Dioxins

In recent years, a growing number of initiatives and legislative actions have been adopted worldwide for environmental pollution control. The need for disposable systems or tools for environmental monitoring encouraged the development of new technologies and suitable methodologies for monitoring environmental pollutants such as pesticides, dioxins, toxins, metals, and so forth (Rodriguez-Mozaz et al. 2006). Pesticides are chemical and biological substances that play an important role in food production by controlling rodents, insects, weeds, organisms responsible for plant diseases, and other pests. A wide range of compounds are used as pesticides such as organochlorines, organophosphates, and carbamates. Pesticide residues are the very small amounts of pesticides that can remain in the crop after harvesting or storage and make their way into the food chain. Most pesticides create some degree of risk to harm humans, animals, or the environment because they are inherently designed to adversely affect or kill selected living organisms or target species. In the global food trade, one should comply with the MRLs of pesticide residues prescribed or restricted by importing countries or with international standards. Hence, quantification of pesticide residues is essential. The conventional methods of determination of pesticide residues are time-consuming, require sample pretreatment, and lack specificity and accuracy. Ghosh et al. (2006) developed an amperometric biosensor by immobilizing acetylcholinesterase and choline oxidase in a polymeric porous network directly on the working electrode for the determination of organophosphorus pesticide like monocrotophos. Zhang et al. (2005) developed a biosensor test based on disposable screen-printed electrodes for the monitoring of organophosphate and carbamate residues in foods with increased fat contents such as milk with a special combination of wild-type acetylcholinesterase with three engineered variants to enhance sensitivity. Nikolelis et al. (2005) reported an electrochemical flow injection amperometric inhibition sensor for the analysis of carbofuran in foods. Air-stable lipid films supported on a methylacrylate polymer incorporating an acetylcholinesterase enzyme were used for the determination of the degree of inhibition caused by the presence of pesticide. Kumar et al. (2006) developed an optical biosensor in which whole cells of *Flavobacterium* spp. were immobilized by trapping in glass fiber filter paper for the detection of methyl parathion pesticide. The lower detection limit was estimated as 0.3 μM.

Dioxin is a generic name for two groups of compounds, polychlorinated dibenzofurans and polychlorinated dibenzo-p-dioxins. They are environmental pollutants and are of concern because of their highly toxic potential. Certain dioxin-like polychlorinated biphenyls (PCBs) with similar toxic properties are also included under the term *dioxins*, which are mainly by-products of industrial processes but can also result from natural processes, such as volcanic eruptions and forest fires. In terms of dioxin release into the environment, waste incinerators (solid waste and hospital waste) are often the worst culprits, due to incomplete burning. More than 90% of human exposure to dioxins is through the food supply, mainly meat, milk and milk products, and fish and shellfish (WHO 2007). Mascini et al. (2005) reported that gold-based quartz crystals were modified with synthetic oligopeptides, designed as biomimetic traps, to obtain piezoelectric sensors selective to the dioxins. The crossreactivity of the system was quantified by spiking the milk samples with commercial PCBs and a mixture of PCBs and dioxins.

9.3.4   BIOSENSORS FOR DETECTION OF MICROBES
AND THEIR METABOLITES IN DAIRY PRODUCTS

Food safety and food biosecurity is always a major concern to consumers, food manufacturers, and regulatory agencies. The safety of milk and milk products from farm-to-plate through the supply chain continuum must be ensured to protect consumers from foodborne infections, intoxications, and toxi-infections. Food safety management systems like Hazard Analysis and Critical Control Points (HACCP) ensure safety and call for the availability of rapid, sensitive, and accurate detection methods. Conventional microbiological methods for enumerating the suspected pathogens or toxins are powerful, error-proof, and dependable; but these lengthy, cumbersome methods are often ineffective because they are not compatible with the perishable nature of milk and milk products and the speed at which such products are manufactured and distributed in the global trade. With the advancement in biosensors, several biosensor tools belonging to the categories of optical, electrochemical, and mass-based tools have been applied in the dairy industry for detection of foodborne pathogens and their toxic metabolites.

9.3.4.1 Pathogenic Microorganisms

Biosensors for bacterial detection generally involve a biological recognition component such as receptors, nucleic acids, or antibodies in intimate contact with an appropriate transducer. Based on the type of detection, they are classified as direct or indirect biosensors. Direct detection biosensors are designed in such a way that the biospecific reaction directly determines in real time by measuring the physical changes induced by the complex formation. Indirect detection biosensors are those in which a preliminary biochemical reaction takes place and the products of that reaction are detected by a sensor (Ivnitski et al. 1999). Dairy animals could be the important reservoirs of foodborne pathogens. Hence, milk may carry *Salmonella, Campylobacter, Listeria,* or *Escherichia coli* O157:H7 organisms (Bhunia 2008). The presence of pathogens in ready-to-eat milk products such as soft cheeses made with unpasteurized milk, ice cream, butter, and so forth is a serious concern because these products generally do not receive any further treatment before consumption. Waswa et al. (2007) used an SPR-based biosensor for direct detection of *Escherichia coli* O157:H7 in milk and reported that the sensitivity of the assay was 10^2–10^3 colony forming units (cfu)/mL. Radke and Alocilja (2005) developed a high-density microelectrode array biosensor fabricated from silicon and consisting of an interdigitated gold electrode array for the detection of *Escherichia coli* O157:H7. A QCM biosensor was developed for the detection of *Listeria monocytogenes* by immobilizing the antibody on the gold surface of the crystals (Minunni et al. 1996). Chang et al. (2006) developed a remote monitoring system with a quartz crystal sensor for determining the bacterial population in raw milk. Immersion of the electrodes in a cell culture with bacteria inoculums resulted in a change of frequency caused by the impedance change due to the microbial metabolism and the adherence of bacteria on the surface of the electrodes. The calibration curve of bacteria density showed a linear correlation over the range of 70×10^6 cfu/mL. The fermentation process of milk inoculated with bacteria was monitored amperometrically using an L-lactate biosensor in the flow injection analysis system. The effect of *Enterococcus faecalis, Bacillus coagulans, Enterobacter sakazakii, Staphylococcus aureus,* and *Bacillus sphericus* was investigated (Skladal et al. 1993).

Biosensor technologies such as acoustic wave devices and optical and thickness-shear mode resonators offer sensitivity but are limited by their need for direct physical contacts. Magnetoelastic sensors don't need direct physical contacts to obtain a sensor response, which prohibit their usage in sealed containers and thus are suitable for the remote detection of pathogens (Guntupalli et al. 2007). Magnetoelastic materials are amorphous ferromagnetic alloys that usually include a combination of iron, nickel, molybdenum, and boron. They exhibit a physical resonance when it is subjected to a time-varying magnetic field that can be monitored by using a pickup coil eliminating the need for direct physical connections. Guntupalli et al. (2007) used a magnetoelastic sensor immobilized with polyclonal antibody for the detection of *Salmonella typhimurium* in fat-free milk. The detection limit was reported as 5×10^3 cfu/mL. Lakshmanan et al. (2007) developed a magnetoelastic sensor immobilized with bacteriophage for the detection of *Salmonella typhimurium* in fat-free milk. Upon exposure to the target pathogen, the response of the phage-immobilized magnetoelastic sensor changed, which was quantified by the shift in the sensor's resonance frequency. The remote-query (wireless, passive) magnetoelastic sensor platform was used for direct, real-time detection and monitoring of *Staphylococcus aureus* in milk (Huang et al. 2008). Using this technique *S. aureus* concentrations were quantified up to 10^3 to 10^7 cells mL^{-1} in milk.

9.3.4.2 Microbial Toxins

Staphylococcus aureus is the predominant microbe involved in causing mastitis in dairy cattle. It is a major human pathogen involved in production of numerous toxins. Real-time detection is important in the diagnostics and detection of foodborne toxins, especially during the food production process. Rasooly and Rasooly (1999) developed a real-time detection system for the detection of Staphylococcal enterotoxin A (SEA) in milk and other foods using an evanescent wave biosensor. The approach used two antibodies. The toxin binds initially to a capturing antibody that is covalently bound on the surface of the biosensor detector. The second antibody binds to the captured toxin. It was reported that the measured signal was proportional up to 10 ng SEA addition. Homola et al. (2002) developed a wavelength modulation-based SPR biosensor for the detection of Staphylococcal enterotoxin B (SEB) in milk and reported that the method is capable of directly detecting concentrations of SEB up to 0.5 ng/mL in milk. Aflatoxins are a group of mycotoxins formed by common fungi such as *Aspergillus flavus*, *A. parasiticus*, *A. nomius*, and so forth. They are found in peanuts, corn, cottonseed, and other nuts, grains, and spices. Aflatoxin B1 in animal feed coverts in part to a hydroxylated compound, aflatoxin M1, which appears in the milk of lactating cows and is highly stable during normal milk-processing operations (Carlson et al. 2000). A *Bacillus megaterium* spore-based biosensor was developed for the detection of aflatoxin M1 in milk (NDRI 2009). The detection limit of aflatoxin M1 was reported to be 0.5 to 4.0 ppb. Cucci et al. (2007) presented a compact fluorometric sensor with a light-emitting diode (LED) source and a highly sensitive photomultiplier tube (PMT) detector for the selective detection of native fluorescence of aflatoxin M1 in liquid solutions.

9.4   CONCLUSION

Milk and milk products provide essential nutrients for all age groups. Hence, ensuring the quality and safety of dairy products offered to consumers is a mandatory objective of the dairy industry. The perishable nature of milk demands periodic evaluation of its quality with respect to the constituents considering the biochemical and microbiological changes throughout the supply chain continuum from production to consumption. Quality control and evaluation are carried out to check the adulterants and contaminants of milk and milk products and quantify the various food additives used in the manufacturing of dairy products for food labeling purposes. The conventional techniques used for detection and determination of quality such as titration, spectrophotometry, chromatography, electrophoresis, amd microbiological methods are slow, expensive, and need well-trained analytical instrument operators. Further, the procedures are not easy to monitor and require sample preparation. Biosensors are small analytical devices that combine a transducer with a biologically active substance (sensing component) that produce a signal related to a single analyte or group of analytes. They are very sensitive and selective due to the extraordinary nature of the biological interaction. Biosensors have been successfully applied to milk and milk products to determine various milk components, namely, lactose and other minor carbohydrates, caseins and other minor proteins, vitamins, cholesterol, estimation of adulterants, detection of antibiotics and pesticide residues, pathogenic bacteria, and toxins. With newer techniques such as magnetoelastic sensors, it is now possible for nondestructional analysis of hermetically sealed foods such as UHT milk and milk products. Although various (hand-held) biosensor devices are developed for online and at-line applications, their commercial applications are limited due to cost considerations and some key technical barriers. These barriers include suitability for automation, miniaturization, and system integration for multiple tasks. The developments in the area of nanotechnology may be of some help to overcome these challenges. Therefore, development of biosensors must be aimed for versatility to support interchangeable biorecognition of elements and feasibility to allow automation and ease of operation at a competitive cost.

ACKNOWLEDGMENT

The authors are grateful to Professor Mehmet Mutlu, editor of this book, for his constant encouragement in preparing this chapter.

REFERENCES

Adanyi, N. 2010. Biosensors. In *Handbook of dairy foods analysis*, ed. L.M.L. Nollet and F. Toldra, 307–327. Boca Raton, FL: CRC Press.

Adanyi, N., and M. Varadi. 2003. Development of organic phase amperometric biosensor for measuring cholesterol in food samples. *European Food Research and Technology* 218:99–104.

Adanyi, N., E.E. Szabo, and M. Varadi. 1999. Multi-enzyme biosensors with amperometric detection for determination of lactose in milk and dairy products. *European Food Research Technology* 209:220–226.

Adrain, J., S. Pasche, and J.M. Diserens et al. 2009. Wave-guide interrogated optical immunosensor (WIOS) for detection of sulphonamide antibiotics in milk. *Biosensors and Bioelectronics* 24:3340–3346.

Akyilmaz, E., and O. Kozgus. 2009. Determination of calcium in milk and water samples by using catalase enzyme electrode. *Food Chemistry* 115:347–351.

Amarita, F., C.R. Fernandez, and F. Alkorta. 1997. Hybrid biosensors to estimate lactose in milk. *Analytica Chimica Acta* 349:153–518.

Arya, S.K., M. Datta, and B.D. Malhotra. 2008. Recent advances in cholesterol biosensor. *Biosensors and Bioelectronics* 23:1083–1100.

Babkina, S.S., and N.A. Ulakhovich. 2004. Amperometric biosensor based on denatured DNA for the study of heavy metals complexing with DNA and their determination in biological, water and food samples. *Bioelectrochemistry* 63:261–265.

Bansal, P., and N. Bansal. 1997. Synthetic milk—Genesis, current status and options. *Current Science* 23:904–905.

Baxter, G.A., J.P. Ferguson, M.C. O'Connor, and C.T. Elliot. 2001. Detection of streptomycin residues in whole milk using an optical immunobiosensor. *Journal of Agricultural and Food Chemistry* 49:3204–207.

Bergstrom, C., A. Sternesjo, P. Bjurling, and S. Lofar. 1999. Design and use of a general capturing surface in optical biosensor analysis of sulphamethazine in milk. *Food and Agricultural and Immunology* 11:329–338.

Bhunia, A.K. 2008. Biosensors and bio-based methods for the separation and detection of food borne pathogens. *Advances in Food and Nutrition Research* 54:1–44.

Blake, C.J. 2007. Analytical procedures for water-soluble vitamins in foods and dietary supplements: A review. *Analytical and Bioanalytical Chemistry* 389:63–76.

Boettcher, P.J., and B.M.A.O. Perera. 2007. Improving the reproductive management of small holder dairy cattle and the effectiveness of artificial insemination: A summary. In *Improving the Reproductive Management of Dairy Cattle Subjected to Artificial Insemination* IAEA-TECDOC-1533, 1–7. Vienna, Austria: International Atomic Energy Agency.

Cacciatore, G., M. Petz, S. Rachid, R. Hakenbeck, and A.A. Bergwerff. 2004. Development of an optical biosensor assay for detection of β-lactam antibiotics in milk using the penicillin-binding protein 2x*. *Analytica Chimica Acta* 520:105–115.

Caelen, I., A. Kalman, and L. Wahlstrom. 2004. Biosensor-based determination of riboflavin in milk samples. *Analytical Chemistry* 76:137–143.

Cao, L., H. Lin, and V.M. Mirsky. 2007. Surface plasmon resonance biosensor for enrofloxacin based in deoxyribonucleic acid. *Analytica Chimica Acta* 589:1–5.

Carlo, D.M., M. Nistor, D. Compagnone, B. Mattiasson, and E. Csoregi. 2006. Biosensors for food quality assessment. In *Food biotechnology*, ed. K. Shetty, G. Paliyath, A. Pometto, and R.E. Levin, 1567–1593. Boca Raton, FL: CRC Press.

Carlson, M.A., C.B. Bargeron, R.C. Benson et al. 2000. An automated handheld biosensor for aflatoxin. *Biosensors and Bioelectronics* 14:841–848.

Caselunghe, M.B., and J. Lindeberg. 2000. Biosensor-based determination of folic acid in fortified food. *Food Chemistry* 70:523–532.

Centonze, D., C.G. Zambonin, and F. Palmisano. 1997. Determination of glucose in non-alcoholic beverages by a biosensor coupled with microdialysis fiber samplers. *Journal of AOAC International* 80:829–833.

Chang, K.-S., H.-D. Jang, C.-F. Lee, Y.-G. Lee, C.-J. Yuan, and S.-H. Lee. 2006. Series quartz crystal sensor for remote bacteria population monitoring in raw milk via the internet. *Biosensors and Bioelectronics* 21:1581–1590.

Choi, M.M.F. 2005. Application of a long shelf-life biosensor for the analysis of L-lactate in dairy products and serum samples. *Food Chemistry* 92:575–581.

Claycomb, R.W. and M.J. Delwiche. 1998. Biosensor for on-line measurement of bovine progesterone during milking. *Biosensors and Bioelectronics* 13:1173–1180.

Corzo, N., A. Olano, and I. Martinez-Castro. 2010. Carbohydrates. In *Handbook of dairy foods analysis*, ed. L.M.L. Nollet and F. Toldra, 139–168. Boca Raton, FL: CRC Press.

Crooks, S.R.H., B. McCarney, I.M. Traynor, C.S. Thompson, S. Floyd, and C.T. Elliott. 2003. Detection of levamisole residues in bovine liver and milk by immunobiosensor. *Analytica Chimica Acta* 483:181–186.

Cucci, C., A.G. Mignani, C. Dall'Asta, R. Pela, and A. Dossena. 2007. A portable fluorometer for the rapid screening of M1 aflatoxin. *Sensors and Actuators B: Chemical* 126:467–472.

Curulli, A., S. Kelly, C.O. O'Sullivan, G.G. Guilbault, and G. Palleschi. 1998. A new interference-free lysine biosensor using a non-conducting polymer film. *Biosensors and Bioelectronics* 13:1245–1250.

Delves-Broughton, J. 2005. Nisin as a food preservative. *Food Australia* 57:525–527.

Dewdney, J.M., L. Maes, J.P. Raynaud et al. 1991. Risk assessment of antibiotic residues of β-lactams and macrolides in food products with regard to their immune-allergic potential. *Food and Chemical Toxicology* 29:477–483.

Dupont, D., and S. Muller-Renaud. 2006. Quantification of proteins in dairy products using an optical biosensor. *Journal of AOAC International* 89:843–848.

Dupont, D., O. Rolet-Repecaud, and S. Muller-Renaud. 2004. Determination of the heat treatment undergone by milk by following the denaturation of α-lactalbumin with a biosensor. *Journal of Agricultural and Food Chemistry* 52:677–681.

Eshkenazi, I., E. Maltz, B. Zion, and J. Rishpon. 2000. A three-cascaded-enzymes biosensor to determine lactose concentration in raw milk. *Journal of Dairy Science* 83:1939–1945.

Ferguson, J., A. Baxter, P. Young et al. 2005. Detection of chloramphenicol and chloramphenicol glucuronide residues in poultry muscle, honey, prawn and milk using surface plasmon resonance biosensor and Qflex kit chloramphenicol. *Analytica Chimica Acta* 529:109–113.

Fodey, T., G. Murilla, A. Cannavan, and C. Elliott. 2007. Characterization of antibodies to chloramphenicol, produced in different species by enzyme-linked immunosorbent assay and biosensor technologies. *Analytica Chimica Acta* 592:51–57.

Fox, P.F., and P.L.H. McSweeney. 1998. *Dairy chemistry and biochemistry.* London: Blackie Academic and Professional.

Gao, Y., F. Guo, S. Gokavi, A. Chow, Q. Sheng, and M. Guo. 2008. Quantification of water-soluble vitamins in milk-based infant formulae using biosensor-based assays. *Food Chemistry* 110:769–776.

Garjonyte, R., V. Melvydas, and A. Malinauskas. 2009. Amperometric biosensors for lactic acid based on baker's and wine yeast. *Microchim Acta* 164:177–183.

Gaudin, V., and P. Maris. 2001. Development of a biosensor-based immunoassay for screening of chloramphenicol residues in milk. *Food and Agricultural Immunology* 13:77–86.

Gaudin, V., J. Fontaine, and P. Maris. 2001. Screening of penicillin residues in milk by surface plasmon resonance-based biosensor assay: Comparison of chemical and enzymatic sample pre-treatment. *Analytica Chimica Acta* 436:191–198.

Ghosh, D., K. Dutta, D. Bhattacharyay, and D. Sarkar. 2006. Amperometric detection of pesticides using polymer electrodes. *Environment Monitoring and Assessment* 119:481–489.

Gillis, E.H., J.P. Gosling, J.M. Sreenan, and M. Kane. 2002. Development and validation of a biosensor-based immunoassay for progesterone in bovine milk. *Journal of Immunological Methods* 267:131–138.

Goktug, T., M.K. Sezginturk, and E. Dinckaya. 2005. Glucose oxidase-β-galactosidase hybrid biosensor based on glassy carbon electrode modified with mercury for lactose determination. *Analytica Chimica Acta* 551:51–56.

Guntupalli, R., R.S. Lakshmanan, M.L. Johnson et al. 2007. Magnetoelastic biosensor for the detection of *Salmonella typhimurium* in food products. *Sensing and Instrumentation for Food Quality and Safety* 1:3–10.

Gustavsson, E., and A. Sternesjo. 2004. Biosensor analysis of β-lactams in milk: Comparison with microbiological, immunological and receptor-based screening methods. *Journal of AOAC International* 87:614–620.

Gustavsson, E., P. Bjurling, and A. Sternesjo. 2002. Biosensor analysis of penicillin G in milk based on the inhibition of carboxypeptidase activity. *Analytica Chimica Acta* 468:153–159.

Gustavsson, E., J. Degelaen, P. Bjurling, and A. Sternesjo. 2004. Determination of β-lactams in milk using surface plasmon resonance-based biosensor. *Journal of Agricultural and Food Chemistry* 52:2791–2796.

Haasnoot, W., E.E.M.G. Loomans, G. Cazemier et al. 2002. Direct versus competitive biosensor immunoassays for the detection of (dihydro) streptomycin residues in milk. *Food and Agricultural Immunology* 14:15–27.

Haasnoot, W., G. Cazemier, M. Koets, and A.V. Amerongen. 2003. Single biosensor immunoassay for the detection of five aminoglycosides in reconstituted skimmed milk. *Analytica Chimica Acta* 488:53–60.

Haasnoot, W., N.G.E. Smits, A.E.M. Kemmers-Voncken, and M.G.E.G. Bremer. 2004. Fast biosensor immunoassays for the detection of cow's milk in the milk of ewes and goats. *Journal of Dairy Research* 71:322–329.

Herrero, A.M., T. Requena, A.J. Reviejo, and J.M. Pingarron. 2004. Determination of L-lactic acid in yoghurt by a bienzyme amperometric graphite-Teflon composite biosensor. *European Food Research Technology* 219:556–559.

Homola, J., J. Dostalek, S. Chen, A. Rasooly, S. Jiang, and S.S. Yee. 2002. Spectral surface plasmon resonance biosensor for detection of *Staphylococcal enterotoxin* B in milk. *International Journal of Food Microbiology* 75:61–69.

Honensinner, V., I. Maier, and F. Pittner. 2007. A 'gold cluster-linked immunosorbent assay': Optical near-field biosensor chip for the detection of allergenic β-lactoglobulin in processed milk matrices. *Journal of Biotechnology* 130:385–388.

Hoyle, D. 2001. *ISO 9000 quality systems handbook*. Oxford: Butterworth-Heinemann.

Huang, S., S. Ge, L. He, Q. Cai, and C.A. Grimes. 2008. A remote-query sensor for predictive indication of milk spoilage. *Biosensors and Bioelectronics* 23:1745–1748.

Immonen, N., and M. Karp. 2007. Bioluminescence-based bioassays for rapid detection of nisin in food. *Biosensors and Bioelectronics* 22:1982–1987.

Indyk, H.E. 2009. Development and application of an optical biosensor immunoassay for α-lactalbumin in bovine milk. *International Dairy Journal* 19:36–42.

Indyk, H.E., and E.L. Filonzi. 2005. Determination of lactoferrin in bovine milk, colostrum and infant formulas by optical biosensor analysis. *International Dairy Journal* 15:429–438.

Indyk, H.E., and E.L. Filonzi. 2004. Direct optical biosensor analysis of folate-binding protein in milk. *Journal of Agricultural and Food Chemistry* 52:3253–3258.

Indyk, H.E., E.A. Evans, M.C.B. Caselunghe et al. 2000. Determination of biotin and folate in infant formula and milk by optical biosensor-based immunoassay. *Journal of AOAC International* 83:1141–1148.

Indyk, H.E., E.L. Filonzi, and L.W. Gapper. 2006. Determination of minor proteins of bovine milk and colostrum by optical biosensor analysis. *Journal of AOAC International* 89:898–902.

Indyk, H.E., J.W. Williams, and H.A. Patel. 2008. Analysis of denaturation of bovine IgG by heat and high pressure using an optical biosensor. *International Dairy Journal* 18:359–366.

Indyk, H.E., I.J. McGrail, G.A. Watene, and E.L. Filonzi. 2007. Optical biosensor analysis of the heat denaturation of bovine lactoferrin. *Food Chemistry* 101:838–844.

Indyk, H.E., B.S. Persson, M.C.B. Caselunghe, A. Moberg, E.L. Filonzi, and D.C. Woollard. 2002. Determination of vitamin B12 in milk products and selected foods by optical biosensor protein-binding assay: Method comparison. *Journal of AOAC International* 85:72–81.

Ingawale, M.V., and R.L. Dhoble. 2004. Buffalo reproduction in India: An overview. *Buffalo Bulletin* 23:4–9.

Ivekovic, D., S. Milardovic, and B.S. Grabaric. 2004. Palladium hexacyanoferrate hydrogel as a novel and simple enzyme immobilization matrix for amperometric biosensors. *Biosensors and Bioelectronics* 20:872–878.

Ivnitski, D., I. Abdel-Hamid, P. Atanasov, and E. Wilkins. 1999. Biosensors for detection of pathogenic bacteria. *Biosensors and Bioelectronics* 14:599–624.

Janssen, C.D., and A.L. Hart. 1996. Measurement of soluble L-lactate in dairy products using screen-printed sensors in batch mode. *Biosensors and Bioelectronics* 11:1041–1149.

Jenkins, D.M., and M.J. Delwiche. 2003. Adaptation of a manometric biosensor to measure glucose and lactose. *Biosensors and Bioelectronics* 18:101–107.

Jenkins, D.M., J. Michael, and M.J. Delwiche. 2002. Manometric biosensor for on-line measurement of milk urea. *Biosensors and Bioelectronics* 17:557–563.

Kappel, N.D., F. Proll, and G. Gauglitz. 2007. Development of a TIRF-based biosensor for sensitive detection of progesterone in bovine milk. *Biosensors and Bioelectronics* 22:2295–2300.

Kelly, S.C., P.J. O'Connell, C.K. O'Sullivan, and G.G. Guilbault. 2000. Development of an interference free amperometric biosensor for determination of L-lysine in food. *Analytica Chimica Acta* 412:111–119.

Kumar, J., S.K. Jha, and S.F. D'Souza, 2006. Optical microbial biosensor for detection of methyl parathion pesticide using *Flavobacterium* sp. whole cells adsorbed on glass fiber filters as disposable biocomponent. *Biosensors and Bioelectronics* 21:2100–2105.

Lakshmanan, R.S., R. Guntupalli, J. Hu, V.A. Petrenko, J.M. Barbaree, and B.A. Chin. 2007. Detection of *Salmonella typhimurium* in fat free milk using a phage immobilized magnetoelastic sensor. *Sensors and Actuators B: Chemical* 126:544–550.

Liu, H., H. Li, T. Ying, K. Sun, Y. Qin, and D. Qi. 1998. Amperometric biosensor sensitive to glucose and lactose based on co-immobilization of ferrocene, glucose oxidase, β-galactosidase and mutarotase in β-cyclodextrin polymer. *Analytica Chimica Acta* 358:137–144.

Lukacheva, L.V., A.A. Zakemovskaya, E.E. Karyakina et al. 2007. Determination of glucose and lactose in food products with the use of biosensors based on Berlin blue. *Journal of Analytical Chemistry* 62:388–393.

Lundback, H., and B. Olsson. 1985. Amperometric determination of galactose, lactose and dihydroxyacetone using galactose oxidase in a flow injection system with immobilized enzyme reactors and on-line dialysis. *Analytical Letters* 18:871–889.

Mahoney, R.R. 1998. Galactosyl-oligosaccharide formation during lactose hydrolysis: A review. *Food Chemistry* 63:147–154.

Malhotra, B.D., R. Singhal, A. Chaubey, S.K. Sharma, and A. Kumar. 2005. Recent trends in biosensors. *Current Applied Physics* 5:92–97.

Mannino, S., M.S. Cosio, and S. Buratti. 1999. Simultaneous determination of glucose and galactose in dairy products by two parallel amperometric biosensor. *Italian Journal of Food Science* 1:57–65.

Mannino, S., O. Brenna, S. Buratti, and M.S. Cosio. 1997. Microdialysis-bioreactor for on-line monitoring of glucose in food samples. *Electroanalysis* 9:1337–1340.

Marrakchi, M., S.V. Dzyadevych, F. Lagarde, C. Martelet, and N. Jaffrezic-Renault. 2008. Conductometric biosensor based on glucose oxidase and beta-galactosidase for specific lactose determination in milk. *Material Science and Engineering*: C 28:872–875.

Mascini, A.M., A. Macagnano, G. Scortichini et al. 2005. Biomimetic sensors for dioxins detection in food samples. *Sensors and Actuators B: Chemical*. 111–112:376–384.

Mayer, M., M. Genrich, W. Kunnecke, and U. Bilitewski, 1996. Automated determination of lactulose in milk using an enzyme reactor and flow analysis with integrated dialysis. *Analytica Chmica Acta* 324:37–45.

Medyantseva, E.P., M.G. Vertlib, G.K. Budnikov, and M.P. Tyshelk. 1998. A new approach to selective detection of 2,4-diclorophenoxyacetic acid with a cholinesterase amperometric biosensor. *Applied Biochemistry and Microbiology* 34:202–205.

Mellgren, C., and A. Sternesjo. 1998. Optical immunobiosensor assay for determining enrofloxacin and ciprofloxacin in bovine milk. *Journal of AOAC International* 81:394–397.

Mello, L.D., and L.T. Kubota, 2002. Review of the use of biosensors as analytical tools in the food and drink industries. *Food Chemistry* 77:237–256.

Minunni, M., M. Mascini, R.M. Carter, M.B. Jacobs, G.J. Lubrano, and G.G. Guilbault. 1996. A quartz crystal microbalance displacement assay for *Listeria monocytogenes*. *Analytica Chimica Acta* 325:169–174.

Moeller, N., E. Mueller-Seitz, O. Scholz, W. Hillen, A.A. Bergwerff, and M. Petz. 2007. A new strategy for the analysis of tetracycline residues in food stuffs by a surface plasmon resonance biosensor. *European Food Research Technology* 224:285–292.

Mohan, K., V. Kumar, M. Sarkar, and B.S. Prakash. 2010. Temporal changes in endogenous estrogens and expression of behaviours associated with estrus during the periovulatory period in Heatsynch treated Murrah biffaloes (*Bubalus bubalis*). *Tropical Animal Health and Production* 42:21–26.

Montagne, M., and J.L. Marty. 1995. Bi-enzyme amperometric D-lactate sensor using macromolecular NAD+. *Analytica Chimica Acta* 315:297–302.

Moscone, D., R.A. Bernardo, E. Marconi, A. Amine, and G. Palleschi. 1999. Rapid determination of lactulose in milk by microdialysis and biosensors. *Analyst* 124:325–329.

Muller-Renaud, S., D. Dupont, and P. Dulieu. 2003. Quantification of κ-casein in milk by an optical immunosensor. *Food and Agricultural Immunology* 15:265–277.

Muller-Renaud, S., D. Dupont, and P. Dulieu. 2004. Quantification of β-casein in milk and cheese using an optical immunosensor. *Journal of Food and Agricultural Chemistry* 52:659–664.

NDRI. 2008. Development of fiber optic biosensor for detecting urea in milk. In *Annual Report 2007–08 of National Dairy Research Institute*. 51 Karnal (India).

NDRI. 2009. Development of biosensors and micro-techniques for analysis of pesticide residues, aflatoxin, heavy metals and bacterial contamination in milk. In *Annual Report 2008–2009 of National Dairy Research Institute*. 44 Karnal (India).

Nikolelis, D.P., M.G. Simantirakia, C.G. Siontoroua, and K. Toth. 2005. Flow injection analysis of carbofuran in foods using air stable lipid film based acetylcholinesterase biosensor. *Analytica Chimica Acta* 537:169–177.

Nygren, L., A. Sternesjo, and L. Bjorck. 2003. Determination of folate-binding proteins from milk by optical biosensor analysis. *International Dairy Journal* 13:283–290.

Olano, A., M.M. Calvo, and N. Corzo. 1989. Changes in the carbohydrate fraction of milk during heating processes. *Food Chemistry* 31:259–265.

Palmisano F., M. Quinto, R. Rizzi, and P.G. Zambonin. 2001. Flow injection analysis of L-lactate in milk and yoghurt by on-line microdialysis and amperometric detection at a disposable biosensor. *Analyst* 126:866–870.

Pandya, A.J., and M.M.H. Khan. 2006. Buffalo milk utilization for dairy products. In *Handbook of milk of non-bovine mammals*, ed. Y.W. Park and G.F.W. Haenlein, 215–256. Oxford: Blackwell Publishing Ltd.

Paradkar, M.M., R.S. Singhal, and P.R. Kulkarni. 2000. An approach to the detection of synthetic milk in dairy milk: 1. Detection of urea. *International Journal of Dairy Technology* 53:87–91.

Parodi, P.W. 1997. Cow's milk folate binding protein: Its role in folate nutrition. *Australian Journal of Dairy Technology* 52:109–18.

Pena, N., G. Ruiz, A.J. Reviejo, and J.M. Pingarron. 2001. Graphite-Teflon composite bienzyme electrodes for the determination of cholesterol in reversed micelles. Application to food samples. *Analytical Chemistry* 73:1190–1195.

Pilloton, R., and M. Mascini. 1990. Flow analysis of lactose and glucose in milk with an improved electrochemical biosensor. *Food Chemistry* 36:213–222.

Radke, S.M., and E.C. Alocilja. 2005. A high density microelectrode array biosensor for detection of *E. coli* O157:H7. *Biosensors and Bioelectronics* 20:1662–1667.

Rajendran, V., and J. Irudayaraj. 2002. Detection of glucose, galactose and lactose in milk with a microdialysis-coupled flow injection amperometric sensor. *Journal of Dairy Science* 85:1357–1361.

Ram, M.K., P. Bertoncello, H. Ding, S. Paddeu, and C. Nicolini. 2001. Cholesterol biosensors prepared by layer-by-layer technique. *Biosensors and Bioelectronics* 16:849–856.

Rasooly, L., and A. Rasooly. 1999. Real time biosensor analysis of *Staphylococcal enterotoxin A* in food. *International Journal of Food Microbiology* 49:119–127.

Raz, S.R., M.G.E.G. Bremer, M. Giesbers, and W. Norde. 2008. Development of a biosensor microarray towards food screening, using imaging surface plasmon resonance. *Biosensors and Bioelectronics* 24:552–557.

Rinken, T., and H. Riik, 2006. Determination of antibiotic residues and their interaction in milk with lactate biosensor. *Journal of Biochemical and Biophysical Methods* 66:13–21.

Rodriguez-Mozaz, S., M.J. Lopez-de-Alda, and D. Barcelo. 2006. Biosensors as useful tools for environmental analysis and monitoring. *Analytical and Bioanalytical Chemistry* 386:1025–1041.

Sarkar, P., I.E. Tothiel, S.J. Setford, and A.P.F. Turner. 1999. Screen-printed amperometric biosensors for the rapid measurement of L- and D-amino acids. *Analyst* 124:865–70.

Schmidt, A., C. Standfuss-Gabisch, and U. Bilitewski, 1996. Microbial biosensor for free fatty acids using an oxygen electrode based on thick film technology. *Biosensors and Bioelectronics* 2:1139–1145.

Sekine, Y., and E.H. Hall. 1998. A lactulose sensor based on coupled enzyme reactions with a ring electrode fabricated from tetrathiafulven-tetracyanoquinodimatane. *Biosensors and Bioelectronics* 13:995–1005.

Serra, B., M.D. Morales, A.J. Reviejo, E.H. Hall, and J.M. Pingarron. 2005. Rapid and highly sensitive electrochemical determination of alkaline phosphates using a composite tyrosinase biosensor. *Analytical Biochemistry* 336:289–294.

Sharma, S.K., R. Singhal, B.D. Malhotra, N. Sehgal, and A. Kumar. 2004. Lactose biosensor based on Langmuir-Blodgett films of poly(3-hexyl thiophene). *Biosensors and Bioelectronics* 20:651–657.

Skladal, P., M. Mascini, C. Salvadori, and G. Zannoni. 1993. Detection of bacterial contamination in sterile UHT milk using an L-lactate biosensor. *Enzyme and Microbial Technology* 15:508–512.

Sternesjo, A., C. Mellgren, and L. Bjorck. 1995. Determination of sulfamethazine residues in milk by a surface plasmon resonance-based biosensor assay. *Analytical Biochemistry* 226:175–181.

Stredansky, M., A. Pizzariello, S. Stredonska, and S. Miertus. 1999. Determination of D-fructose in food stuffs by an improved amperometric biosensor based on a solid binding matrix. *Analytical Communications* 36:57–61.

Takhistov, P. 2006. Biosensors technology for food processing, safety, and packaging. In *Handbook of food science, technology and engineering*, vol. 3, ed. Y.H. Hui, 128-1–128–120. Boca Raton, FL: CRC Press.

Thavarungkul, P., S. Dawan, P. Kanatharana, and P. Asawatreratanakul. 2007. Detecting penicillin G in milk with impedimetric label-free immunosensor. *Biosensors and Bioelectronics* 23:688–694.

Toldra, F., and M. Reig. 2006. Methods for rapid detection of chemical and veterinary drug residues in animal foods. *Trends in Food Science and Technology* 17:482–489.

Trivedi, U.B., D. Lakshminarayana, I.L. Kothari et al. 2009. Potentiometric biosensor for urea determination in milk. *Sensors and Actuators B: Chemical* 140:260–266.

Verma, N., and M. Singh. 2003. A disposable microbial based biosensor for quality control in milk. *Biosensors and Bioelectronics* 18:1219–1224.

Vidal, J.C., E. Garcia-Ruiz, and J.R. Castillo. 2000. Strategies for the improvement of an amperometric cholesterol biosensor based on electropolymerization in flow system: Use of charge-transfer mediators and platinization of the electrode. *Journal of Pharmaceutical and Biomedical Analysis* 24:51–63.

Waswa, J., J. Irudayaraj, and C. DebRoy. 2007. Direct detection of *E.coli* O157:H7 in selected food systems by a surface plasmon resonance biosensor. *LWT-Food Science and Technology* 40:187–192.

WHO (World Health Organization). 2010. Dioxins and their effects on human health. *Fact Sheet No. 225*. WHO. Available at: http://www.who.int/mediacentre/factsheets/fs225/en/index.html (accessed September 1, 2010).

Yadav, H., S. Jain, and P.R. Sinha. 2007. Formation of oligosaccharides in skim milk fermented with mixed dahi cultures, *Lactococcus lactis* ssp. *diacetylactis* and probiotic strains of lactobacilli. *Journal of Dairy Research* 74:154–159.

Yman, I.M., A. Eriksson, M.A. Johnson, and K.E. Hellenas. 2006. Food allergen detection with biosensor immunoassays. *Journal of AOAC International* 89:856–861.

Zaydan, R., M. Dion, and M. Boujtita. 2004. Development of a new method, based on a bioreactor coupled with an L-lactate biosensor, toward the determination of a nonspecific inhibition of L-lactic acid production during milk fermentation. *Journal of Agricultural and Food Chemistry* 52:8–14.

Zhang, Y., S.B. Muench, H. Schulze et al. 2005. Disposable biosensor test for organophosphate and carbamate insecticides in milk. *Journal of Agricultural and Food Chemistry* 53:5110–5115.

10 Electrochemical Biosensors as a Tool for the Determination of Phenolic Compounds and Antioxidant Capacity in Foods and Beverages

Montserrat Cortina-Puig, Thierry Noguer, Jean-Louis Marty, and Carole Calas-Blanchard

CONTENTS

10.1 INTRODUCTION

Free radicals are highly reactive molecules with unpaired electrons that can be generated *in vivo* during metabolic processes. Reactive oxygen species (ROS) represent the most important class of free radicals generated in living systems. It is accepted that ROS play different roles *in vivo*. Some are positive and are related to their involvement in energy production, phagocytosis, regulation of cell growth and intercellular signaling, and synthesis of biologically important compounds (Halliwell 1997). However, ROS may be very damaging, since they can attack lipids in cell membranes, proteins in tissues or enzymes, carbohydrates, and DNA to induce oxidations, which cause membrane damage, protein modification (including to enzymes), and DNA damage. This oxidative damage is believed to play a causative role in aging and several degenerative diseases associated with it, such as heart disease, cataracts, cognitive dysfunctions, and cancer (Wattanapitayakul and Bauer 2001; Valcour and Shiramizu 2004; Marchetti et al. 2006; Roessner et al. 2008). Humans have evolved with antioxidant systems to protect against free radicals. These systems include some antioxidants produced in the body and others obtained from the diet. Due to the incomplete efficiency of the endogenous defense system and the existence of some environmental or behavioral stressors (pollution, sunlight exposure, cigarette smoking, excessive alcohol consumption, etc.), in which ROS are produced in excess, dietary antioxidants are needed for diminishing the cumulative effects of oxidative damage (Halliwell 1994). It has been suggested that the intake of fruits and vegetables is associated with a low risk of cancer and cardiovascular diseases (Knekt et al. 1996).

The determination of free radicals and antioxidants has been widely investigated in food technology. Traditional techniques such as spectrophotometry, fluorescence, and gas or liquid chromatography are being replaced by other innovative technologies. In this direction, electrochemical biosensors are promising tools because they exhibit advantages such as minimal sample preparation, selectivity, sensitivity, reproducibility, relatively low cost, rapid time of response, and simple use for continuous on-site analysis. Whereas in the medical field the main objective is the evaluation of the ability of some compounds to scavenge free radicals, in food science, research aims to detect and quantify them. In this sense, two different kinds of biosensors are reported in the antioxidant domain. On the one hand, several amperometric biosensors for the detection of mono- and polyphenols (the main antioxidant compounds in food) have been developed on the basis of enzymes such as tyrosinase, laccase, or peroxidase. On the other hand, biosensors for the assessment of the antioxidant capacity are based on the free-radical scavenging activity. This review is focused on both electrochemical biosensors.

In the following sections, the different biosensors developed for the determination of both the total phenol content and the ROS scavenging capacity in foods and beverages are described.

10.2   BIOSENSORS FOR DETERMINATION OF THE TOTAL PHENOL CONTENT

Polyphenolic compounds, which are widely distributed in plant-derived foods including wine, tea, cacao, and fruit, recently attracted much attention because of their possible health benefits arising from their antioxidant activity, such as free-radical scavengers and inhibition of lipoprotein oxidation (Sánchez-Moreno et al. 1999; Lu and Yeap Foo 2000). They can be broadly divided into two categories, flavonoid and nonflavonoid polyphenols (Ratnam et al. 2006). Flavonoids, the target class of polyphenols, may be divided into different subclasses according to the degree of oxidation of the heterocyclic ring: flavonols, flavanones, flavones, anthocyanidins, catechins, and isoflavones. On the other hand, the nonflavonoids (phenolic acids) may be classified into hydrobenzoic acids and hydroxycinnamic acids.

Due to their antioxidant properties, the scientific community has shown a great interest in the development of analytical methods allowing the estimation of the *total phenol content*. Different electrochemical biosensors based on polyphenol oxidases (tyrosinase or laccase) or peroxidase have been proposed for the detection of phenolic compounds. They are essentially monoenzymatic systems, although some bi- or trienzymatic sensors combining the previously cited enzymes have also been described. Table 10.1 summarizes the developed biosensors for the determination of phenolic compounds in foods or beverages.

10.2.1 POLYPHENOL OXIDASE-BASED BIOSENSORS

Both tyrosinase and laccase enzymes, commonly referenced as polyphenol oxidases, are copper-containing oxidoreductases, which catalyze the oxidation of various mono- and polyphenolic compounds. In the presence of oxygen, tyrosinase catalyzes the oxidation of monophenols and *o*-diphenols to quinones, whereas laccase catalyzes the oxidation of a larger variety of aromatic compounds, such as substituted mono- and polyphenols, aromatic amines and thiol compounds, with subsequent production of radicals. These active species can be converted to quinones in the second stage of oxidation.

10.2.1.1 Tyrosinase-Based Biosensors

Tyrosinase is the most common enzyme used for the determination of phenolic compounds in food samples with biosensors. The enzyme is predominantly immobilized according to two different procedures based on its physical entrapment in a matrix or on its covalent cross-linking immobilization with glutaraldehyde. Tyrosinase-based biosensors can be classified according to the detection system involved in the analytical device. Some of them are based on the amperometric detection of the oxygen consumption using a Clark-type electrode, whereas others consist on the reversible electrochemical reduction of the *o*-quinone formed from phenols in the enzymatic reaction. In both cases, the measured signal is proportional to the polyphenol concentration in solution.

By using the first detection strategy, Campanella and coworkers (Campanella et al. 1999, 2008) developed a biosensor for the phenol analysis in olive oils capable of operating in organic solvents, the tyrosinase being immobilized by entrapment in a kappa-carrageenan gel. On one hand, they evaluated the progressive rancidification of three different types of olive oil (olive oil, virgin olive oil, and extra virgin olive oil) by simultaneously using two different indicators: a classic one consisting of the peroxide number, and an innovative one consisting of the progressive decrease in the content of polyphenols (Campanella et al. 1999). It was demonstrated that this biosensor could be used to supplement the classical measurement of the peroxide number as a new indicator of an oil's stability to oxidation. On the other hand, the polyphenols' deterioration of extra virgin olive oils was also studied during heating in an

TABLE 10.1

Biosensors for the Detection of Polyphenols in Foods and Beverages

Bioelement	Immobilization	Detection	Electrode	Sample	Reference
Tyrosinase	Physical entrapment	O_2 consumption	Clark-type electrode	Olive oil	(Campanella et al. 1999)
Tyrosinase	Covalent cross-linking	O_2 consumption	Clark-type electrode	Olive oil	(Capannesi et al. 2000)
Tyrosinase	Physical entrapment	O_2 consumption	Clark-type electrode	Red and white wines	(Campanella et al. 2004b)
Tyrosinase	Covalent cross-linking	O_2 consumption	Clark-type electrode	Tea	(Abhijith et al. 2007)
Tyrosinase	Physical entrapment	O_2 consumption	Clark-type electrode	Dry spices	(Bonanni et al. 2007)
Tyrosinase	Physical entrapment	O_2 consumption	Clark-type electrode	Extra virgin olive oil	(Campanella et al. 2008)
Tyrosinase	Physical entrapment	Quinone reduction	CPE	Beer	(Eggins et al. 1997)
Tyrosinase	Covalent cross-linking	Quinone reduction	Graphite SPE	Green tea	(Romani et al. 2000)
Tyrosinase	Physical entrapment	Quinone reduction	CPE	Wine	(Jewell and Ebeler 2001)
Tyrosinase	Covalent cross-linking	Quinone reduction	nAu-GCE	Wine	(Carralero Sanz et al. 2005)
Tyrosinase	Covalent cross-linking	Oxidation of ferrocene mediator	Ferrocene-modified carbon SPE	Wine	(Montereali et al. 2005)
Laccase	Covalent cross-linking	Quinone reduction	GCE	Wine	(Gamella et al. 2006)
Laccase	Physical retention	Quinone reduction	Pt	Red wine	(Gomes et al. 2004)
Laccase	Physical retention	Quinone reduction	CPE	Plant extracts	(Franzoi et al. 2009)
Laccase	Covalent cross-linking	Quinone reduction	Sonogel-carbon electrode	Beer	(ElKaoutit et al. 2008)
Laccase + Tyrosinase	Covalent cross-linking	Quinone reduction	Sonogel-carbon electrode	Beer	(Elkaoutit et al. 2007)
Peroxidase	SAM	Quinone reduction	Gold electrode	Wine, tea	(Imabayashi et al. 2001)
Peroxidase	Physical cross-linking	Quinone reduction	Silica-titanium CPE	Vegetable extracts	(Mello et al. 2003)
Peroxidase	Physical cross-linking	Quinone reduction	Silica-titanium CPE	Tea	(Mello et al. 2005)

Note: SPE: Screen-printed electrode; CPE: carbon paste electrode; GCE: glassy carbon electrode; nAu: gold nanoparticles; SAM: self-assembled monolayer.

oxidizing atmosphere (Campanella et al. 2008). This tyrosinase-based biosensor was also applied to the quantification of polyphenols in other different matrices such as dry spices (Bonanni et al. 2007) and red and white wines (Campanella et al. 2004b).

The same strategy, based on the Clark-type electrode, was used by Capannesi et al. (2000) for the monitoring of the polyphenols in oil samples. In this case the enzyme was covalently cross-linked with glutaraldehyde and immobilized on a cellophane membrane. This biosensor was applicable both in batch conditions and in a flow injection analysis system. Additionally, another research group developed a comparable sensor for evaluating the tea quality in terms of polyphenol content (Abhijith et al. 2007).

In the second proposed detection strategy, a more direct approach is considered. In the presence of tyrosinase, catechol and its derivatives are oxidized by dissolved oxygen into the corresponding quinones, which are electrochemically reduced on the surface of the working electrode polarized at the appropriate potential. Using this principle, several biosensors have been developed in order to detect polyphenolic compounds in natural complex matrices. Romani et al. (2000) determined the total polyphenolic content of green tea, which was in accordance with the values determined using high-performance liquid chromatography (HPLC) and differential pulse voltammetry. However, they could not test other natural extracts (olive and grape skin extracts) because these anthocyanic fractions were unstable at neutral pH, necessary for optimum tyrosinase performances. In this case, tyrosinase was covalently immobilized with glutaraldehyde on graphite screen-printed electrodes. By using the same immobilization technique, other researchers modified the carbon electrodes with electrodeposited gold nanoparticles (Carralero Sanz et al. 2005) or with a ferrocene mediator (Montereali et al. 2005). Both biosensors were applied to the measurement of polyphenols index in wines.

Considering that tyrosinase can be found in various fruits and vegetables, some biosensors were developed incorporating natural plant tissues in carbon paste electrodes. Taking advantage of this characteristic, Eggins et al. (1997) used banana-, potato-, and apple-based sensors for the determination of total flavanols in commercial beers. The same approach was employed by Jewell and Ebeler (2001) to evaluate the polyphenolic content in red and white wines.

10.2.1.2 Laccase-Based Biosensors

One of the most important enzymes in terms of applicability and versatility in the food industry is laccase because it can catalyze the oxidation not only of phenol and o-diphenols but also m- and p-diphenols to the corresponding quinones or radical species while oxygen is reduced to water. For this reason laccase-based biosensors are especially appropriate for the determination of the total phenol content.

In contrast to tyrosinase-based biosensors, the only detection method involved in laccase-based biosensors is the reduction of the quinones formed during the catalytic oxidation of the phenols.

Several biosensors have been developed based on different techniques for the immobilization of the enzyme. Gomes et al. (2004) developed a biosensor based on the physical retention of the enzymatic solution on a Pt electrode by a polyethersulfone membrane. This laccase-based sensor was applied to the determination of polyphenolic compounds in a red wine. By applying two different potentials, this biosensor was able to discriminate between catechin and caffeic acid when testing artificial mixtures. However, due to the complexity of the red wine matrix, its direct application on commercial samples was unsuccessful. Consequently, a preliminary step of solid phase extraction was required to improve the detection, slightly overcoming some interference. Laccase was also immobilized onto a glassy carbon electrode by cross-linking with glutaraldehyde (Gamella et al. 2006). This biosensor exhibited a good analytical performance allowing a convenient estimation of the polyphenol index in white, rosé, and red wines. Recently, another biosensor was constructed using laccase immobilized in carbon paste mixed with ionic liquids for determination of rosmarinic acid in plant extracts (Franzoi et al. 2009). The concentrations determined for lemon balm were in good agreement with those obtained using capillary electrophoresis.

On the other hand, Elkaoutit et al. (2008) developed three different enzymatic electrodes based on laccase, tyrosinase, or peroxidase for the polyphenol monitoring in beers. The immobilization step was accomplished by doping a sonogel-carbon electrode with a mixture of the individual enzyme and Nafion ion exchanger as additive-protective. After comparing their analytical performances, the laccase-based biosensor was selected as better achieving the polyphenol index determination in commercial lager beers. The same immobilization procedure was followed to develop a biosensor based on the coimmobilization of both laccase and tyrosinase enzymes in order to estimate the total polyphenol index in some beer samples (Elkaoutit et al. 2007). The analytical performances of this bienzymatic biosensor were enhanced when compared with the monoenzymatic one, although in both cases the obtained results were correlated to those obtained by the classical Folin-Ciocalteu method.

10.2.2 PEROXIDASE-BASED BIOSENSORS

Horseradish peroxidase (HRP) biosensors are based on the well-known fact that phenolic compounds can act as electron donors for peroxidase in the catalytic reduction of peroxide. The amount of phenols can then be detected as the reduction current of the oxidized phenols generated by the enzyme reaction cycle with H_2O_2. Considering the higher electron-donating ability of polyphenols compared with phenols, it is expected that polyphenol compounds can be detected using an HRP-immobilized biosensor system. In contrast to tyrosinase- and laccase-based biosensors, these biosensors require the external addition of H_2O_2 as a cosubstrate for the catalytic reaction.

Two different kinds of HRP biosensors have been developed based on the different HRP immobilization technique. On one hand, Imabayashi et al. (2001) developed an amperometric biosensor using HRP covalently immobilized on a self-assembled monolayer (SAM)-modified gold electrode for the polyphenol detection in wine and tea samples. The total amounts of polyphenols were well correlated with those determined by the Folin-Ciocalteu method. On the other hand, Mello et al. (2003) presented a biosensor based on HRP immobilized onto a silica-titanium carbon paste electrode and its application for measuring the polyphenolic compounds in teas and coffees. This biosensor showed a good suitability when compared with the Folin-Ciocalteu traditional method. Furthermore, a good correlation was obtained between the total phenol content determined with this biosensor and the total antioxidant activity investigated by the 1,1-diphenyl-2-picrylhydrazyl (DPPH) radical-scavenging method (Mello et al. 2005).

10.3 BIOSENSORS FOR DETERMINATION OF THE REACTIVE OXYGEN SPECIES (ROS) SCAVENGING CAPACITY

The presence of antioxidant substances in a sample where radicals are generated involves their decomposition. Therefore, evaluation of the antioxidant capacity of different compounds can be determined based on the variation of the ROS concentration in the reaction medium. In living systems, superoxide radicals ($O_2^{\cdot-}$) and hydroxyl radicals (OH$^\bullet$) are the main ROS produced in normal metabolism (Valentao et al. 2002). There are several processes to produce them.

$O_2^{\cdot-}$ is mainly produced as an intermediate of the oxidation reaction of xanthine or hypoxanthine to uric acid in the presence of the enzyme xanthine oxidase (XOD) (Fridovich 1978):

$$\text{xanthine} + O_2 + H_2O \xrightarrow{\text{XOD}} \text{uric acid} + 2H^+ + O_2^{\cdot-} \tag{10.1}$$

Hydroxyl radicals (OH$^\bullet$) can be generated by the Fenton reaction, where reduced transition metal ions, such as Fe(II), Cu(I), or Cr (II), react with H_2O_2 in a one-electron redox reaction producing OH$^\bullet$ and hydroxide anion:

$$Me^n + H_2O_2 \rightarrow Me^{n+1} + OH^- + OH^\bullet \tag{10.2}$$

The addition of a reducing agent increases the radical generation rate. Alternatively, the transition metal can be reduced by the application of an appropriate electrode potential (Fojta et al. 2000).

OH$^\bullet$ can also be produced during TiO_2 photocatalysis. The radical generation starts with the absorption of light of a wavelength higher than its band gap by TiO_2, which results in the transition of an electron from the valence band (VB) to the conduction band (CB), leaving a hole behind (Equation 10.3). Then, adsorbed water or hydroxide ions are trapped by holes to produce OH$^\bullet$ (Equation 10.4 and Equation 10.5). Subsequently, electrons are trapped by the reaction with adsorbed O_2 to produce superoxide radical ($O_2^{\cdot-}$) (Equation 10.6), which then forms more OH$^\bullet$ (Equation 10.7) (Nagaveni et al. 2004):

$$TiO_2 + h\nu \rightarrow h_{VB}^+ + e_{CB}^- \tag{10.3}$$

$$h_{VB}^+ + H_2O_{(ads)} \rightarrow OH^\bullet + H^+ \tag{10.4}$$

$$h_{VB}^+ + 2OH_{(ads)}^- \rightarrow OH^- + OH^\bullet \tag{10.5}$$

$$e_{CB}^- + O_2 \rightarrow O_2^{\cdot-} \tag{10.6}$$

$$O_2^{\cdot-} + 2H_2O \rightarrow 2OH^\bullet + 2OH^- + O_2 \tag{10.7}$$

With the aim to assess antioxidant capacity based on the measurement of $O_2^{\bullet-}$ concentration, two main types of biosensors have been developed, using cytochrome c (cyt c) or superoxide dismutase (SOD) enzyme. $O_2^{\bullet-}$ determination using a cyt c-based sensor lacks in selectivity, since this heme protein is not specific for $O_2^{\bullet-}$; moreover, its inherent property as a peroxidase, able to reduce H_2O_2 endogenously coexisting in biological systems, greatly limits its application for detection of $O_2^{\bullet-}$ in real samples. SOD-based biosensors, on the contrary, are much more specific and sensitive.

Another method to determine the antioxidant capacity is by measuring the damage produced to DNA by free radicals. In this case, the presence of antioxidants involves a decrease in DNA alterations.

Many electrochemical biosensors for the assessment of the antioxidant capacity, based on the determination of the free-radical scavenging activity, have been developed. However, only few of them have been applied in foods or beverages. Table 10.2 summarizes the biosensors based on the determination of both $O_2^{\bullet-}$ and OH$^{\bullet}$ radicals that have been applied to evaluate the antioxidant capacity in the food industry.

10.3.1 CYT c-BASED BIOSENSORS

The detection principle of these biosensors is based on the redox reaction between cyt c-(Fe^{3+}) and $O_2^{\bullet-}$ followed by the further oxidation of the cyt c-(Fe^{2+}) at the electrode surface, leading to an oxidation current proportional to the radical concentration in solution (Tammeveski et al. 1998). To avoid interference from H_2O_2, generated as a final product of the enzymatic reaction or by the spontaneous dismutation of $O_2^{\bullet-}$ (Equation 10.8), catalase enzyme is sometimes added to the reaction media. The addition of antioxidants reduces the radical concentration and thus the oxidation current, allowing the quantification of the antioxidant capacity.

$$2O_2^{\bullet-} + 2H^+ \rightarrow O_2 + H_2O_2 \tag{10.8}$$

Usually, the electron transfer between the electrode and the redox protein is extremely slow. Consequently, traditional electrochemical methods cannot detect it (Krylov et al. 2004). An elegant way to modify gold electrodes, most commonly used in this area, is the formation of SAMs. Short-chain alkanethiols show a high efficiency of communication between cyt c and the electrode (Manning et al. 1998; Tammeveski et al. 1998). However, they cannot form dense films and thus do not effectively block the electrode from interfering substances. To eliminate electroactive interferences, such as H_2O_2 and uric acid, cyt c has been immobilized on long-chain alkanethiols-modified electrodes. These cyt c-based biosensors can be used for $O_2^{\bullet-}$ detection and for the analysis of antioxidant activities (Lisdat et al. 1999a,b; Ignatov et al. 2002). To increase the electron transfer rate of cyt c, Ge and Lisdat (2002) employed mercaptoundecanoic acid/mercaptoundecanol mixed SAM-modified gold electrodes. This procedure had already been described and applied by Gobi and Mizutani (2000, 2001) to the $O_2^{\bullet-}$ sensing. However, they still used short-chain modifiers, which limited the selectivity. These mixed SAM-modified biosensors showed more sensitivity to $O_2^{\bullet-}$ (Beissenhirtz et al. 2003).

TABLE 10.2

Biosensors Applied to the Determination of the ROS Scavenging Capacity

Bioelement	Immobilization	Detection Probe	Electrode	Sample	Reference
Cyt c	SAM	$O_2^{\cdot-}$	Gold wire electrode	Chinese tonifying herbs	(Beissenhirtz et al. 2004)
Cyt c/XOD	SAM	$O_2^{\cdot-}$	Gold SPE	Garlic samples	(Cortina-Puig et al. 2009a)
Cyt c/XOD	SAM	$O_2^{\cdot-}$	Gold SPE	Orange juices	(Cortina-Puig et al. 2009b)
SOD	Physical adsorption	H_2O_2	H_2O_2 electrode	Red and white wines	(Campanella et al. 2004b)
SOD	Physical adsorption	H_2O_2	H_2O_2 electrode	Plant extracts	(Campanella et al. 2001b)
SOD	Physical adsorption	H_2O_2	H_2O_2 electrode	Tea samples and herbal products	(Campanella et al. 2003b)
SOD	Physical adsorption	H_2O_2	H_2O_2 electrode	Dry spices	(Bonanni et al. 2007)
SOD	Physical adsorption	H_2O_2	H_2O_2 electrode	Aromatic herbs, olives, and fresh fruit	(Campanella et al. 2003a)
SOD	Physical adsorption	H_2O_2	H_2O_2 electrode	Algae samples	(Campanella et al. 2005)
SOD	Physical adsorption	H_2O_2	H_2O_2 electrode	Phytotherapeutic diet integrators	(Campanella et al. 2004a)
SOD	Physical adsorption	H_2O_2	H_2O_2 electrode	Papaya fruit and papaya-based food integrators	(Campanella et al. 2009)
SOD	Physical adsorption	O_2	Clark-type electrode	Extra virgin olive oil	(Amati et al. 2008)
dsDNA	Physical adsorption	$OH^{\bullet}$	Carbon SPE	Plant extracts	(Mello et al. 2006)
dsDNA/ $[Co(phen)_3]^{3+}$	Physical adsorption	$OH^{\bullet}$	Carbon SPE	Yeast polysaccharides	(Bučková et al. 2002)
dsDNA/ $[Co(phen)_3]^{3+}$	—	$OH^{\bullet}$	GCE	Plant extracts	(Labuda et al. 2002)
Guanine/ adenine	—	$OH^{\bullet}$	GCE	Flavored waters	(Kamel et al. 2008)

Note: SPE: Screen-printed electrode; CPE: carbon paste electrode; GCE: glassy carbon electrode; nAu: gold nanoparticles; SOD: superoxide dismutase.

Recently, we developed an enzymatic biosensor based on the coimmobilization of both cyt c and XOD on a mercaptoundecanoic acid/mercaptoundecanol mixed SAM-modified gold electrode. By using this approach, better sensitivities were achieved because cyt c and XOD were very close and the generated $O_2^{\cdot-}$ radicals could quickly react with cyt c, avoiding its spontaneous dismutation. This biosensor has been successfully applied to the determination of the antioxidant capacity of different orange juices and both alliin and allicin extracted from garlic bulbs (Cortina-Puig et al. 2009a,b).

Apart from *in vitro* analysis, cyt *c*–based biosensors allow the measurement of the $O_2^{\bullet-}$ produced *in vivo*, for example, during ischemia and reperfusion injury. Beissenhirtz et al. (2004) compared the *in vitro* $O_2^{\bullet-}$ scavenging activity and the *in vivo* antioxidant potential of methanolic extracts prepared from 10 Chinese tonifying herbs. Results did not show quantitative correlation. However, for 8 out of 10 samples a similar tendency was found.

10.3.2 Superoxide Dismutase-Based Biosensors

SOD biosensors are shown as a promising alternative to cyt *c* biosensors for the evaluation of the antioxidant capacity. Compared with cyt *c*, SOD is more specific for $O_2^{\bullet-}$, and the rate of the reaction between $O_2^{\bullet-}$ and SOD is several orders of magnitude higher. SOD has been well addressed for the dismutation of the $O_2^{\bullet-}$ radical into O_2 and H_2O_2 with strong activity and great specificity.

$$2O_2^{\bullet-} + 2H^+ \xrightarrow{\text{SOD}} O_2 + H_2O_2 \qquad (10.9)$$

First-generation SOD-based sensors are mainly based on the detection of the H_2O_2 oxidation at the electrode surface. Nevertheless, this oxidation reaction occurs in a high potential (>0.5 V vs. Ag/AgCl), which results in interference problems, limiting the practical applications of these biosensors. Strategies to improve the selectivity of the developed sensors include the use of an H_2O_2-impermeable Teflon membrane (Song et al. 1995) or two-channel sensors (Lvovich and Scheeline 1997). The latter method allows the simultaneous determination of $O_2^{\bullet-}$ and H_2O_2.

Campanella and coworkers (Campanella et al. 2001b, 2003a,b, 2004a,b, 2005, 2009; Bonanni et al. 2007) have been working on the development of different SOD-based biosensors for assessing the antioxidant capacity of several compounds. Most of their work is based on the immobilization of SOD in a kappa-carrageenan gel and the amperometric detection of H_2O_2. The gel containing the enzyme is sandwiched between a cellulose acetate membrane, which improves the selectivity of the electrode by blocking the access to possible electroactive interferences, and an external dialysis membrane. The addition of a sample possessing antioxidant properties produces a decrease in the signal strength as the antioxidant species react with the $O_2^{\bullet-}$ radical, thus reducing its concentration in solution. There is a consequent decrease in the H_2O_2 released and thus also in the intensity of the amperometric signal. This biosensor has been used to evaluate red and white wines (Campanella et al. 2004b), fresh aromatic herbs, olives and fresh fruit, bulbs and vegetables, plant products sold by herbalists and/or pharmacies, tea (Campanella et al. 2001b, 2003a,b), dry spices (Bonanni et al. 2007), algae (Campanella et al. 2005), phytotherapeutic diet integrators (Campanella et al. 2004a), and papaya fruit and papaya-based food integrators (Campanella et al. 2009).

However, it is known that a large number of molecules with interesting scavenging properties are difficult to determine because of their very low solubility in water. This latter point leads the same authors to the development of a biosensor capable of operating in organic solvents, in which these compounds are more soluble. The developed O_2^- biosensor, successfully used in a dimethyl sulfoxide (DMSO) solution, is based on SOD entrapped in a cellulose triacetate layer (sandwiched between two gas-permeable membranes) or in a kappa-carrageenan gel layer (sandwiched between an external gas-permeable membrane and an internal cellulose acetate membrane), coupled to an O_2 amperometric transducer (Campanella et al. 2001a). This biosensor has been applied to check the free-radical concentration of extra virgin olive oil (Amati et al. 2008).

These biosensors provide a reliable method to measure the antioxidant capacity, as it has been demonstrated by comparing results with other detection methods, for example, cyclic and pulse voltammetry, spectrophotometry (N,N-dimethyl-p-phenylenediamine-FeCl$_3$ method) and fluorimetry (oxygen radical absorbance capacity, ORAC, method), the last one often considered as the reference method. Moreover, the developed biosensors are robust, easily miniaturizable, can perform analyses *in situ*, and do not require expensive or sophisticated equipment.

10.3.3 DNA-BASED BIOSENSORS

Nucleic acids can be oxidized by various oxidants and in particular by OH$^\bullet$ radicals. Cellular DNA oxidation by ROS always results in DNA bases being damaged, which has been acknowledged as a significant source of mutations leading to cancer, premature aging, and other degenerative diseases (Portugal and Waring 1987; Jaruga and Dizdaroglu 1996; Evans and Cooke 2004).

Taking advantage of this property, several DNA-based sensors have been developed for the measurement of the antioxidant capacity of different compounds. In many cases the Fenton reaction is used as the method of inducing damage by generating OH$^\bullet$ radicals.

The simplest strategy is based on the immobilization of double-stranded DNA (dsDNA), usually from *calf thymus*, on screen-printed carbon electrodes by simple adsorption, and the detection of the guanine oxidation peak by square wave voltammetry (SWV). Since the peak current intensity is proportional to the guanine concentration, the immersion of the DNA-modified electrode into a Fenton solution produces a signal decrease in the peak current intensity. The introduction of antioxidants into the Fenton solution results in the scavenging of the OH$^\bullet$. Consequently, after immersion of the electrode into this solution, the peak current intensity is very close to the original one, which demonstrates the DNA integrity. The results demonstrated that the DNA-based biosensor was suitable as a rapid screening test for the evaluation of antioxidant properties of different plant extracts (Mello et al. 2006).

Another way to detect the DNA damage is to use an electrochemical label able to interact with dsDNA. This interaction is based on an intercalation phenomenon (predominantly at high ionic strength) or on an electrostatic interaction (predominantly at low ionic strength). The amount of redox label bound to the DNA layer, which can be measured by differential pulse voltammetry (DPV), decreases proportionally to the concentration of radicals present in the sample (Labuda et al. 1999; Korbut et al. 2001). As before, the presence of antioxidants recovers the electrochemical signal. By using this strategy, Bučková et al. (2002) used [Co(phen)$_3$]$^{3+}$ as a redox marker to evaluate the antioxidant capacity of different yeast polysaccharides. The same approach was used by Labuda et al. (2002) to assess the antioxidant activity of plant extracts containing phenolic compounds.

Recently, Kamel et al. (2008) electrochemically immobilized guanine and adenine as DNA bases on glassy carbon electrodes to determine the antioxidant capacity of flavored water samples. The method relies on monitoring the changes of the intrinsic anodic response of the surface-confined guanine and adenine species, resulting from its interaction with free radicals from Fenton-type reaction in the absence and presence of antioxidant.

DNA sensors are promising devices to perform simple tests for the routine evaluation of the antioxidant capacity of samples in an easy way. Moreover, the choice of screen-printed carbon as an electrode material leads to disposable and low-cost analysis tools.

10.4 CONCLUSION

This chapter describes in detail the electrochemical biosensors for the detection of antioxidants in the food and beverage industries. Two different kinds of biosensors are described: those based on the total phenol content determination and those based on the antioxidant capacity measurement.

Biosensors based on polyphenol oxidase or peroxidase have been proposed for the detection of phenolic compounds, which are widely distributed in foods. The most sensitive phenolic biosensors are based on tyrosinase. However, these devices usually present short lifetimes due to the low stability of this enzyme. As an alternative to tyrosinase, laccase was used. Laccase-based biosensors present rather broad substrate specificity, although they are characterized by detection limits higher than those achieved with tyrosinase biosensors. On the other hand, peroxidase-based biosensors have been proposed to be sensitive to a wider range of phenolic compounds compared with the polyphenol oxidase biosensors, because some phenols without the active structure to polyphenol oxidase cannot be monitored by the latter. However, in contrast to tyrosinase and laccase-based biosensors, peroxidase sensors require the external addition of H_2O_2 as a substrate for the catalytic reaction.

Biosensors based on the antioxidant capacity measurement deserve special attention because they provide information about the real antioxidant capacity of a compound, enabling the exploitation of its beneficial properties. Many biosensors based on the determination of both $O_2^{\cdot-}$ and $OH^\bullet$ have been applied to evaluate the antioxidant capacity of different compounds. For the determination of $O_2^{\cdot-}$, usually generated through the xanthine/XOD enzymatic system, cyt c and SOD biosensors are commonly used. Whereas all cyt c biosensors are based on the direct electron transfer between the immobilized redox protein and the electrode surface, promoted by SAMs, only first-generation SOD sensors have been applied to evaluate the antioxidant capacity. Nevertheless, SOD-based biosensors are more sensitive and, in principle, more selective. In order to determine $OH^\bullet$, mainly generated by the Fenton reaction, DNA-based sensors have been developed. These sensors are based on the damage induced to DNA by free radicals, and they are shown to be an attractive alternative to assess antioxidant capacity.

The applicability of electrochemical biosensors for the detection of antioxidants in the food and beverage industries is demonstrated. There is a growing interest in the development of such devices, as demonstrated by the numerous configurations that have recently appeared in the literature. Their advantages in terms of simplicity, rapidity, and low cost with respect to traditional techniques promote their investigation and exploitation. Nevertheless, further work is required to avoid the interferences problem and to consolidate them as practical and current antioxidant assessment tools.

ACKNOWLEDGMENT

The authors greatly acknowledge the European Commission for financial support through the project Nutra-Snacks (FOOD-CT-2005–023044).

REFERENCES

Abhijith, K.S., P.V. Sujith Kumar, M.A. Kumar, and M.S. Thakur. 2007. Immobilised tyrosinase-based biosensor for the detection of tea polyphenols. *Analytical and Bioanalytical Chemistry* 389:2227–2234.

Amati, L., L. Campanella, R. Dragone, A. Nuccilli, M. Tomassetti, and S. Vecchio. 2008. New investigation of the isothermal oxidation of extra virgin olive oil: Determination of free radicals, total polyphenols, total antioxidant capacity, and kinetic data. *Journal of Agricultural and Food Chemistry* 56:8287–8295.

Beissenhirtz, M., F. Scheller, and F. Lisdat. 2003. Immobilized cytochrome *c* sensor in organic/aqueous media for the characterization of hydrophilic and hydrophobic antioxidants. *Electroanalysis* 15:1425–1435.

Beissenhirtz, M.K., R.C.H. Kwan, K.M. Ko, R. Renneberg, F.W. Scheller, and F. Lisdat. 2004. Comparing an *in vitro* electrochemical measurement of superoxide scavenging activity with an *in vivo* assessment of antioxidant potential in Chinese tonifying herbs. *Phytotherapy Research* 18:149–153.

Bonanni, A., L. Campanella, T. Gatta, E. Gregori, and M. Tomassetti. 2007. Evaluation of the antioxidant and prooxidant properties of several commercial dry spices by different analytical methods. *Food Chemistry* 102:751–758.

Bučková, M., J. Labuda, J. Sandula, L. Krizkova, I. Stepanek, and Z. Durackova. 2002. Detection of damage to DNA and antioxidative activity of yeast polysaccharides at the DNA-modified screen-printed electrode. *Talanta* 56:939–947.

Campanella, L., A. Bonanni, and M. Tomassetti. 2003b. Determination of the antioxidant capacity of samples of different types of tea, or of beverages based on tea or other herbal products, using a superoxide dismutase biosensor. *Journal of Pharmaceutical and Biomedical Analysis* 32:725–736.

Campanella, L., E. Martini, and M. Tomassetti. 2005. Antioxidant capacity of the algae using a biosensor method. *Talanta* 66:902–911.

Campanella, L., A. Bonanni, D. Bellantoni, and M. Tomassetti. 2004a. Biosensors for determination of total antioxidant capacity of phytotherapeutic integrators: Comparison with other spectrophotometric, fluorimetric and voltammetric methods. *Journal of Pharmaceutical and Biomedical Analysis* 35:303–220.

Campanella, L., A. Bonanni, G. Favero, and M. Tomassetti. 2003a. Determination of antioxidant properties of aromatic herbs, olives and fresh fruit using an enzymatic sensor. *Analytical and Bioanalytical Chemistry* 375:1011–1016.

Campanella, L., A. Bonanni, E. Finotti, and M. Tomassetti. 2004b. Biosensors for determination of total and natural antioxidant capacity of red and white wines: Comparison with other spectrophotometric and fluorimetric methods. *Biosensors and Bioelectronics* 19:641–651.

Campanella, L., G. Favero, M. Pastorino, and M. Tomassetti. 1999. Monitoring the rancidification process in olive oils using a biosensor operating in organic solvents. *Biosensors and Bioelectronics* 14:179–186.

Campanella, L., G. Favero, L. Persi, and M. Tomassetti. 2001b. Evaluation of radical scavenging properties of several plants, fresh or from a herbalist's, using a superoxide dismutase biosensor. *Journal of Pharmaceutical and Biomedical Analysis* 24:1055–1064.

Campanella, L., T. Gatta, E. Gregori, and M. Tomassetti. 2009. Determination of antioxidant capacity of papaya fruit and papaya-based food and drug integrators, using a biosensor device and other analytical methods. *Monatshefte für Chemie* 140:965–972.

Campanella, L., A. Nuccilli, M. Tomassetti, and S. Vecchio. 2008. Biosensor analysis for the kinetic study of polyphenols deterioration during the forced thermal oxidation of extra-virgin olive oil. *Talanta* 74:1287–98.

Campanella, L., S. De Luca, G. Favero, L. Persi, and M. Tomassetti. 2001a. Superoxide dismutase biosensors working in non-aqueous solvent. *Fresenius' Journal of Analytical Chemistry* 369:594–600.

Capannesi, C., I. Palchetti, M. Mascini, and A. Parenti. 2000. Electrochemical sensor and biosensor for polyphenols detection in olive oils. *Food Chemistry* 71:553–562.

Carralero Sanz, V., M.L. Mena, A. González-Cortés, P. Yáñez-Sedeño, and J.M. Pingarrón. 2005. Development of a tyrosinase biosensor based on gold nanoparticles-modified glassy carbon electrodes: Application to the measurement of a bioelectrochemical polyphenols index in wines. *Analytica Chimica Acta* 528:1–8.

Cortina-Puig, M., X. Muñoz-Berbel, C. Calas-Blanchard, and J.L. Marty. 2009a. Electrochemical characterization of a superoxide biosensor based on the co-immobilization of cytochrome c and XOD on SAM-modified gold electrodes and application to garlic samples. *Talanta* 79:289–294.

Cortina-Puig, M., X. Muñoz-Berbel, R. Rouillon, C. Calas-Blanchard, and J.L. Marty. 2009b. Development of a cytochrome c-based screen-printed biosensor for the determination of the antioxidant capacity of orange juices. *Bioelectrochemistry* 76:76–80.

Eggins, B.R., C. Hickey, S.A. Toft, and D.M. Zhou. 1997. Determination of flavanols in beers with tissue biosensors. *Analytica Chimica Acta* 347:281–288.

Elkaoutit, M., I. Naranjo-Rodriguez, K.R. Temsamani, M.D. De La Vega, and J.L.H.-H. De Cisneros. 2007. Dual laccase-tyrosinase based sonogel-carbon biosensor for monitoring polyphenols in beers. *Journal of Agricultural and Food Chemistry* 55:8011–8018.

Elkaoutit, M., I. Naranjo-Rodriguez, K.R. Temsamani, M.P. Hernández-Artiga, D. Bellido-Milla, and J.L.H.-H. De Cisneros. 2008. A comparison of three amperometric phenoloxidase-sonogel-carbon based biosensors for determination of polyphenols in beers. *Food Chemistry* 110:1019–1024.

Evans, M.D., and M.S. Cooke. 2004. Factors contributing to the outcome of oxidative damage to nucleic acids. *Bioessays* 26:533–542.

Fojta, M., T. Kubicarova, and E. Palecek. 2000. Electrode potential-modulated cleavage of surface-confined DNA by hydroxyl radicals detected by an electrochemical biosensor. *Biosensors and Bioelectronics* 15:107–115.

Franzoi, A.C., J. Dupont, A. Spinelli, and I.C. Vieira. 2009. Biosensor based on laccase and an ionic liquid for determination of rosmarinic acid in plant extracts. *Talanta* 77:1322–1327.

Fridovich, I. 1978. The biology of oxygen radicals. *Science* 201:875–880.

Gamella, M., S. Campuzano, A.J. Reviejo, and J.M. Pingarron. 2006. Electrochemical estimation of the polyphenol index in wines using a laccase biosensor. *Journal of Agricultural and Food Chemistry* 54:7960–7967.

Ge, B., F. Lisdat. 2002. Superoxide sensor based on cytochrome *c* immobilized on mixed-thiol SAM with a new calibration method. *Analytica Chimica Acta* 454:53–64.

Gobi, K.V., and F. Mizutani. 2000. Efficient mediatorless superoxide sensors using cytochrome *c*-modified electrodes: Surface nano-organization for selectivity and controlled peroxidase activity. *Journal of Electroanalytical Chemistry* 484:172–181.

Gobi, K.V., and F. Mizutani. 2001. Amperometric detection of superoxide dismutase at cytochrome *c*-immobilized electrodes: Xanthine oxidase and ascorbate oxidase incorporated biopolymer membrane for *in-vivo* analysis. *Analytical Sciences* 17:11–15.

Gomes, S.A.S.S., J.M.F. Nogueira, and M.J.F. Rebelo. 2004. An amperometric biosensor for polyphenolic compounds in red wine. *Biosensors and Bioelectronics* 20:1211–1216.

Halliwell, B. 1994. Free radicals, antioxidants, and human disease: Curiosity, cause, or consequence? *Lancet* 344:721–724.

Halliwell, B. 1997. Antioxidants and human disease: A general introduction. *Nutrition Reviews* 55:S44–S52.

Ignatov, S., D. Shishniashvili, B. Ge, F.W. Scheller, and F. Lisdat. 2002. Amperometric biosensor based on a functionalized gold electrode for the detection of antioxidants. *Biosensors and Bioelectronics* 17:191–199.

Imabayashi, S.-I., Y.-T. Kong, and M. Watanabe. 2001. Amperometric biosensor for polyphenol based on horseradish peroxidase immobilized on gold electrodes. *Electroanalysis* 13:408–412.

Jaruga, P., and M. Dizdaroglu. 1996. Repair of products of oxidative DNA base damage in human cells. *Nucleic Acids Research* 24:1389–1394.

Jewell, W.T., and S.E. Ebeler. 2001. Tyrosinase biosensor for the measurement of wine polyphenolics. *American Journal of Enology and Viticulture* 52:219–222.

Kamel, A.H., F.T.C. Moreira, C. Delerue-Matos, and M.G.F. Sales. 2008. Electrochemical determination of antioxidant capacities in flavored waters by guanine and adenine biosensors. *Biosensors and Bioelectronics* 24:591–599.

Knekt, P., R. Jarvinen, A. Reunanen, and J. Maatela. 1996. Flavonoid intake and coronary mortality in Finland: A cohort study. *British Medical Journal* 312:478–481.

Korbut, O., M. Buckova, P. Tarapcik, J. Labuda, and P. Grundler. 2001. Damage to DNA indicated by an electrically heated DNA-modified carbon paste electrode. *Journal of Electroanalytical Chemistry* 506:143–148.

Krylov, A.V., W. Pfeil, and F. Lisdat. 2004. Denaturation and renaturation of cytochrome *c* immobilized on gold electrodes in DMSO-containing buffers. *Journal of Electroanalytical Chemistry* 569:225–231.

Labuda, J., M. Bučková, L. Heilerová et al. 2002. Detection of antioxidative activity of plant extracts at the DNA-modified screen-printed electrode. *Sensors* 2:1–10.

Labuda, J., M. Bučková, M. Vaníčková, J. Mattusch, and R. Wennrich. 1999. Voltammetric detection of the DNA interaction with copper complex compounds and damage to DNA. *Electroanalysis* 11:101–107.

Lisdat, F., B. Ge, R. Reszka, and E. Kozniewska. 1999b. An electrochemical method for quantification of the radical scavenging activity of SOD. *Fresenius' Journal of Analytical Chemistry* 365:494–498.

Lisdat, F., B. Ge, E. Ehrentreich-Forster, R. Reszka, and F.W. Scheller. 1999a. Superoxide dismutase activity measurement using cytochrome *c*-modified electrode. *Analytical Chemistry* 71:1359–1365.

Lu, Y., and L. Yeap Foo. 2000. Antioxidant and radical scavenging activities of polyphenols from apple pomace. *Food Chemistry* 68:81–85.

Lvovich, V., and A. Scheeline. 1997. Amperometric sensors for simultaneous superoxide and hydrogen peroxide detection. *Analytical Chemistry* 69:454–462.

Manning, P., C.J. Mcneil, J.M. Cooper, and E.W. Hillhouse. 1998. Direct, real-time sensing of free radical production by activated human glioblastoma cells. *Free Radical Biology and Medicine* 24:1304–1309.

Marchetti, M.A., W. Lee, T.L. Cowell, T.M. Wells, H. Weissbach, and M. Kantorow. 2006. Silencing of the methionine sulfoxide reductase A gene results in loss of mitochondrial membrane potential and increased ROS production in human lens cells. *Experimental Eye Research* 83:1281–1286.

Mello, L.D., M.D.P.T. Sotomayor, and L.T. Kubota. 2003. HRP-based amperometric biosensor for the polyphenols determination in vegetables extract. *Sensors and Actuators B: Chemical* 96:636–645.

Mello, L.D., A.A. Alves, D.V. Macedo, and L.T. Kubota. 2005. Peroxidase-based biosensor as a tool for a fast evaluation of antioxidant capacity of tea. *Food Chemistry* 92:515–519.

Mello, L.D., S. Hernandez, G. Marrazza, M. Mascini, and L.T. Kubota. 2006. Investigations of the antioxidant properties of plant extracts using a DNA-electrochemical biosensor. *Biosensors and Bioelectronics* 21:1374–1382.

Montereali, M.R., W. Vastarella, L. Della Seta, and R. Pilloton. 2005. Tyrosinase biosensor based on modified screen printed electrodes: Measurements of total phenol content. *International Journal of Environmental Analytical Chemistry* 85:795–806.

Nagaveni, K., M.S. Hegde, N. Ravishankar, G.N. Subbanna, and G. Madras. 2004. Synthesis and structure of nanocrystalline TiO_2 with lower band gap showing high photocatalytic activity. *Langmuir* 20:2900–2907.

Portugal, J., and M.J. Waring. 1987. Hydroxyl radical footprinting of the sequence-selective binding of netropsin and distamycin to DNA. *FEBS letters* 225:195–200.

Ratnam, D.V., D.D. Ankola, V. Bhardwaj, D.K. Sahana, and M.N.V.R. Kumar. 2006. Role of antioxidants in prophylaxis and therapy: A pharmaceutical perspective. *Journal of Controlled Release* 113:189–207.

Roessner, A., D. Kuester, P. Malfertheiner, and R. Schneider-Stock. 2008. Oxidative stress in ulcerative colitis-associated carcinogenesis. *Pathology—Research and Practice* 204:511–524.

Romani, A., M. Minunni, N. Mulinacci et al. 2000. Comparison among differential pulse voltammetry, amperometric biosensor, and HPLC/DAD analysis for polyphenol determination. *Journal of Agricultural and Food Chemistry* 48:1197–11203.

Sánchez-Moreno, C., J. A. Larrauri, and F. Saura-Calixto. 1999. Free radical scavenging capacity and inhibition of lipid oxidation of wines, grape juices and related polyphenolic constituents. *Food Research International* 32:407–412.

Song, M.I., F.F. Bier, and F.W. Scheller. 1995. A method to detect superoxide radicals using Teflon membrane and superoxide dismutase. *Bioelectrochemistry and Bioenergetics* 38:419–422.

Tammeveski, K., T.T. Tenno, A.A. Mashirin, E.W. Hillhouse, P. Manning, and C.J. McNeil. 1998. Superoxide electrode based on covalently immobilized cytochrome c: Modelling studies. *Free Radical Biology and Medicine* 25:973–978.

Valcour, V., and B. Shiramizu. 2004. HIV-associated dementia, mitochondrial dysfunction, and oxidative stress. *Mitochondrion* 4:119–129.

Valentao, P., E. Fernandes, F. Carvalho, P.B. Andrade, R.M. Seabra, and M.L. Bastos. 2002. Antioxidative properties of cardoon (*Cynara cardunculus L.*) infusion against superoxide radical, hydroxyl radical, and hypochlorous acid. *Journal of Agricultural and Food Chemistry* 50:4989–4993.

Wattanapitayakul, S.K., and J.A. Bauer. 2001. Oxidative pathways in cardiovascular disease: Roles, mechanisms, and therapeutic implications. *Pharmacology & Therapeutics* 89:187–206.

11 Neural Networks
Their Role in the Field of Sensors

José S. Torrecilla

CONTENTS

11.1 INTRODUCTION

The almost-perfect examples of sensors are in the human body, with the senses continuously supplying real-time data to the brain. By using our sensory elements (eyes, ears, skin, nose, and tongue), all perceptible information contained in our environment can be obtained. Then, by specific and ideal signal transducers, this information is selected, filtered, and processed in the most wonderful natural computer, the human brain. Finally, depending on the results, humans react in a short period of time.

    The two most important parts of this marvelous system are the five sensors and certainly our natural "fast computer," but in using these senses not all signals can be detected (e.g., radiation out of the visible range, adulterations, low concentration of impurities), nor can the natural computer work using all formats of signals (pressure, electrical signals, etc.). The real meaning of these limitations teaches us to adapt the design to the property to be measured and the type of signal to be processed. In the technical field, to quantify the desired properties, the appropriate physicochemical characteristics should be found; and the mathematical algorithms used should work using adequate information and format.

In the commercial and research fields, there is an oversupply of sensors. Among this large number of sensors, an extensive family is formed by biosensors. Their history started with the first reference to these types of sensors appearing in the 1960s (Clark and Lyons 1962). Biosensors and nanobiosensors are measurement systems for the detection of an analyte that combines a biological component (enzymes, cell receptors, protein, peptide, oligonucleotide, etc.) with a physico-chemical detector. These types of sensors are capable of continuous measurement of analytes in biological media such as blood serum, urine, and so forth (Torrecilla et al. 2008b). They can also be used to measure online biomoleculars and/or monitor biological processes.

Focusing on chemometric tools, there are many references where linear and nonlinear algorithms treat output of the sensors, and their calculations are used to determine/quantify compounds. Although linear algorithms use a lower number of parameters than nonlinear algorithms—because the latter show statistical performance—these are more widely used in the chemical field. Recently, coupling biosensor responses with computation strategies based on neural networks (NNs) have been growing in importance because of their application to multicomponent analysis (Lovanov et al. 2001; Torrecilla et al. 2009a) or for interpreting experimental data in the determination of pesticides (Trojanowick 2002), phenolic compounds (Gutes et al. 2005; Torrecilla et al. 2007a), neuroactive species (Ziegler 2000), ethanol–glucose mixtures (Lovanov et al. 2001), and the like.

In this chapter, different successful approaches using neural networks and different types of sensors and biosensors have been studied. In particular, five applications of supervised and unsupervised NNs in the quantification of the concentrations of chemicals in five different chemical systems are described: (1) two ionic liquids (ILs) and two hydrocarbons in their quaternary mixtures; (2) lycopene and β-carotene in food samples; (3) polyphenolic compounds in olive oil mill wastewater; (4) glucose, uric acid, and ascorbic acid in biological mixtures; and finally, (5) the chapter describes the results achieved of two unsupervised NNs in the identification of edible oils and in detecting adulterations of extra virgin olive oil (EVOO) with other cheaper oils.

11.2 PRINCIPAL TOOLS

11.2.1 Neural Networks

Depending on the methods used to optimize neural networks, there are two broad types of NNs, namely, supervised and unsupervised neural networks. In the first group, in order to optimize their performances, input and output data are required. In the latter, given that these types of NNs are able to optimize themselves, only input data is required. These two types will be briefly described here.

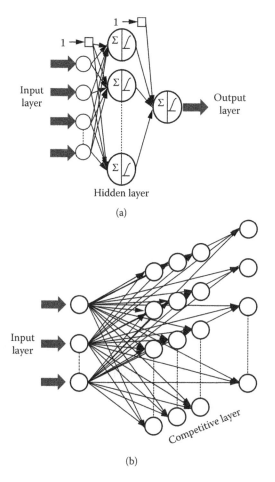

Input
layer

Output
layer

Hidden layer

(a)

Input
layer

Competitive layer

(b)

FIGURE 11.1 Schematic diagram of a multilayer perceptron (supervised neural network: a) and a self-organizing map model (unsupervised neural network: b).

The *supervised* NNs used in the first four applications described are based on a multilayer perceptron (MLP) consisting of several neurons arranged in three layers: input, hidden, and output layers (Figure 11.1). The topology of the NN is given by the number of layers and number of neurons in each layer. The input layer is used to input data into the NN; the nonlinear calculations are carried out in the other two layers. The calculation process in each neuron of the hidden and output layers consists of transfer and activation functions. The activation function, Equation (11.1), means that the input data to each neuron are multiplied by a self-adjustable parameter called weight, w; the result, x_k, is then fed into a transfer function. The two most common transfer function algorithms used are the sigmoid and hyperbolical tangent functions, Equations (11.2) and (11.3), respectively. The calculated value, y_k,

is the output of the considered neuron. The NNs used were designed by MATLAB®
Version 7.01.24704 software (Demuth et al. 2005).

$$x_k = \sum_{j=1} w_{jk} \cdot y_j \qquad (11.1)$$

$$y_k = f(x_k) = \left(\frac{1}{1+e^{-x_k}} \right) \qquad (11.2)$$

$$y_k = f(x_k) = \left(\frac{1-e^{-2x_k}}{1+e^{-2x_k}} \right) \qquad (11.3)$$

The learning process, which updates the weights to improve the predictive capacity
of the NNs, is carried out by minimizing the error prediction, Equation (11.4), using
the back-propagation model (Torrecilla et al. 2009a). Once the NN model has been
optimized, it should be internally and externally validated (OECD 2007). More details
regarding this type of NN can be found in the literature (Arbib 2003; Fine 1999).

$$E_k = \frac{1}{2} \sum_k (r_k - y_k)^2 \qquad (11.4)$$

The *unsupervised* NN potential applicability is outstanding, although the number
of applications of this type of NN is notably less than the supervised NN. Self-
organizing maps (SOMs), or Kohonen neural networks, and learning vector quantization (LVQ) networks are some of the most important NNs from this group of
algorithms. More details about these unsupervised NNs can be found in the literature (Arbib 2003). SOMs are the most interesting in the competitive neural network
field (Kohonen 1987). SOM models can learn to detect irregularities and correlations
in their input and adapt their future responses to that input accordingly; that is, they
are able to recognize groups with similar characteristics (Kohonen 1987; Demuth
et al. 2005). The architecture of SOM models is shown in Figure 11.1. Every circle
and arrow represents a neuron and weight, respectively; that is, there are as many
weights as arrows, and the number of neurons is equal to the product of the width
and length of the competitive layer. In this layer, each neuron has as many weights as
the input descriptors. Every neuron is represented by a vector of weights.

Given that SOMs classify input data according to how they are grouped in the
input space, throughout the learning process, in order to adequately represent all
input data, its weights are optimized. As every neuron is represented as a weight
vector, during this process, the neurons look for the best place to represent the whole
input database. The learning process of the SOM involves two steps: ordering and
tuning phases (Demuth et al. 2005). In the former, the ordering phase learning rate
(OLr) and neighborhood distance (ND) are decreased from both that rate and the
maximum ND between two neurons to the tuning phase learning rate (TLr) and
the tuning phase neighborhood distance, respectively. The ordering phase lasts for
a given number of steps (named ordering phase steps). In the latter, the learning rate

is decreased much more slowly than in the ordering phase, while the ND remains constant. Therefore, the number of epochs for the tuning phase of the SOM learning process should be much larger than the number of steps in the ordering phase. Given that the SOM model is a competitive and unsupervised NN, its weight optimization can be summarized in five stages:

1. Assign random values to the weights.
2. A data set from the learning sample is presented to the SOM model.
3. The neuron with the least Euclidean distance between its weights and data set (D) is selected (Equation 11.5). It is called the *winning neuron.*
4. The weights of the selected neuron are optimized so that they become more like the input vector (Equation 11.6).
5. The weights of the neighborhood neurons are also optimized but with proportionally less Euclidean distance to the wining neuron (Equation 11.7).

The process is repeated iteratively. When the whole database has been presented to the SOM, an epoch has finalized (Fonseca et al. 2006). Once the SOM has been trained, it is able to extract the relevant information in order to classify new input vectors, which are interpolated in the learning range.

$$D_j = \|X - W_j\| = \sqrt{\sum_i^S (X_i - W_{ij})^2}; \quad j = 1, 2, \ldots, N \tag{11.5}$$

$$W_j(n) = W_j(n-1) + Lr \cdot \lfloor X(n-1) - W_j(n-1) \rfloor \tag{11.6}$$

$$W_j(n) = (1 - Lr) \cdot W_j(n-1) + Lr \cdot X(n-1) \tag{11.7}$$

In Equations (11.5, 11.6, and 11.7), W, n, X, Lr, N, and S are the weights, iteration of a given epoch, input vector, learning rate, number of weights of the SOM, and number of data sets of the learning sample, respectively (Raju et al. 2006; Demuth et al. 2005). The SOM model used in this work was designed using MATLAB Version 7.01.24704 (R14) (Demuth et al. 2005).

Learning vector quantization networks are based on the combination of supervised and unsupervised NNs. These models can classify not only any set of input vectors but also linearly separable sets of input vectors. The only requirements are that the competitive layer must have enough neurons and each class must be assigned enough competitive neurons (Lloyd et al. 2007). LVQ models classify input vectors into target classes by using a competitive layer to find subclasses of input vectors, and then combining them into the target classes defined by the user. Therefore, LVQ networks consist of two layers: unsupervised (competitive) and supervised (linear) layers (Demuth et al. 2005).

The competitive layer learns to classify input vectors in much the same way as the competitive layers of SOMs (*vide supra*). The linear layer transforms the competitive layer's classes into target classifications defined by the user. The linear layers have one neuron per class (Demuth et al. 2005). The LVQ model used in this work was designed using MATLAB Version 7.01.24704 (R14).

11.2.2 Principal Component Analysis Description

Principal component analysis (PCA) is one of the most valuable results from applied linear algebra. PCA is used abundantly in all forms of analysis (from neuroscience to computer graphics) because it is a simple and nonparametric method of extracting relevant information from confusing data sets. The PCA technique reduces complex data sets to a lower number of dimensions, revealing the underlying simplified structures. It is based on the assumption that most information about classes is contained in the direction along which the variation is the largest. All mathematical steps of the PCA technique are summarized in Figure 11.2. Preserving the information from the original data, its dimensionality can be reduced following the PCA technique. More detailed information can be found in the literature (Wang and Paliwal 2003; Torrecilla et al. 2009d).

11.2.3 Sensors

In general, sensors are able to detect a physical and/or chemical change that varies with the property of interest. For instance, a liquid-in-glass thermometer transforms an expansion/contraction of the fluid on a temperature signal. In this case, there is a linear dependence between changes of volumes of the liquid and its temperature. Regretfully, there is hardly ever a linear relation between characteristics, and the mathematical relation would probably change with the time. Mainly for this reason, there are no ideal sensors. Because of this, when looking for the best possible performance, a sensor is considered as good if it obeys three main rules, namely, the sensor should be sensitive to the measured property and

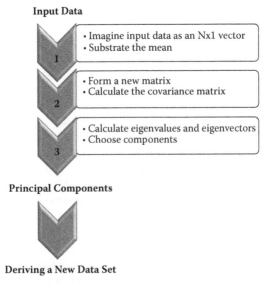

FIGURE 11.2 Summary of the calculation steps of the principal component analysis technique.

insensitive to any other, and it should be influenced by neither the sample nor the measured property. In addition, the mathematical relation between the output signal and the measured property value should be linear. However, although this relation would be mathematically linear, several types of deviations can be observed, which make the measurement process more difficult. This deviation could be based on systematic or random error. Examples of the former deviations are hysteresis, long-term drift, digitalization error, offset, and so forth. These systematic errors can be solved by strategies based on calibration or compensated processes. One of the most important effects of the second deviation group is the noise in the output signal of the sensor, and this can be reduced by processing the signal using a filter (Palancar et al. 1996, 1998; Torrecilla et al. 2008c, 2009b).

In this work, four types of sensors have been used: two commercial ultraviolet-visible (UV-vis) spectrophotometers (Varian Cary 1E UV-vis, Torrecilla et al. 2009a; and Pharmacia Ultrospec 4000 UV/vis, Torrecilla et al. 2008a), a laccase biosensor (Campuzano et al. 2002), and a gold nanoparticle enzyme biosensor (Cai et al. 2001; Mena et al. 2005). Finally, nuclear magnetic resonance spectroscopy (^{1}H-NMR and ^{31}P-NMR) has been used to identify edible and vegetable oils and also to detect adulteration of EVOO (Torrecilla et al. 2009c).

11.3 PRINCIPAL RESULTS AND DISCUSSIONS

One of the principal problems in accurately quantifying concentrations of chemicals in complex mixtures is that the chemical signals overlap. In general, three methods can be used to overcome this: the design of a specific measurement system, the application of powerful mathematical algorithms on the sensor output signals (Torrecilla et al. 2009a), and depending on the system, both techniques mentioned can be applied simultaneously (Torrecilla et al. 2008b). Here, systems based on the latter will be studied. The application of NNs on the five aforementioned chemical systems is shown here.

11.3.1 DETERMINATION OF TWO IONIC LIQUIDS, HEPTANE, AND TOLUENE CONCENTRATIONS

IL is a class of solvent or material composed of an organic cation and an inorganic anion. This product is interesting because of its high dissolving power, negligible vapor pressure, low flammability, thermomechanical and electrochemical stability, wide range in a liquid state, solvating properties for diverse kinds of material, ability to act as an acid or base, high ionic conductivity, wide electrochemical window, and so forth. On the other hand, in order to reach specific properties, the ions that constitute the IL can be selected. For instance, this feature could control its miscibility with water or with organic solvent, liquid range or other characteristics. In recent years, these solvents have gained popularity as being suitable in industrial and research fields (Wasserscheid and Welton 2007).

Recently, although ILs are being measured using interpolation in physicochemical properties (density, viscosity, refractive index, etc.), proton nuclear magnetic resonance, gas chromatography, and so forth, these are not adequate to measure/control online chemical processes (extraction, distillation, etc.) because of the time required to prepare samples. Given the importance of these processes, an analytical technique with a sample preparation time less than the sampling time of the process and a reliable algorithm are necessary.

In order to validate the NN algorithms as a powerful chemometric tool, the system based on low concentrations (less than 15 ppm) of toluene, heptanes, 1-ethyl-3-methylimidazolium ethylsulfate ([emim][EtSO$_4$]), and 1-butyl-3-methylimidazolium methylsulfate ([bmim][MeSO$_4$]) ILs in acetone was selected. Because of the sample preparation time required by UV-vis spectroscopy, it was selected as the analytical technique. Given that the imidazolium rings of both ILs and toluene are UV-active in the same region, the NN algorithms can be reliably tested to solve overlapping effects of quaternary mixtures on line.

An NN/UV-vis approach has been optimized and validated using samples with toluene, heptanes, [emim][EtSO$_4$] and [bmim][MeSO$_4$] IL concentrations between 0 and 15 ppm. As a result of an application of principal component analysis to UV-vis absorbance values between 190 and 900 nm wavelengths, seven principal components have been selected as the main variables. Using these seven variables, the NN algorithm's parameters are optimized by an experimental design. The values of the optimized parameters are shown in Table 11.1. Then, the optimized NN was externally validated by the application of this NN to estimate the concentration of 25 new samples distinct from those used in the learning process of the model. In this process, the mean prediction error (MPE, Equation 11.8) was less than 2.5%, and the mean correlation coefficient was higher than 0.95.

$$MPE = \frac{1}{N} \sum_n \frac{|r_n - y_n|}{r_n} \cdot 100 \qquad (11.8)$$

TABLE 11.1

Parameters of the Supervised NN Models Used

	UV-Vis Spectroscopy		Biosensors	
	Hydrocarbons and ILs Determination	Carotenoids Determination	Cathecol and Caffeic Acid Determination	Glucose, Ascorbic Acid, and Uric Acid Determination
Transfer function	Sigmoid	Sigmoid	Sigmoid	Sigmoid
Training function	Bayesian regulation	Bayesian regulation	Bayesian regulation	Bayesian regulation
Input nodes	7	2	1	11
Hidden neurons	20	5	7	13
Output neurons	4	2	2	3
Learning coefficient	0.5	0.32	1	0.001
Learning coefficient decrease	0.018	0.67	0.879	1
Learning coefficient increase	51	57	117	100

In Equation (11.8), N, y_n, and r_n are the number of observations, model estimation, and real value, respectively. Therefore, the PCA/NN/UV can be adapted to deconvolute the contribution of each chemical in the quaternary mixture studied. As a result, in the IL field, this approach is interesting for further applications to digital control and to measurement devices (Torrecilla et al. 2007b, 2009a).

11.3.2 Determination of Carotenoid Concentrations in Foods

Lycopene and β-carotene chemicals belong to the carotenoids family, which is widespread in nature. This is the main group of pigments with important metabolic functions. Due to its antioxidant activity, this group of chemicals shows a strong correlation between carotenoid intake and a reduced risk of some diseases and health problems, such as cancer, atherogenesis, bone calcification, eye degeneration, neuronal damages, and others.

Given that the lycopene and β-carotene are active in the same region of UV-vis spectroscopy, their determination by linear algorithms is not suitable (Torrecilla et al. 2008a). In order to use this fast, simple analytical technique, a nonlinear algorithm based on NN algorithm has been applied on the UV absorbance data at 446 and 502 nm wavelengths. Using these absorbance values of 25 binary mixtures composed of lycopene and β-carotene with concentrations ranging between 0.4 and 3.2 μg mL^{-1} and their respective concentrations distributed following an experimental design, the NN was optimized (Table 11.1). Once the NN model was optimized, NN/UV-vis spectroscopy was applied to determine the concentration of both chemicals in food samples such as tomato concentrate, tomato sauce, ketchup, tomato juice, and tomato puree. The mean prediction error value was less than 1.5%, and the correlation coefficient was higher than 0.99. The mean prediction error is 50 times lower than when a linear model is used in place of nonlinear algorithms. This improvement in the results is extremely valuable for its application to a fast and reliable lycopene and β-carotene evaluation in food samples. Because of this, algorithms based on NNs could be applied to the online quality control of these types of foods in the industry and/or in homes of consumers.

11.3.3 Determination of Polyphenolic Compounds Concentrations in Olive Oil Mill Wastewater

In the manufacturing of EVOO, waste is produced and has a serious environmental impact due to its high content of organic substances (sugars, tannins, polyphenols, polyalcohols, pectins, lipids, etc.) It is known that caffeic acid (CA) and catechol (CT) are two of the major contributors to the toxicity of these wastes. Given their electrochemical characteristics, a laccase biosensor is commonly used to determine CA and CT components. Because of the similarities in the oxidized species produced, the amperometric signals overlapping in the reduction of voltammograms is high, and therefore a powerful tool is required to solve this signal. Because of this a nonlinear algorithm based on NNs was tested.

Using voltammogram profiles of 300 samples and their respective concentrations of CA and CT, an NN was optimized. Once the NN model had been optimized, it was externally validated using real concentrations taken from three different olive oil mills in Spain (Almendralejo, Badajoz; Martos, Jaén; and Villarejo de Salvanes, Madrid). The mean prediction error was less than 0.5%, and the correlation coefficient was higher than 0.99. These statistical results are even better and more selective than those when using other nonportable commercial analytical equipment. Therefore, the integrated NN/laccase biosensor system is an adequate approach to estimate both hazardous chemicals in olive oil mill wastewater.

11.3.4 Determination of Glucose, Uric Acid, and Ascorbic Acid in Biological Mixtures

The importance of the determination of glucose, uric acid, and ascorbic acid in human samples is well known. Here, an amperometric biosensor based on a colloidal gold–cysteamine–gold disk electrode with an enzyme glucose oxidase and a redox mediator, tetrathiafluvalene, coimmobilized atop the modified electrode was used for the simultaneous determination of glucose and ascorbic and uric acids in ternary mixtures. The major obstacle for the amperometric detection of glucose in real samples is the interference arising from electro-oxidizable substances such as ascorbic and uric acids existing in a measured system. Because of this, a nonlinear algorithm based on an NN has been tested.

As a consequence of an experimental design, 125 cyclic voltammograms of ternary mixtures and their respective concentrations were used to optimize the NN model. The concentrations of these chemicals were between 0 and 1 mM. The optimized parameters values are shown in Table 11.1. Then, the optimized NN was externally validated. The mean prediction error was less than 1.74%, and the correlation coefficient was higher than 0.99. In the light of these results, the NN model is able to solve the interferences between glucose and ascorbic and uric acids without any chemical pretreatment. These promising results open the door to apply these types of algorithms to real samples from humans.

11.3.5 Identification of Edible and Vegetable Oils and Detection of Extra Virgin Olive Oil (EVOO) Adulteration

For millennia, the adulteration of food products with cheaper and more available substitutes has been a worldwide problem. Currently, adulteration of foods is more and more prevalent, mainly in those relatively expensive products. Given that the highest quality olive oil (EVOO) is in this category, a large number of cases of adulteration of this oily juice have been detected in recent years. The substitution or adulteration of EVOO with cheaper ingredients is not only an economic fraud but may also on occasion have severe health implications for consumers. An example is the Spanish toxic oil syndrome resulting from the consumption of aniline denaturalized rapeseed oil that involved more than 20,000 people and caused serious illness and even death. To fight against the increase of these fraudulent activities, the chemical compositions of specific olive oils have been qualified and protected by denominations of origin. Technically, mathematical algorithms have been also proposed (Fonseca et al. 2006; Marini et al. 2007; Torrecilla et al. 2009c).

The learning, verification, and validation processes of the models have been developed using data from the literature (Vigli et al. 2003). The learning, verification, and validation samples are composed of data that characterize the classification process. These data consist of values of the acidity; iodine value; ratio of 1,2-diglycerides to the total diglycerides; and the concentrations of total sterols, total diglycerides, 1,2-diglycerides, 1,3-diglycerides, saturated fatty acid (SFA), oleic acid, linolenic acid, and linoleic acid determined by analysis of the respective ^{1}H NMR and ^{31}P NMR spectra (Vigli et al. 2003). These properties have been calculated in nearly 200 samples corresponding to hazelnut, sunflower, corn, soybean, sesame, walnut, rapeseed, almond, palm, groundnut, safflower, coconut, and extra virgin olive oils. In addition, to test the adulteration detection capability of SOM and LVQ models, the aforementioned properties have also been measured and calculated in nearly 30 samples that consisted of the mixture of EVOO/corn oil (6 samples), EVOO/soya (6 samples), EVOO/sunflower oil (6 samples), and EVOO/hazelnut oil (10 samples) (Vigli et al. 2003). Finally, in order to test the competence of the optimized models, an external validation process was carried out using six other bibliographical references (Torrecilla et al. 2009c). In order to select the most important variables, the PCA technique has been applied. Then, SFA (mainly composed of palmitic and stearic acids), oleic acid, and linoleic acid concentrations were selected.

The optimization of parameters of the SOM model has been carried out by Central Composite Design 2^5 + star experimental design. The response of the experimental design was the number of incorrect classifications of the oil samples from the verification sample. In order to reach the least possible number of misclassifications, the optimum parameter values have been fixed at 0.1, 1500, 0.01, 0.5, and 30,000 to OLr, ordering phase, TLr, ND, and the number of epochs necessary in the learning process, respectively. The only LVQ parameter to optimize is the learning rate (Lr). In the optimization process, Lr was tested between 1×10^{-3} to 1 (Demuth et al. 2005). Taking into account that the minimum number of misclassifications is required, the best Lr value was 0.01. Once the models had been optimized, they were externally validated, and less than 5.5% of the samples were misclassified. The adulteration of EVOO with corn, soya, sunflower, and hazelnut oils was detected when their concentration was higher than 10, 5, 5, and 10%, respectively.

In light of these results, both unsupervised models are adequate to classify these samples studied in 13 types of edible oils. Although the results could be improved by specifically designed models for the adulteration databases, the results reached here are promising and open a way to identify vegetable oils or to determine the protected denomination of origin. In addition, the techniques proposed are suitable to detect adulteration at relatively low concentrations. A detailed description of these models and their applications can be found in the literature (Torrecilla et al. 2009c).

11.4 CONCLUSIONS

In order to test the capability of algorithms based on NNs to solve the overlapping effect between chemicals, different types of sensors have been reviewed here. In light of the statistical results, chemometric tools based on NNs are suitable to solve the overlapping effect in the systems reviewed here, without any chemical pretreatment.

And given the short time taken to estimate the concentration, this tool can be applied to calculate the concentration of chemicals online. Although every application should be previously tested and readjusted, taking into account chemical and/or physical considerations (where necessary), these successful results are extremely promising for other types of biosensors or sensors. In addition, two unsupervised NNs can also be used to identify edible oils and to detect adulterations in EVOO with cheaper edible oils. This opens the door to other applications in the food sector (denomination of origin, quality control online, etc.).

REFERENCES

Arbib, M.A. 2003. *The handbook of brain theory and neural networks*, 2nd ed. Cambridge: Massachusetts Institute of Technology.

Cai, H., C. Xu, P. He, and Y. Fang. 2001. Colloid Au-enhanced DNA immobilization for the electrochemical detection of sequence-specific DNA. *Journal of Electroanalytical Chemistry* 510:78–85.

Campuzano, S., R. Galvez, M. Pedrero, F. Manuel de Villena, and J. Pingarrón. 2002. Preparation, characterization and application of alkanethiol self-assembled monolayers modified with tetrathiafulvalene and glucose oxidase at a gold disk electrode. *Journal of Electroanalytical Chemistry* 526:92–100.

Clark, L.C.J., and C. Lyons. 1962. Electrode systems for continuous monitoring in cardiovascular surgery. *Annals of the New York Academy of Sciences* 102:29–45.

Demuth, H., M. Beale, and M. Hagan. 2005. Neural network toolbox for use with MATLAB: User's guide, Version 4.0.6. Ninth printing revised for Version 4.0.6 (Release 14SP3). Natick, MA: The Math Works, Inc.

Fine, T.L. 1999. *Feedforward neural network methodology*. New York: Springer-Verlag.

Fonseca, A.M., J.L. Biscaya, J. Aires-de-Sousa, and A.M. Lobo. 2006. Geographical classification of crude oils by Kohonen selforganizing maps. *Analytica Chimica Acta* 556:374–382.

Gutes, A., F. Cespedes, S. Alegret, and M. Valle. 2005. Determination of phenolic compounds by a polyphenol oxidase amperometric biosensor and artificial neural network analysis. *Biosensors and Bioelectronics* 20:1668–1673.

Kohonen, T. 1987. *Self-organization and associative memory*, 2nd ed. Berlin: Springer-Verlag.

Lloyd, G.R., R.G. Brereton, R. Faria, and J.C. Duncan. 2007. Learning vector quantization for multiclass classification: Application to characterization of plastics. *Journal of Chemical Information and Modeling* 47:1553–1563.

Lovanov, A., L. Borisov, S. Gordon, R. Greene, T. Leathers, and A. Reshetilov. 2001. Analysis of ethanol–glucose mixtures by two microbial sensors: Application of chemometrics and artificial neural networks for data processing. *Biosensors and Bioelectronics* 16:1001–1007.

Marini, F., A.L. Magrì, R. Bucci, and A.D. Magrì, 2007. Use of different artificial neural networks to resolve binary blends of monocultivar Italian olive oils. *Analytica Chimica Acta* 599:232–240.

Mena, M., P. Yañez-Sedeño, and J. Pingarrón. 2005. A comparison of different strategies for the construction of amperometric enzyme biosensors using gold nanoparticle-modified electrodes. *Analytical Biochemistry* 336:20–27.

OECD (Organisation for Economic Cooperation and Development). 2007. Guidance document on the validation of (quantitative) structure activity relationship [(Q)SAR] models, No. 69. OECD Series on Testing and Assessment. Paris: OECD. Available at: http://www.oecd.org (accessed October 10, 2009).

Palancar, M.C., J.M. Aragón, and J.S. Torrecilla. 1998. pH-control system based on artificial neural networks. *Industrial and Engineering Chemistry Research* 37:2729–2740.

Palancar, M.C., J.M. Aragón, J.A. Miguens, and J.S. Torrecilla. 1996. Application of a model reference adaptive control system to pH control effects of lag and delay time. *Industrial and Engineering Chemistry Research* 35:4100–4110.

Raju, K.S., D.N. Kumar, and L. Duckstein. 2006. Artificial neural networks and multicriterion analysis for sustainable irrigation planning. *Computers and Operations Research* 33:1138–1153.

Torrecilla, J.S., A. Fernández, J. García, and F. Rodríguez. 2007b. Determination of 1-ethyl-3-methylimidazolium ethylsulfate ionic liquid and toluene concentration in aqueous solutions by artificial neural network/UV spectroscopy. *Industrial and Engineering Chemistry Research* 46:3787–3793.

Torrecilla, J.S., A. Fernández, J. García, and F. Rodríguez. 2008c. Design and optimisation of a filter based on neural networks. Application to reduce noise in experimental measurement by TGA of thermal degradation of 1-ethyl-3-methylimidazolium ethylsulfate ionic liquid. *Sensors and Actuators B-Chemical* 133:426–34.

Torrecilla, J.S., J. García, E. Rojo, and F. Rodríguez. 2009d. Estimation of toxicity of ionic liquids in leukemia rat cell line and acetylcholinesterase enzyme by principal component analysis, neural networks and multiple lineal regressions. *Journal of Hazardous Materials* 164:182–194.

Torrecilla, J.S., M.L. Mena, P. Yáñez-Sedeño, and J. García. 2007a. Quantification of phenolic compounds in olive oil mill wastewater by artificial neural network/laccase biosensor. *Journal of Agricultural and Food Chemistry* 55:7418–7426.

Torrecilla, J.S., M.L. Mena, P. Yáñez-Sedeño, and J. García. 2008b. A neural network approach based on gold-nanoparticle enzyme biosensor. *Journal of Chemometrics* 22:46–53.

Torrecilla, J.S., E. Rojo, J.C. Domínguez, and F. Rodríguez. 2009b. Chaotic parameters and their role in quantifying noise in the output signals from UV, TGA and DSC apparatus. *Talanta* 79:665–668.

Torrecilla, J.S., M. Cámara, V. Fernández-Ruiz, G. Piera, and J.O. Caceres. 2008a. Solving the spectroscopy interference effects of β-carotene and lycopene by neural networks. *Journal of Agricultural and Food Chemistry* 56:6261–6266.

Torrecilla, J.S., E. Rojo, J. García, M. Oliet, and F. Rodríguez, 2009a. Determination of toluene, n-heptane, [emim][EtSO₄], and [bmim][MeSO₄] ionic liquids concentrations in quaternary mixtures by UV-vis spectroscopy. *Industrial and Engineering Chemistry Research* 48:4998–5003.

Torrecilla, J.S., E. Rojo, M. Oliet, J.C. Domínguez, and F. Rodríguez. 2009c. Self-organizing maps and learning vector quantization networks as tools to identify vegetable oils. *Journal of Agricultural and Food Chemistry* 57:2763–2769.

Trojanowick M. 2002. Determination of pesticides using electrochemical enzymatic biosensors. *Electroanalysis* 14:1311–1328.

Vigli, G., A. Philippidis, A. Spyros, and P. Dais. 2003. Classification of edible oils by employing 31P and 1H NMR spectroscopy in combination with multivariate statistical analysis. A proposal for the detection of seed oil adulteration in virgin olive oils. *Journal of Agricultural and Food Chemistry* 51:5715–5722.

Wang, X., and K.K. Paliwal. 2003. Feature extraction and dimensionality reduction algorithms and their applications in vowel recognition. *Pattern Recognition* 36:2429–2439.

Wasserscheid, P., and T. Welton. 2007. *Ionic liquids in synthesis,* 2nd ed. Weinheim: Wiley-VCH.

Ziegler, C., A. Harsch, and W. Goepel. 2000. Natural neural networks for quantitative sensing of neurochemicals: An artificial neural network analysis. *Sensors and Actuators B-Chemical* 65:160–162.

12 Trends in Biosensing and Biosensors

Frank Davis and Séamus P.J. Higson

CONTENTS

12.1 INTRODUCTION

The purpose of this chapter is to detail many of the recent advances in biosensing and to attempt to predict future trends. The detection of a wide variety of compounds is necessary in today's world. Many medical conditions can be detected by the presence of various biomarkers in medical samples such as blood or breath. One obviously widespread example of this is that diabetics have at any time a reliable and convenient method to easily monitor their own blood glucose levels with the blood glucose biosensor. Biological markers exist for many other conditions, and the detection of these markers can provide a diagnosis of the condition and assess its severity and the efficacy of any treatment program. The detection of low levels of many other compounds within living systems is also crucial, with benefits such as allowing us to diagnose and follow cases of accidental poisoning.

In addition to health applications, environmental monitoring is also a widespread application. Many common chemicals and biological materials can have potentially devastating effects if released into the environment, posing dangers to humans, livestock, and wild animals. Damage to the environment can also occur both to crops and other plant species, as well as cause contamination of food and water supplies. Releases can be due to industrial accidents such as the release of methyl isocyanate at Bhopal, India, or to deliberate actions as exemplified by the sarin gas attacks in Tokyo. Besides these relatively simple chemicals, the use of biologically derived poisons such as ricin or bacteriological agents such as anthrax has caused a demand for the reliable detection of these types of agents. Other industrial or agricultural use of chemicals can also lead to potential contamination such as the use of pesticides and fertilizers that can run off the land treated, leading to contamination of local groundwater—and potentially leading to the contamination of food and water supplies. Natural deterioration of food with time can also lead to bacteriological contamination, with reliable analysis providing early warning of any problems as well as possibly helping to prevent exposure of people or animals to these toxins. Adulteration of food samples has also proved to be of concern such as the recent deliberate addition of melamine to milk in China or the giving of antibiotics to farm animals, leading to their presence in meat and milk.

The nature of the sample to be analyzed will inevitably influence the choice of analytical technique to be used. Samples can exist in a wide variety of forms, including, for example, blood, plasma, urine, saliva, breath, meat, milk, water, soil, and air. Often, conventional methods of testing these materials may require time-consuming extraction steps or removal of interfering materials. The levels of analyte required to be measured may be extremely low and need to be determined accurately. These requirements mean that many current analytical procedures are time-consuming, can only be performed in the laboratory, and require the use of expensive equipment in conjunction with the services of highly trained staff. Biosensors offer an alternative to this. The high sensitivity and selectivity of many biosensors potentially allow the accurate determination of low levels of analytes in complex media, and this has led to intensive research and development of these systems.

12.2 ELECTROCHEMICAL SENSING

Since the development of early glucose biosensors in the 1960s (Clark and Lyons 1962), there has been widespread interest in electrochemical biosensors; and within this section we will attempt to detail some of the most recent advances. There are several advantages to electrochemical sensing, with the equipment required often being quite simple and portable, lending itself to operation in the field. Also, the availability of cheap, screen-printed electrodes allows the use of these systems as disposable, one-shot sensors, thereby avoiding the problems of sensor regeneration and fouling of the sensor by interferents. However, there can be a problem of communication between the bioactive moieties used in the sensor and the electrode. Various methods have been utilized to address this problem, such as with the commercial glucose electrodes, mediators such as ferrocene compounds are used to shuttle electrons between the electrode and the immobilized enzymes. This approach does have some limitations; antibody–antigen reactions, for example, do not produce electrons. To enhance sensitivity, direct "wiring" of the sensing moiety to the electrode using species such as conducting polymers or carbon nanotubes has been actively investigated.

12.2.1 CONDUCTING POLYMERS

Conducting polymers were first discovered in the 1970s, and the 2000 Nobel Prize for chemistry was awarded for this advance. There are a variety of conductive organic polymers, but the most commonly used in sensing are polyanilines, polypyrroles, and polythiophenes, along with their derivatives (Figure 12.1). Much interest in these systems has arisen because they can be deposited electrochemically onto a variety of conductive electrodes, thereby facilitating their synthesis as thin, even-conductive films. Species of a biological nature can be either entrapped within these films during the polymerization process or grafted onto the film using a variety of chemical or affinity methods. There have been a number of reviews on conducting polymers in biosensors (Gerard et al. 2002; Cosnier 2005; Geetha et al. 2006; Peng et al. 2009; Sadik et al. 2009), so within this chapter we will present just an overview.

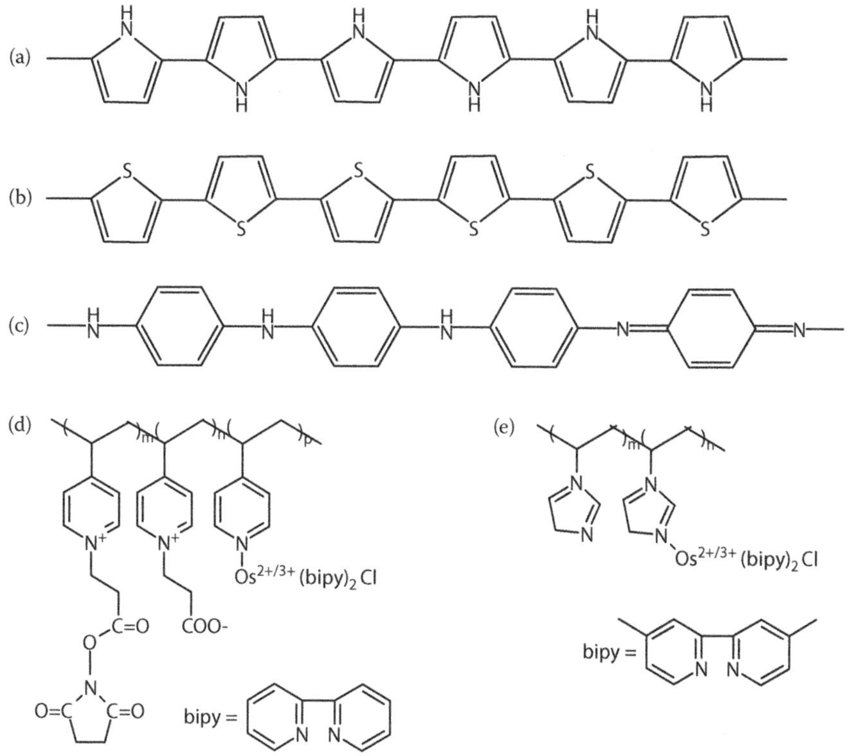

FIGURE 12.1 Structures of (a) polypyrrole, (b) polythiophene, and (c) polyaniline and redox-active polymers based on osmium bipyridyl complexes substituted onto (d) polyvinyl pyridine and (e) polyvinyl imidazole.

The close association between the conductive polymer and the biomolecule can potentially facilitate rapid electron transfer between the active species and an electrode surface. Alternatively, the electronic structure of conductive polymers is highly sensitive to the environment and conformation of the polymer chains. Often a binding event will give rise to conformational changes that affect the film, leading to a measurable change in its electrochemical or optical properties. Conducting polymers have been used as a basis for many sensors, for example, within DNA sensors. In early work polypyrrole containing entrapped single-stranded oligonucleotides could be deposited and shown to give specific responses toward the counterstrand (Wang et al. 1999). Polyaniline and polydiaminobenzene have also been successfully utilized as hosts for DNA (Davis et al. 2004).

Although entrapment does make films capable of specific recognition, there are problems in that the conditions and potentials required for polymer deposition can damage DNA and other biomolecules. Also there are problems in that target molecules may not be able to access the entrapped recognition species. Therefore covalent attachment of these species to conductive polymer films has been widely studied. This can be done by substituting the biomolecule of interest with a polymerizable group and depositing it onto an electrode. For example, oligonucleotides substituted with pyrrole groups were synthesized and electrodeposited as a copolymer with pyrrole onto silicon microelectrode arrays and applied to the genotyping of hepatitis C virus in blood samples (Livache et al. 1998). However, a much more widely used approach is to synthesize a conducting polymer film bearing either chemically reactive groups or biotin units and then grafting it on the biomolecule of choice.

In a typical example of this approach, a pyrrole substituted with a good leaving group, N-hydroxyl phthalimide was copolymerized with pyrrole-3-acetic acid, and then this film reacted with an amino-terminated oligonucleotide to give a conductive polymer with grafted oligonucleotide sidechains. A variation of this approach involved copolymerizing pyrrole along with a pyrrole containing butanoic acid sidechains; these acid groups could then be reacted with an amino-substituted oligonucleotide using a carbodiimide catalyst (Peng et al. 2005). In an interesting variation on this theme, one group has successfully electrodeposited polythiophene films containing arylsulphonamide sidegroups (Gautier et al. 2005). This group can be electrochemically cleaved and then further reacted to give a reactive sulphonyl chloride group to which an oligonucleotide could be grafted. The electrochemical cleavage gives rise to the possibility of putting down the polymers onto microelectrode arrays and then individually cleaving each electrode coating separately and grafting with different oligonucleotides, thereby creating a DNA microarray. Other workers have also widely used conducting polymer films containing reactive species such as n-hydroxy succinimide, which then react with groups such as amines (contained within many enzymes and antibodies), thereby covalently immobilizing them on the polymer surface (Cosnier 2005).

Chemical reactions offer one method of attaching biomolecules to a surface; another is to make use of affinity reactions (Cosnier 2005). A popular method involves the extremely strong interactions between the avidin families of proteins toward biotin units. For example, one method involved depositing biotin-functionalized polypyrrole and then utilizing the strong biotin-avidin interaction to immobilize first a layer of avidin followed by a layer of biotinylated antihuman immunoglobulin G (IgG) (Ouerghi et al. 2002) to give an immunosensor toward human IgG, which could be interrogated by alternating current (AC) impedance and had a linear dynamic range of 10–80 ng ml^{-1} of antigen and a detection limit of 10 pg ml^{-1}. Many other groups have also utilized the biotin–avidin interaction for binding biomolecules to conducting polymers (Cosnier 2005). Within our group we have made a variety of immunosensors based on this interaction. Polyaniline could be deposited onto screen-printed carbon electrodes and then chemically treated to introduce biotin moieties. These films were then coated with a layer of avidin and then used to immobilize a series of biotinylated antibodies. Using this method, immunosensors for the antibiotic ciprofloxacin (Garifallou et al. 2007, Tsekenis et al. 2008a), the heart drug digoxin (Barton et al. 2009a), and myelin basic protein (Tsekenis et al. 2008b) were constructed that could detect concentrations in the range of 1 ng ml^{-1} of the target.

Affinity methods have also been used to immobilize DNA onto conducting polymers; a biotinylated pyrrole monomer, for example, was deposited as a copolymer with pyrrole onto gold microelectrodes. This was then treated with avidin and finally a biotinylated molecule such as glucose oxidase or a single-stranded oligonucleotide (Bidan et al. 2004). These systems were then shown to be capable of binding complementary DNA strands by a variety of electrical and optical techniques.

Most of the sensors described within are based on simple planar films; however, we have also pioneered the use of conducting polymer microarrays. We found that a nonconducting polymer, poly(1,2-diaminobenzene), could be electrochemically deposited onto any conductive surface and then sonochemically ablated to give a random array of micropores in the film, which exposed the underlying conductive layer (Figure 12.2). It then became possible to electrodeposit polyaniline within

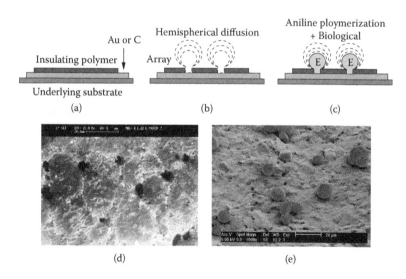

FIGURE 12.2 Schematic of the formation of polyaniline microarrays: (a) deposition of insulating layer, (b) sonochemical formation of pores, (c) polymerization of aniline; scanning electron microscopy (SEM) pictures of (d) pores and (e) polyaniline protrusions.

these pores to give conductive polymer protrusions containing entrapped moieties such as glucose oxidase (Barton et al. 2004). Other work developed sensors for pesticides with detection limits of 10^{-17} mol l^{-1} (Law and Higson 2005). We then compared the performance of our arrays with entrapped anti–bovine serum albumin (anti-BSA) with similar planar films and showed up to a thousandfold increase in sensitivity (Barton et al. 2009a). Within the same work we also investigated planar films containing entrapped antibodies to digoxin compared with electrodes where the antibody had been grafted using an affinity protocol and found large increases in sensitivity when the affinity protocol was used.

This led us to combine the two approaches, firstly by constructing polyaniline microelectrode arrays and then grafting on antibodies using an affinity protocol. This gave immunosensors with a very high sensitivity. Limits of detection for prostate-specific antigen were 1 ng ml^{-1} when simple entrapment was used to immobilize the antibody but 1 pg ml^{-1} when an affinity protocol was used (Barton et al. 2008a). Similar protocols were used to construct immunosensors for the stroke marker protein neuron-specific enolase and S100[β] with similar high sensitivities (Barton et al. 2008b, 2009b). These sensors all showed high sensitivity, low levels of interference from other proteins, and good storage capabilities.

A great deal of research lately has been devoted to the use of AC impedance techniques as opposed to the more conventional cyclic voltammetry and other voltammetric and direct current (DC) methods. AC impedance has several advantages over DC techniques; for instance, it only uses a small excitation amplitude, which causes minimal perturbation of the system, and also it can obtain data on both electrode capacitance and charge transfer processes. Equivalent circuits for these systems can be derived using established theories. AC impedance was used to detect antibody–antigen reactions within our work as described above and also used to detect DNA hybridization. For example, DNA hybridization has been detected at modified polythiophene electrodes, with binding of the second strand leading to decreases in impedance (Peng et al. 2007; Gautier et al. 2007) caused by facilitation of cation movement into and out of the film. The application of impedance spectroscopy to DNA detection along with the application of many different conducting polymers as a basis for these types of sensors has been reviewed elsewhere (Peng et al. 2009).

Conducting polymers conduct electrons along the polymer backbone; however, other polymers have been utilized that instead shuttle electrons between electroactive groups bound along the polymer chain. An early example is shown by this composite material containing polypyridine and osmium 2,2-bipyridine (Figure 12.1d). Reactive groups such as succinimide could be incorporated to covalently immobilize enzymes (Heller 1990). This gave rise to the construction of films of up to 1 μm thick, which gave a strong electrochemical response to glucose. Pyruvate sensors have also been constructed using these systems (Gajovic et al. 1999). Osmium-modified polyvinyl imidazole (Figure 12.1e) as a composite with a polyethylene glycol–based cross-linker and Nafion allowed constructions of oxidase-based sensors with linear ranges of 6–30 mmol l^{-1} (glucose) and 4–7 mmol l^{-1} (lactate) with only a negligible response being observed to common interferents (Ohara et al. 1994). It should be noted that the commercial Freestyle Navigator continuous glucose monitoring system utilizes osmium-based mediators (Freeman and Lyons 2008).

Other examples include a layered enzyme electrode with a polyvinyl pyridine/osmium polymer being used to "wire" both glucose oxidase and bilirubin oxidase to a glassy carbon electrode with concentrations of glucose as low as 2 fmol l^{-1} being detected in the presence of atmospheric oxygen (Mano and Heller 2005). Other redox-active polymers include ferrocene-containing hydrogels which, combined with photolithographic techniques, were utilized to immobilize glucose oxidase on gold electrodes (Sirkar and Pishko 1998) to produce patterned sensors. Glucose sensors were also made by plasma-polymerizing vinyl ferrocene onto a needle-type electrode to give a redox layer as a suitable substrate for the immobilization of glucose oxidase (Hiratsuka et al. 2005).

Single-stranded DNA was bound at an electrode surface along with an osmium-containing polymer and hybridization with a probe DNA combined with horseradish peroxidase labeling, allowed detection of DNA down to levels of just 3,000 copies (Zhang et al. 2003).

12.2.2 Electrochemical Microarrays

Many of the biosensors described above can only detect one analyte; however, a sensing platform that can simultaneously detect a number of analytes has much wider applications. This has led to the widespread investigation of microarrays where many small amounts of differing sensing molecules can be immobilized onto a chip and their individual responses measured. One mature application of this technology is in DNA microarrays where many spots of different oligonucleotide sequences are immobilized onto glass or plastic substrates, treated with labeled oligonucleotides in solution, and then the resultant binding visualized by fluorescence microscopy. Many reviews have been written on this subject such as Sassolas et al. (2008). As an alternative to fluorescence imaging, the use of electrochemically addressed microarrays with their inherent advantages of relative low cost, portability, and ability to be used in turbid samples has become more common.

Various approaches to making these arrays have been reviewed (Sadik et al. 2009) and include the use of boron-doped diamond microdisk arrays (Wang 2006). Commercial arrays have been developed such as the CombiMatrix arrays containing 12,544 electrodes cm^{-1}, each electrode being 44 μm in diameter. Different oligonucleotides were immobilized on the array and could be used along with microfluidic technology to detect hybridization at levels as low as 7.5×10^{-13} M of target (Ghindilis et al. 2007). Nucleic acid assays have been developed for genotyping a variety of pathogens including *Bacillus anthracis*, *Yersinia pestis*, *Escherichia coli*, and *Bacillus subtilis*. A schematic of this apparatus is shown in Figure 12.3.

An interesting alternative to electrodeposition was reported where glucose, glucose oxidase, and pyrrole were combined in solution (Ramanavicius et al. 2005). This led to the enzymatic production of hydrogen peroxide, which served as the initiator for the polymerization of pyrrole with the resultant formation of polypyrrole/glucose oxidase nanoparticles (Figure 12.4). Casting of the solution on carbon electrodes followed by glutaraldehyde cross-linking led to formation of glucose sensors with enhanced stability compared with controls manufactured from cross-linked enzyme only.

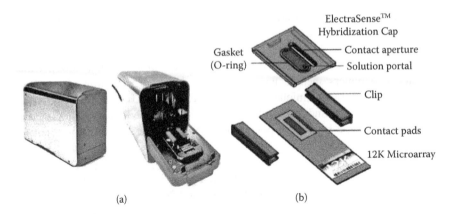

FIGURE 12.3 ElectraSense™ Reader (a) and ElectraSense™ 12K microarray with hybridization cap (b). (From Ghindilis, A.L., M.W. Smith, K.R. Scharztkopf et al., 2007, *Biosensors and Bioelectronics* 22:1853–1860. With permission.)

12.2.3 Incorporation of Nanosized Objects into Electrochemical Systems

Later in this chapter we will discuss nanobiosensors; however, various nanosized objects have been incorporated into standard types of electrode systems.

It has become possible to synthesize a wide variety of metal nanoparticles with exquisite control of size, composition, and surface coatings. Because these particles often have similar sizes as many of the biological species used within this field, attempts have been made to incorporate them into composites where the small size of the nanoparticles permits intimate contact within the composite. The metal nanoparticles are often conducting in nature, allowing "wiring" of the enzyme.

Because of its relative inertness and capability to be substituted with thiols, gold has been widely studied. For example, gold electrodes can be modified with a monolayer of gold nanoparticles, which are then used as a substrate to covalently attach glucose oxidase. Nanoparticle-containing sensors demonstrated enhanced electron transfer between enzyme and electrode and also enhanced stability for up to 30 days (Zhang et al. 2005). Another study where glucose oxidase was cross-linked onto films of colloidal gold immobilized on gold and carbon electrodes also demonstrated

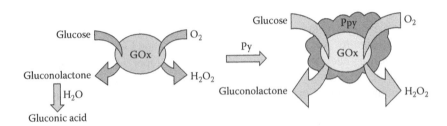

FIGURE 12.4 Synthesis of polypyrrole encapsulated glucose oxidase nanoparticles. (From Ramanavicius, A., A. Kausaite, and A. Ramanaviciene, 2005, *Sensors and Actuators B: Chemical* 111:532–539. With permission.)

higher sensitivity and stability compared electrodes without the colloid (Mena et al. 2005).

Silver nanoparticles have also been used along with glucose oxidase within a cross-linked polymer film immobilized on a platinum electrode, with again, much-enhanced electron transfer demonstrated by increased current responses in comparison to systems without nanoparticles (Ren et al. 2005). As an alternative, platinum nanoparticles encapsulated within polymeric dendrimers can be deposited along with glucose oxidase onto platinum electrodes to give glucose sensors (Zhu et al. 2007).

Since many of the sensors developed so far use carbon-based electrodes, once carbon nanotubes became available in usable quantities, their use in these applications became inevitable. The high aspect ratio, the high surface-area-to-volume ratio, their ability to undergo chemical substitution by a variety of procedures, and their electrical conductivity make them attractive for use in biosensors. The structures of single-walled carbon nanotubes (SWNTs) and multiwalled carbon nanotubes (MWNTs) are shown in Figure 12.5. These will be discussed in more detail later in the chapter, but a few brief examples will be given here.

A glassy carbon electrode could be modified with a cast composite of glucose oxidase, chitosan, and MWNTs to give a glucose sensor that displayed direct electron transfer (Liu et al. 2005). The same enzyme could be entrapped along with carboxylic acid–modified MWNTs in an electrodeposited film of polypyrrole to give a hydrogen peroxide sensor capable of detecting the product of the enzymatic reaction of glucose with a linear range up to 50 mmol l^{-1} (Wang and Musameh 2005). When polypyrrole and a single-stranded oligonucleotide were codeposited onto carbon nanotube–modified electrodes, the resultant biosensor could detect 10^{-6} mol l^{-1} of the counterstrand—and was also found to be capable of differentiating between the counterstrands and other oligonucleotides with one, two, and three base mismatches (Cai et al. 2003). Combinations of metal nanoparticles and carbon nanotubes have also been utilized; SWNTs and platinum nanoparticles, for example, could be codispersed within Nafion

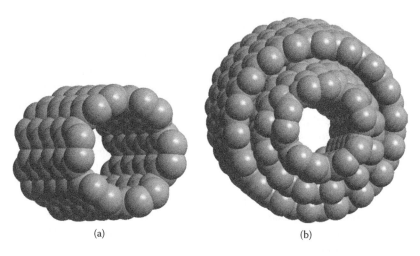

(a)   (b)

FIGURE 12.5   Structures of carbon nanotubes: (a) single-walled and (b) multiwalled.

along with glucose oxidase to give sensors with enhanced sensitivity over sensors modified with platinum (Pt) or SWNTs alone (Hrapovic et al. 2004).

12.2.4 PRACTICAL APPLICATIONS OF ELECTROCHEMICAL SENSORS

We will discuss here some of the practical applications of electrochemical sensors. The classical example of a biosensor is the glucose biosensor, which has greatly improved life expectancy and quality for millions of diabetics. Current techniques still require the withdrawal of blood by piercing the skin, a process that many diabetics find painful and distasteful. A much better method would be continuous monitoring, in which a sensor is implanted subcutaneously and relays regular reports back to a monitoring device. Implantable sensors suitable for *in vivo* glucose monitoring would be required to be extremely small and show long-term stability (with minimal drift, thereby removing the need for frequent calibration), display no oxygen dependency, and also show high biocompatibility. Such *in vivo* glucose sensors have been developed; however, as yet they have demonstrated only limited lifetimes mainly because of biofouling due to protein deposition on the sensor or formation of fibrous tissue. Also, if used intravascularly, there is the potential of immune rejection or thrombus formation, which could degrade device performance and potentially harm the user. The field of *in vivo* glucose monitoring has been extensively reviewed (Pickup et al. 2005), so only a brief history and the most recent advances will be given here.

One potential application utilized a needle-based electrochemical probe coated with Nafion (Moussy et al. 1993) that is just 0.5 mm in diameter and can be inserted subcutaneously through an 18-gauge needle, allowing *in vivo* measurements over periods of 2 weeks. A $2 \times 4 \times 0.5$ mm biochip containing two platinum microarray working electrodes onto which glucose and lactate oxidase were immobilized has been reported as a potential implantable biosensor (Guiseppi-Elie et al. 2005). A ribbon-like magnetoelastic sensor could be coated with a pH sensitive polymer and a layer of glucose oxidase (GOx) and catalase (Pang et al. 2007). The enzyme-catalyzed oxidation of glucose to gluconic acid causes a local pH change, leading to shrinkage of the pH-sensitive polymer and a decrease in the sensor mass loading. Application of a magnetic field causes vibration of the sensor, which can be measured by a pickup coil and is dependent on the mass of the sensor. This potentially could be used as an *in vivo* glucose sensor without any physical connection between a sensor and a monitoring device (Pang et al. 2007). As an alternative to implanted sensors, flexible and wearable glucose oxidase–based sensors on polydimethylsiloxane have been reported with the potential to measure glucose in sweat or tears (Kudo et al. 2006).

Implantable glucose sensors are now available commercially, usually consisting of an implanted sensor combined with a pocket-size monitoring and logging device. A subcutaneously implantable device, the Guardian® Real-Time Continuous Glucose Monitoring System, has been commercialized by Medtronic (www.minimed.com). The glucose biosensor probe is inserted subcutaneously, usually in the abdomen, and is capable of providing a glucose reading every 5 minutes for up to 72 hours. These and other data such as meal times and exercise periods can also be recorded and all data then downloaded to a personal computer. Other systems include the Dexcom Seven® Plus (www.dexcom.com), which has Food and Drug Administration (FDA) approval for up to 7 days of continuous use, and the Abbott FreeStyle Navigator®, which is also FDA approved.

Externally worn insulin pumps have also been developed, such as those by Medtronic (www.minimed.com), and remove the need for injections by introducing insulin subcutaneously though a cannula. Coupling this system with a glucose biosensor allows the device to inject insulin "on demand," in effect acting as an artificial pancreas, as exemplified by the Paradigm® Revel system. This was the first device of its type to receive FDA approval (Newman et al. 2004), although unfortunately these at present only have useful lifetimes of a few days without maintenance. A true artificial pancreas would enhance the quality of life for millions of diabetics, especially those with type 1 diabetes. Open microflow methods have also been studied (Ahmed et al. 2005), where delivery of fluid to the sensor is through a tissue-positioned cannula that houses the needle sensor, but with the delivery open-ended into the implant site. Use of polyurethane coatings led to good biocompatibility, low fouling, and low drift during monitoring.

Electrochemical sensors have also been recently utilized for cancer monitoring. We have already described our own work on the detection of prostate cancer markers (Barton et al. 2008a), and other workers have also utilized various assays. Cancer marker genes extracted from human breast tissue (Tansil et al. 2005) could be detected with a detection limit of 1.5 pmol l^{-1}, corresponding to 0.60 ng ml^{-1}. The use of electrochemical immunosensors to detect tumor markers has also been reviewed (Lin and Ju 2005). Multiple antibody functionalized chips have been proposed that can quantificate a number of proteins characteristic of different stages of the disease. Also, the simultaneous detection of carcinoembryonic antigen (CEA) and α-fetoprotein has been reported here (Wilson 2005). Other workers have also detected CEA by attaching the antigen to gold nanoparticles, immobilizing it onto a gold electrode by entrapment in an electrodeposited poly-o-aminophenol film, and interrogating by AC impedance (Tang et al. 2007).

Suspensions of cancer cells could be placed on multielectrode arrays and the consumption of oxygen by these cells measured. Treatment of the cells with various toxins and drugs determined their susceptibility towards these agents, measured as a decrease in metabolic rate. This opens up the possibility of taking biopsy samples from individual patients and determining the potential susceptibility of the cancers toward various treatment regimes (Andreescu and Sadik 2005).

Other uses for these sensors include the detection of food pathogens, and these have been reviewed (Ahmed et al. 2008; Sadik et al. 2009). For example, a gold microelectrode array combined with a microfluidic device could be combined with magnetic nanoparticle/antibody conjugates to give an electrochemical biosensor that could detect pathogens such as *E. coli* in ground beef (Gomez-Sjoberg et al. 2005).

12.3 OPTICAL BIOSENSORS

As an alternative to electrochemical approaches, a wide variety of biosensors have been developed using optical methods. Optical biosensors display several advantages over electronic devices. They are much less sensitive to electronic interference— because the information is carried as photons rather than electrons. The use of glass chips or fiber-optic cable fibers minimizes the weight and the size of the sensors, and also glass sensors usually display a high chemical stability, that is, they are not corroded easily and are usually unaffected by organic solvents as well as displaying good thermal and mechanical stability.

Much of the initial work in the field of optical biosensing involved the use of laboratory-based equipment, such as surface plasmon resonance (SPR). The field of optical biosensors is dominated by the SPR method with commercial SPR systems such as those manufactured by GE Healthcare under the name Biacore being the major systems provider at the time of this writing. SPR is a method that combines optical and electrochemical phenomena at a metal surface and is a powerful technique to measure biomolecular interactions in real time in a label-free environment; detailed reviews beyond the scope of this chapter have been published elsewhere (Ivarsson and Malmqvist 2002; Karlsson 2004; Ligler 2009). The wide range of commercial instruments has led to SPR becoming one of the most popular techniques used for studying intramolecular interactions. Several application areas are emerging including food analysis, a variety of drug detection applications, immunosensors, and the study of protein interactions.

Much recent work on SPR has gone into methods of increasing sensitivity by the incorporation of nanoparticles into the system. For example, gold nanoparticles have been deposited onto the surface of SPR chips (Chen et al. 2004). Both the gold surface and the nanoparticles produce plasmons on excitation, thereby enhancing the local electromagnetic field. This effect can be controlled by varying the size and volume fraction of embedded gold nanoparticles, offers a 10-fold improvement in resolution compared with conventional SPR biosensors, and has been applied to the detection of the surface coverage of biomaterials.

Samples for SPR are usually in the form of a solution or dispersion, meaning that a major drawback of the technique will be interference by species other than the analyte adsorbing to the surface of the optical chip. This is especially significant when attempts are being made to analyze biological fluids such as blood from which protein may deposit upon almost any type of surface. Cleanup methods such as centrifugation or filtration are often utilized to remove as many interferents as possible. Several methods also exist for coating SPR chips with a variety of materials in an attempt to minimize nonselective deposition such as coating with BSA, although these approaches are often not sufficient to prevent interference.

There has been a great deal of research on nanostructured surface coatings, and there is potential for their application in preventing such interactions, such as films based on p-xylylene where the hydrophobicity can be controlled by surface chemistry and roughness (Boduroglu et al. 2007). For example, surfaces can be manufactured that are thermally switchable between hydrophilic and hydrophobic states, which could facilitate the cleaning of SPR chips (Chunder et al. 2009).

SPR is not the only optical technique used for biosensing. The small size of optical fibers (diameters of 3–10 μm) enables bundles of thousands of fused optical fibers to be easily assembled into optical arrays as reviewed by Epstein and Walt (2003). If we combine the ability of optical fibers to be easily integrated with many different sensing schemes with their ease of assembly into high-density arrays (2×10^7 sensors cm^{-2}), the possibility of utilization of these systems for the preparation of a multitude of sensors including artificial noses and high-density oligonucleotide arrays becomes apparent. Selective etching of individual fiber cores is possible, enabling the formation of a high-density microwell array that can serve as a platform for an array of microsensors. The transmission from individual bundles allow for the monitoring of individual locations on a sensor chip. Various spotting techniques could be used to immobilize different receptors at these various locations; combining this with the fast response and versatility of this method enables not only monitoring of single analytes but analysis of the components of more complex mixtures. A number of sensing schemes and applications are described by Epstein and Walt (2003).

Security concerns and the need to "clear up" land mines and other unexploded ordnances have made the reliable detection of explosive compounds an intensively researched field in which many advances have been recently made. Trinitrotoluene (TNT) and other aromatic nitro compounds are the active part of many of these devices, and sensors capable of detecting low levels of these compounds have been developed.

TNT and other nitro compounds contain highly electron-deficient aromatic moieties that are known to form strong $\pi-\pi$ bonds to other aromatic species. Polyarylethynylenes (Figure 12.6) are highly fluorescent polymers in the solid state and have been utilized in attempts to detect TNT (Juan and Swager 2005). The polymer fluorescence is dramatically quenched when exposed to small quantities of TNT vapors due to electron donation from the polymer to the analyte. Not only is there high affinity between the polymer and analyte, but also the conjugated structure of the polymer makes it act like a molecular wire. When a single molecule of TNT is bound, the conjugation is broken and the entire chain ceases to fluoresce. This makes this technique especially sensitive, since the quenching of the whole polymer

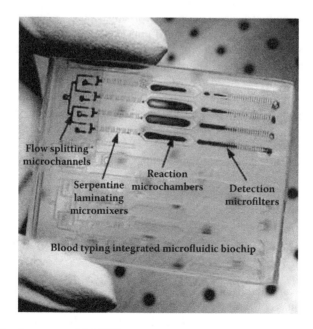

FIGURE 12.6   Structures of (a) TNT and (b) polyarylethynylenes.

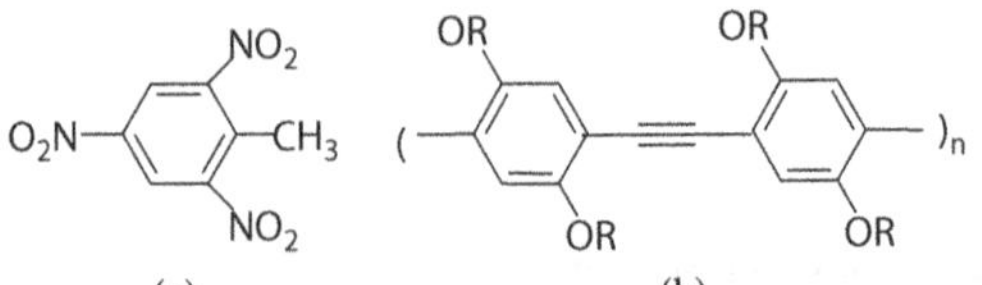

FIGURE 12.7 Blood typing microfluidic biochip. (From Kim D.S., S.H. Lee, C.H. Ahn, J.Y. Lee, and T.H. Kwon, 2006a, *Lab on a Chip* 6:794–802 With permission.)

chain in effect amplifies the effect of each TNT molecule. A commercial product in the Fido® line (www.icxt.com) weighs about a kilogram and can detect a range of nitroaromatics, with sensitivities claimed to be as low as part per quadrillion. Potentially these types of polymers could be coupled with biological recognition elements to produce optical biosensors with extreme sensitivity. Work is ongoing in the detection of a variety of chemical and biological targets.

In attempts to miniaturize optical sensors, chips are being integrated with various microfluidic devices. A typical example (Ligler et al. 2002) involves the integration of microfluidic flow devices with optical wave guides, and this platform proved capable of being utilized to develop immunoassays with sensitivities of 30–50 pg ml^{-1} for antigens such as *Staphylococcal enterotoxin B*. Other workers used wave guides to develop optical biosensors that demonstrated the label-free detection of BSA (White et al. 2006). Microfluidic channels and optical wave guides can also be fabricated in a single substrate by photopatterning of glass (Applegate et al. 2007). This type of technology has, for example, enabled the construction of a disposable microfluidic biochip (Kim et al. 2006a) that can determine blood type within 3 minutes, gives a response visible to the human eye, and requires just a single drop of blood (Figure 12.7). Integration of the detection mechanisms of optical sensors can also be incorporated into the chips by the utilization of organic photodiodes (Hofmann et al. 2005).

As an alternative method, SPR imaging where the SPR behavior of a metal surface can be imaged has been utilized. This is a much more versatile technique than simple SPR because individual areas of an SPR chip can be spotted with various antibodies, oligonucleotides, and so forth, and then any binding visualized in a single experiment. This has been used to screen milk samples for seven different antibiotics simultaneously (Raz et al. 2009). Antibiotics were immobilized as spots onto commercial SPR chips, and then a competitive assay utilized where the chip was then exposed to antibodies to these materials. Levels of detection well below the maximum residue limit were established for most of these compounds in buffer and 10-fold diluted milk.

12.4 QUARTZ CRYSTAL MICROBALANCE AND SURFACE ACOUSTIC WAVE SENSORS

The quartz crystal microbalance (QCM) is a highly sensitive method for studying changes in the mass of thin films. A QCM consists of a thin disk cut from a single crystal of piezoelectric quartz onto which electrodes (usually gold) have been plated. When an oscillating electric field is applied across the plate, a resultant acoustic wave is induced throughout the crystal in a direction perpendicular to the crystal surface with a frequency dependent on the mass of the crystal, including any thin surface layers deposited on it. Although the QCM itself is a sensitive detector for mass changes, it has no inherent specificity. However, deposition of a thin sensing layer onto the QCM is possible and therefore the resultant device should combine the sensitivity of the QCM to mass changes with the specificity conferred by the coating.

A wide variety of antibodies or single-stranded oligonucleotides have been immobilized onto QCM surfaces, and many applications of this technology have been reviewed, along with the various methods used to immobilize the antibodies onto the QCM plate (O'Sullivan and Guilbault 1999). The species that have been detected include drugs such as cocaine, quinine, and atropine. Also biological targets such as the HIV virus and specific single strands of DNA have been successfully detected.

QCM has also found application in detecting microorganisms, as reviewed in Dickert et al. (2003). Exposing clean QCM crystals to water streams will cause the formation of biofilms that cause a mass change and are easily detected. Although this process will respond to any microbes that colonize the surface, the immobilization of antibodies on the surface allows selectivity. As detailed in the review, early work developed specific sensors for *Candida albicans* with response times of just 30 minutes (Muramatsu et al. 1986), and sensors have also been developed for *Salmonella* (Deng et al. 1996), *E. coli* (Qu et al. 1998), foot and mouth virus (Gajendragad et al. 2001), and plant viruses (Eun et al. 2002) that could be detected in crude plant sap. As an alternative technique, biofilms were allowed to grow on QCM crystals, and these films were then exposed to solutions of various antibodies. The selective binding of antibodies to cells in the biofilm causes a mass change, thereby allowing determination of which microorganisms are present in the biofilm (Dickert et al. 2003).

Another acoustic technique used for the construction of biosensors is the surface acoustic wave (SAW) method. These are similar to QCM devices in that a SAW is generated using a piezoelectric substrate and surface electrodes and can be used to detect, among other parameters, mass changes. There are a wide number of formats for SAW generation, different electrode patterns, and so forth, and a detailed description of these is beyond the remit of this work but is reviewed elsewhere (Rocha-Gaso et al. 2009).

As in the case of QCM, the SAW technique is non-elective in itself; however, functionalization of the SAW chip with biorecognition elements is used to construct biosensors. Among many recent examples, a SAW sensor was coated with a polystyrene film (Deobagkar et al. 2005) that was further modified with a polyclonal antibody layer. This proved capable of detection of *E. coli* O157:H7 at levels from 0.4 to 100 cells μl^{-1} in water. In other work, Berkenpas et al. (2006) coated a SAW device with Neutravidin and then used this to attach a biotinylated antibody for *E. coli* O157:H7; and although minimal responses were found against the bacterium in water due to coupling between the liquid and the SAW chip, these authors successfully developed a dip and dry method that proved capable of detection of the bacterium between 10^3 and 10^6 cells μl^{-1}. These SAW devices proved capable of being integrated into microfluidic devices (Lange et al. 2006) as shown in Figure 12.8. Early experiments with BSA demonstrated their capability to be utilized in sensing applications.

Simultaneous detection of two bacteria, *E. coli* and *Legionella*, proved possible using a SAW with two channels, and could detect both species at levels between 10^3

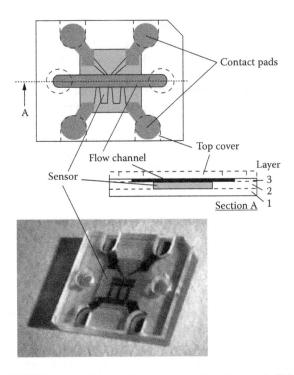

FIGURE 12.8   SAW biosensor chip: outline (top) and photo (bottom). Chip area: 8 mm ×
10 mm. (From Lange, K., G. Blaess, A. Voigt, R. Gotzen, and M. Rapp, 2006, *Biosensors and
Bioelectronics* 22:227–232. With permission.)

and 10^6 cells μl^{-1} in 3 hours (Howe and Harding 2000). Multipurpose immunosensors
have also been developed (Moll et al. 2008) by attachment of antibodies to a variety
of species onto a SAW chip. The resultant sensors have been shown to be capable of
detection of bacteria, viruses, and proteins. Other workers (Bisoffi et al. 2008) have
demonstrated rapid detection of Coxackie virus B4 and Sin Nombre virus with very
rapid (within seconds) detection times. Although these biosensors are all laboratory
sensors, there has been some commercial development of SAW biosensors, and a
product S-sens has been developed by the Centre of Advanced European Study and
Research (Bonn, Germany) that has been utilized within several research applications
(Rocha-Gaso et al. 2009).

12.5   MICRO- AND NANOBIOSENSORS

One of the major fields within biosensor research concerns the miniaturization
of already-developed techniques to produce biosensors that are either portable or
implantable. The great advances in construction and characterization of small,
well-defined structures by methods such as photolithography and atomic force
microscopy have furthered this research. The field of micro- and nanobiosensors has
been recently reviewed (Urban 2009).

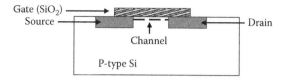

FIGURE 12.9 Schematic of an ISFET.

12.5.1 Miniaturized Transducers

There has been research into miniaturization of the transducers used in biosensors such as exemplified, for example, in the development of enzyme field effect transistors (ENFETs), which are ISFETs (ion-sensitive field effect transistors) with immobilized enzymes. A schematic of an ISFET is shown in Figure 12.9 and basically consists of p-type silicon, with two n-type regions, the source and the drain, implanted within the bulk p-type silicon, and coated with a silicon oxide insulator gate layer. Applying a positive voltage leads to depletion of holes close to the surface between the source and the drain. Electrons are attracted to this region and bridge the potential gap between the source and the drain, leading to formation of a conducting channel. This current is influenced by the potential at the SiO_2/aqueous solution interface, which can be related to ion concentration within the solution. If, for example, glucose oxidase is immobilized at the surface of an ISFET, it will catalyze the formation of gluconic acid, which will reduce the local pH and affect the ISFET. Recent examples of this are reported by Luo et al. (2004), who developed a glucose oxidase–based ENFET. Multiple ISFETs can also be used, for example, in a sensor for glucose and lactate (Hanazato et al. 1989). However, these devices are extremely affected by any changes in local pH, so no commercial biosensors have been produced.

Some microsensors containing hydrogels have also been produced. Microlenses can be produced from hydrogels composed of poly(N-isopropylacrylamide-co-acrylic acid) that was modified with biotin groups and coated with antibodies to biotin (Kim et al. 2006b). The antibody was then also bound photochemically to the hydrogel using a photochemical protocol to cross-link the surface layers. Exposure to free biotin disrupted the cross-linking, changing the swelling of the hydrogel, and this could be observed by a simple bright-field optical microscopy technique. As an alternative, hydrogels could be incorporated into microelectromechanical systems (Strong et al. 2002), sandwiched between a diaphragm and a baseplate. If the hydrogels were substituted, for example, with enzymes or antibodies, to make them bioresponsive, then these devices would have potential to act as microbiosensors.

12.5.2 Microarrays

One method of simultaneous analyzing for a number of analytes is to use array technology, where a range of biological agents is spotted onto a solid surface, usually a glass slide or chip. One field in which this technology has been especially important is that of DNA microarrays. One of the major biotechnology success stories of recent times has been the sequencing of the human genome. Early methods of detection of specific DNA sequences involved chemical cleavage of DNA strands and identification of the individual bases. This process was intensively laborious, expensive, and time-consuming, and the appearance of DNA arrays that allow multiple sequence detection on small volumes of DNA solution quickly and with high specificity and rapid response times greatly simplified this process. A wide variety of instruments for construction of DNA microarrays and their interrogation are now commercially available. The recent history and construction of DNA microarrays has been extensively reviewed (Campas and Katakis 2004) and will only be summarized here.

A large number of single-stranded oligonucleotides can either be spotted or even be synthesized onto glass slides. Upon exposure to a mixture of DNA strands in solution, they will bind their counterparts. If the counterparts have been labeled with fluorescent groups, they can be visualized by a fluorescence microscope. The construction of high-density DNA microarrays requires high densities and small spot sizes. Pin deposition is often utilized, where small pins are immersed into solutions of the DNA probes, withdrawn, and touched against the solid substrate to transfer a drop of solution. Spot sizes of 50–360 µm can be obtained by this method. Other alternatives include ink-jet printing technology, which is a noncontact technique, thus reducing the risk of cross-contamination between spots and can give spot densities of up to 2500 spots cm^{-2}. Both of these utilize preformed oligonucleotides; however, photolithographic techniques (first introduced by Affymetrix) involve the *in situ* synthesis of oligonucleotides with specific structure and location on a solid substrate (Fodor et al. 1991). A reactive surface, usually containing amino groups, is passivated with a photolabile protecting group and then exposed to light through a mask to activate certain areas of the substrate. These are then derivatized with a nucleic acid monomer, with no reaction occurring in unactivated areas. Further irradiation steps through a variety of masks then take place, allowing the buildup of different oligonucleotide sequences at every site on the chip addressable by light. Since photolithographic methods have a high resolution, this protocol leads to microarrays with some of the highest spot densities available; for example, commercial arrays with more than 40,000 spots on a glass slide are now available.

These arrays have a variety of uses, for example, comparing gene expression between two related organisms such as wild-type yeast and a mutant. Other uses could include the comparison of pathenogenic and related nonpathenogenic organisms or the differentiation of healthy and cancerous cells. DNA arrays are commonly used in universities and hospitals as research tools. It is widely anticipated that a new generation of DNA-sequencing platforms will become commercially available over the next few years, enabling resequencing of human genomes at a very high throughput and low cost, meaning that obtaining a human genome could be attained for $1,000 (Mardis 2006).

Within living systems, many biological functions are carried out by proteins. This has led to the development of protein-based biochips, similar in structure to DNA microarrays, to allow identification of protein–ligand interactions as reviewed here (Lueking et al. 2005; Stoll et al. 2005). To construct protein biochips, a wide variety of proteins for immobilization is required. These are usually generated by recombinant expression within *E. coli* and purification (Lueking et al. 2005). These proteins can then be immobilized using similar methods to those used to generate DNA microarrays. Similar interrogation methods can also be used.

There are numerous potential uses for protein microarrays. Protein microarrays can be exposed to an antibody solution and the binding monitored to give a measure of the selectivity and cross-reactivity of the antibody. Alternatively of course, a series of antibodies can be immobilized instead, which can then be exposed to a mixture of proteins. This second approach is often preferred because most antibodies have the same basic structure and can be immobilized using the same techniques, whereas different proteins often have widely different structures. As an example, antibody arrays were exposed to serum from a number of patients with a variety of differing diseases (Lueking et al. 2005). It was found that different diseases led to different binding patterns, indicating the possibility of using these chips for diagnosis.

Proteins can also interact with a wide variety of species such as other proteins, DNA, RNA, peptides, oligosaccharides, and other chemical compounds such as drugs. For example, a chip containing 37,000 yeast recombinant proteins (requiring just 10 pg of each material) was used to identify proteins that bound a peptide that is part of the platelet membrane protein integrin (Lueking et al. 1999). This approach demonstrates the high potential of these biochips to detect and identify previously unknown protein interactions.

Protein microarrays can potentially be used to study protein function within biological systems. There are issues with protein arrays; however, proteins are not as easily handled or immobilized as DNA, and the prediction of affinities is much more problematic. However, antibody microarrays have been used to quantify autoantibodies in the sera of patients with autoimmune disease (Joos et al. 2000). This procedure uses an ELISA-type format and is sensitive and highly specific. As little as 40 fg of a protein standard can be detected with minimal cross-reactivity to nonspecific proteins.

Biosensor arrays can also be constructed using living cells; for example, a 1536-well microtiter plate was used as the substrate and a platinum electrode integrated into a well. This could be covered with a hydrogel and then bacterial growth monitored within the well by an AC impedance protocol (Spiller et al. 2006). This allowed monitoring of cell growth and could also distinguish between living and dead cells, allowing it to be used as an assessment of antibiotic resistance of bacterial cells.

12.5.3 Nanobiosensors

The incorporation of nanotechnology into biosensing can be attained in a variety of methods. One is incorporation of nanomaterials as described briefly in Section 12.2.3 and also described in more detail in the following section, and another is the fabrication of nanosized transducers. For example, cantilever-based sensors have been produced in which a cantilever of 100 nm thickness can be used as a substrate for biological recognition agents. These cantilevers can be easily manufactured from silicon using well-developed microengineering techniques. Multiple cantilevers can be arranged in an array (Figure 12.10a). The sensitivity of these systems is very high; in fact, they can potentially detect single molecules if they are suitably labeled, and a combined cantilever/magnetic bead assay, for example, has been proposed with potential single molecular sensitivity and simultaneous multianalyte capacity (Baselt et al. 1998). Ziegler's (2004) review summarizes much of the work on biosensing using these systems.

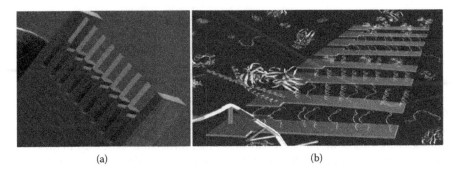

(a)   (b)

FIGURE 12.10 (a) Array of silicon-based cantilevers with individual functionalized surfaces; (b) principle of a cantilever-based biosensor for oligonucleotide detection. (From Urban, G.A., 2009, *Measurement Science and Technology* 20:012001. With permission from the Institute of Physics and Professor Martin Hegner.)

Microcantilevers can be selectively substituted on one side, for example, by evaporating a layer of gold onto one side and then using gold-thiol chemistry to substitute this surface. Biomolecules can then be attached to this by a variety of binding protocols. Binding of an analyte leads to stress within the system and can cause bending of the cantilever, which then can be detected optically by a laser. The specific binding of DNA strands was shown (Fritz et al. 2000) where an array of cantilevers was fabricated with two different 12-mers of single-stranded DNA bound to opposite sides of the cantilever (Figure 12.10b). The sequential binding of complementary strands to first one and then the other oligonucleotide could be clearly detected. Single base mismatches could also be detected, and within the same report, protein–protein interactions were also noted. In similar work 20- or 25-mer thiolated oligonucleotides were bound to gold-coated cantilevers and shown to recognize 10-mer complementary probes with a net positive deflection, while hybridization with targets containing one or two internal mismatches resulted in net negative deflection (Hansen et al. 2001).

Immunochemistry responses have also been studied, and cantilever sensors coated with an antibody to myoglobin were shown to be capable of detection of 85 ng ml^{-1} of the target (Grogan et al. 2001). The same group also demonstrated binding of herbicides to a suitable monoclonal antibody and also biotin–avidin interactions (Raiteri et al. 1999). Other workers (Wu et al. 2001) also reported the detection of two forms of prostate-specific antigen (PSA) in a clinically relevant concentration range from 0.2 ng ml^{-1} to 60 μg ml^{-1} in a background of human serum albumin (HSA) and human plasminogen (HP) at 1 mg ml^{-1}.

Other applications have been demonstrated for these sensors. One group, for example, utilized cantilever sensors to detect mRNA biomarker candidates in total cellular RNA (Zhang et al. 2006). It was possible to observe the differential gene expression of the gene 1–8U, a potential marker for cancer progression or viral infections in a complex background. Detection at the picomolar level without target amplification was possible within minutes, and also the results demonstrated sensitivity to base mismatches. Transcription proteins can also be detected at nanomolar levels by binding double-stranded DNA to cantilever sensors (Huber et al. 2005). Commercial production of these microcantilever sensors, for example, the Cantisens® range of cantilever measurement and functionalization platforms (www.concentris.com), will in all probability increase the amount of research on these systems.

12.6 GOLD–THIOL MONOLAYERS

A variety of methods have been used to attach monolayers of organic compounds to solid substrates. One of the most widely used is that of gold–thiol monolayers, based on binding between "soft" gold and sulphur atoms (Ulman 1991, 1998). Advantages of using gold are that it is relatively easy to clean, it does not oxidize under standard lab conditions, and any weakly physically adsorbed impurities are displaced by the sulphur species. Also, many of the functional groups present in biological species, acids, amines, and so forth are relatively "hard" and do not interact strongly with the gold surface. Early studies looked at simple thiols (Bain et al. 1989) that form stable, well-assembled monolayers that are highly resistant to washing due to the strong chemisorbtion of the sulphur atoms. Molecules such as hexadecanethiol lead to well packed quasi-crystalline monolayers, whereas shorter thiols such as hexanethiol give liquidlike monolayers.

The stability and thinness of these layers rapidly led to these systems being investigated as possible means of attaching biological recognition moieties to gold substrates. The use of a gold substrate also makes the resultant systems easily investigated by QCM, Fourier transform infrared spectroscopy (FTIR), atomic force microscopy (AFM), SPR, and electrochemical methods. Another advantage of these monolayers in a sensing field is that they can be easily printed onto gold using a soft polymeric stamp with resolution down to 30 nm (Xia and Whitesides 1998), opening up the possibility to directly manufacture "circuits" from these monolayers. Numerous reviews have been published on the use of gold–thiol monolayers in biosensors (Davis and Higson 2005; Arya et al. 2009).

Monolayers of bioactive molecules can be immobilized directly on gold if they contain sulphur groups. Alternatively a gold–thiol monolayer can be deposited and then used to adsorb a biomolecule via adsorption, electrostatic attractions, or covalent binding. Large and rigid species such as DNA and proteins may not adsorb as effectively as simple thiols; however, this may improve their suitability as biosensing agents because highly ordered crystalline monolayers may have steric interactions that prevent the binding of substrates or the hybridization of complementary DNA strands.

As an early example, glucose oxidase chemically modified with thiol groups could be adsorbed onto gold microelectrodes (McRipley and Linsenmeier 1996) to give an amperometric glucose sensor with fast (<20 s) response to glucose in the range of 0–50 mmol l^{-1}. Alternatively, mercaptopropionic acid monolayers on gold were used as a point of attachment for glucose oxidase (Gooding et al. 1998) via covalent binding to give an amperometric glucose biosensor. Other work (Darder et al. 2000) involved the adsorption of a difunctional disulphide containing an alkyl chain and an aromatic acid on gold. Membrane-bound type enzymes such as fructose dehydrogenase bound effectively to these layers, up to 3×10^{12} molecules cm^{-2} of enzyme as measured by QCM. Using gold electrodes led to the construction of a voltammetric sensor for fructose in fruit juice with linear response behavior up to 0.7 mmol l^{-1} and detection limit of 0.02 mmol l^{-1}.

An interesting variant on the use of gold–thiol monolayers has been rather than attaching the thiol to the gold electrode; instead, use small colloidal gold particles as the substrate. For example, a gold electrode was modified with cystamine and further reactions to give a reactive thiol surface (Xiao et al. 1999). This surface-bound colloidal gold and horseradish peroxidase could be adsorbed onto to the bound colloid to enable the construction of a peroxide sensor. Polymer nanospheres coated with thiol groups have also been assembled onto gold electrodes and then a layer of colloidal gold bound on top (Xu and Han 2004). Once again, immobilization of horseradish peroxidase allowed the construction of a peroxide biosensor.

Immunosensors have also been constructed utilizing the gold–thiol linkage to immobilize antibodies on gold surfaces. Human cytomegalovirus (Susmel et al. 2000) antibodies could be bound onto QCM crystals in a variety of binding formats, the best being binding to a thiosalicylic acid monolayer via a polylysine coupling layer. An epitope of the virus could be selectively detected via QCM at 1 µg ml^{-1}. Similar work bound thiolated *Salmonella* antibodies directly onto a QCM crystal (Park and Kim 1998) and gave a sensor capable of rapid (<1 hr) detection of *Salmonella* bacteria in the range of 10^6–10^8 colony-forming units (cfu) ml^{-1}.

Since oligonucleotides terminated with thiol groups can be easily made, use of these monolayers as potential DNA sensors comprises a large and varied field. One of the earliest reports in this field (Okahata et al. 1992) describes the immobilization of a thiol-substituted single-stranded DNA onto a QCM crystal (about 6×10^{11}, molecules cm^{-2} corresponding to 8% coverage). This bound complementary DNA and could also determine the number of mismatches. Adsorption could be reversed by heating.

DNA could be bound to gold via a thiol group or electrostatically onto a cationic thiol monolayer (Caruso et al. 1997). QCM and SPR results demonstrated the importance of DNA configuration at the surface. Thiol-bound DNA could bind complementary DNA even though it had a flat orientation on the surface, whereas the electrostatically bound DNA was shown to lie flat, to pack inefficiently, and to not hybridize with their complementary strands. Further work (Herne and Tarlov 1997) showed that mixed layers of thiol DNA with mercaptohexanol displayed higher efficiency (up to 100%), stability, and reversibility of hybridization compared with layers of solely thiolated DNA that did not allow hybridization to occur. Other work on mixed monolayers (Steel et al. 1998) showed good hybridization up to about 4×10^{12} DNA molecules cm^{-2}, but the efficiency fell off at higher surface densities.

Patterning of gold–thiol monolayers (Brockmann et al. 1999) by photochemically removing the thiols, followed by adsorption of an amino-substituted thiol to which DNA was covalently grafted, was used to form a DNA array. This array was then shown by SPR to selectively adsorb DNA-binding proteins onto the DNA-modified portions of the substrate. SPR imaging (Nelson et al. 2001) was also used to detect the hybridization of DNA probes with either complementary oligonucleotides or RNA from *E. coli* down to levels of 2 nmol l^{-1} without any need for labeling.

Biotinylated DNA could bind to a thiol-dextran-streptavidin modified QCM crystal, and has been shown to hybridize with a complementary strand found in pathogenic bacteria (Tombelli et al. 2000). The combination of this technique with *polymerase chain reaction* (PCR) allowed isolation and quantification of bacterial presence in water, vegetable, or human specimens. A similar procedure could make a biosensor capable of detection of the Tay-Sachs mutant gene (Bardea et al. 1999) by AC impedance.

Deposition of a range of thiolated oligonucleotides onto gold microelectrodes (Albers et al. 2003; Nebling et al. 2004) could be used to construct arrays that could then be used to detect a range of viral DNAs. Coupling of the microarray with soluable oligonucleotides led to detection of a variety of viral (e.g., herpes simplex, cytomegalovirus, and Epstein-Barr) DNAs at levels of about 2 nmol l^{-1}.

Much of the recent work has been on the optimization of these monolayers in sensors; for example, biotinylated thiols and biotinylated BSA were both deposited on QCM crystals with the biotin–thiol being shown to give better binding of streptavidin and better capture of biotinylated DNA (Aung et al. 2008). Partially fluorinated disulphides have been used to minimize nonspecific adsorption of proteins (Klein et al. 2008) on gold surfaces and to thereby improve the sensitivity of immunosensors made containing these materials. Besides using the avidin–biotin approach, other methods such as the synthesis of monolayers containing surface aldehyde groups have been utilized for the immobilization of proteins via interactions with the amino groups in the protein (Hahn et al. 2007). Later work by the same group utilized molecules containing a thiol group, a polyethylene glycol chain to minimize nonspecific interactions, and a latent aldehyde function for the binding of biomolecules (Hölzl et al. 2007). Other advantages of these monolayers include their ability to immobilize biomolecules in a defined manner and again improving sensitivity (Arya et al. 2009).

These monolayers have been used for the assembly of enzymatic biosensors; for example, apo-glucose oxidase could be combined with a 1.4-nanometer gold nanocrystal that had been functionalized with the enzyme cofactor flavin adenine dinucleotide (Xiao et al. 2003) and this system immobilized on a gold electrode via an aromatic dithiol. This "wiring" of the enzyme allowed very rapid electron transfer turnover about seven times that observed with molecular oxygen. Mixed samples of thiol-substituted carboxylic acids, which were used to immobilize an alcohol oxidase and a peroxidise enzyme (Hasunuma et al. 2004), were used along with an electron mediator 11-ferrocenyl-1-undecanethiol to modify a gold electrode. The resultant system could be used to measure methanol in plant extracts in real time.

These systems have also been used to greatly enhance the sensitivity and selectivity of DNA sensors, for example, with combinations of DNA-modified SAMs and enzymatic amplification being used to develop DNA assay platforms capable of detecting hybridization of single DNA strands (Li et al. 2008) that were contained in an array of femtoliter-sized reaction wells. DNA chains modified with ferrocenyl groups at the end opposite to the thiol could be used as probe DNA giving extremely sensitive sensors because the hybridization of the target with the probe DNA chain modified the distance between the electrode and the redox-active ferrocene (Fan et al. 2003). The resulting change in electron transfer efficiency is readily measured by cyclic voltammetry at target DNA concentrations as low as 10 pmol l^{-1}. Peptide nucleic acids have also been immobilized to make DNA sensors, and even though they are up to 6–7 nm in length, they are capable of distinguishing complementary chains from those with a single mismatch (Briones et al. 2005).

There has also been much recent work on utilizing SAMs in immunosensors; for example, the ratio of 11-mercaptoundecanoic acid/11-mercaptoundecanol has been shown to be crucial in the performance of immunosensors capable of detecting the antibodies characteristic of autoimmune diabetes (Ayela et al. 2007). Similar mixed SAMs containing polyethylene glycol chains have been shown to allow the construction of immunosensors for *E. coli* with low, nonspecific binding (Subramanian et al. 2006). The immobilization of glycopeptides onto SAMs on top of SPR chips can be utilized to facilitate the detection of carbohydrate binding proteins (Maljaars et al. 2008).

These examples demonstrate the stability, specificity, and versatility of gold–thiol monolayers within these applications. Also for systems requiring the use of very small amounts of material (a typical SAM requires 2×10^{-7} g cm^{-2}) and the potential control and dimensions of the ultrathin films, especially using techniques such as soft lithography to print chips and microfluidic devices with sizes measured in nanometers, means these types of systems could well be the next leap in biosensor technology.

12.7 NANOMATERIALS

We have already discussed within this chapter some of the uses of nanoparticles and carbon nanotubes when utilized to modify such systems as electrodes or SPR chips. However, a wide variety of nanomaterials have also been studied for other biosensing uses. What makes these nanomaterials of such interest is that quite often they are of a similar size as the biological molecules that they are being used in conjunction with. This allows the nano- and biocomponents to be highly dispersed and also aids intimate contact between them. Therefore, there has been a great deal of interest in using nanomaterials to "wire" biological species to various transducers. Also, many nanosystems display interesting optical properties such as shown by quantum dots or novel electrical properties as shown by conducting nanotubes and nanofibers. Finally, their extremely small size allows them to be utilized where larger systems would be unusable, for example, inside individual cells.

We have already discussed how gold nanoparticles can be used to wire enzymes to electrodes; however, other materials also demonstrate this property. Arrays of gold nanotubes could be formulated within the pores of polycarbonate track-etched membranes and used to immobilize glucose oxidase (Delvaux and Champagne 2003) to make sensors with extremely high sensitivity to glucose. By incorporation of a second enzyme, horseradish peroxidase, the potential required for the detection of glucose was lowered to a level where interferences from other electro-oxidizable compounds are minimal (Delvaux et al. 2005). Other materials have also been used, including inorganic nanowires. These have been investigated for their ability to be utilized in the detection of viruses due to their high sensitivity, potentially down to detection of a single virus as reviewed recently (Bentzen et al. 2006).

By use of controlled deposition of metals within track-etched membrane pores, "striped" nanowires of different metals can be synthesized—a metallic equivalent of a bar code. Various receptor molecules can be conjugated to these and a different bar code used for each receptor (Keating and Natan 2003). These can then be incubated with fluorescent labeled targets and the binding of targets detected by simultaneous analysis of reflectance and fluorescence images. The use of striped Au–Ag nanowires conjugated with antibodies gave an immunoassay for a variety of simulated biological warfare agents (Tok et al. 2006). Detection of degree of fluorophore-tagged antigen binding to the nanowires can be obtained after just 2–3 hours.

Nanowire-based field-effects transistors (FETs) have been intensely investigated (Patolsky et al. 2006) as possible platforms for biosensors. For example, silicon nanowires can be deposited onto the gates of FETs and substituted with tyrosine kinase to make biosensors capable of the detection of adenosine triphosphate (ATP) due to accumulation of negative charge (from the ATP). The interference effects of various drugs could also be determined. Similarly Si nanowires could be substituted with peptide nucleic acids and hybridization with the DNA counterpart detected (Hahm and Lieber 2004), allowing identification of fully complementary versus mismatched DNA samples and detection down to 10^{-14} mol l^{-1}.

A multiple array of nanowire FETs could be constructed and modified with antibodies to PSA, carcinoembryonic antigen, and mucin-1 (Zheng et al. 2005). This array proved capable of detecting marker proteins at levels below 10^{-12} pg ml^{-1} with ~100% specificity. PSA could also be detected at levels below 10^{-12} pg ml^{-1} in serum. Finally, this platform could also be used to detect viruses. Modification of a nanowire array with antibodies to the influenza virus and exposure to dilute solutions of the target clearly demonstrates discrete conductance changes characteristic of binding and unbinding of single viruses at the surface of nanowire devices (Patolsky et al. 2004). Optical measurements confirmed these were due to single binding/unbinding events. Devices containing two FETs, one modified with influenza antibody and one with adenovirus antibodies, conclusively showed selective binding of each virus to the correct array. This opens up the possibility of larger arrays capable of detection of multiple viruses in parallel. A number of carbon nanotube–coated FETs have also been reported (Byon et al. 2008).

Nanobiosensors also open up the possibility of investigating reactions that occur actually within a single cell. An Si/SiO$_2$ microneedle with an integrated silicon nanowire sensor at the tip could be constructed as shown in Figure 12.11. A proof of principle of the construction of this novel intracellular sensor and its responses to different pHs has already been published (Park et al. 2007). We can surmise that with suitable chemical modification, biosensors could be developed for localized biochemical

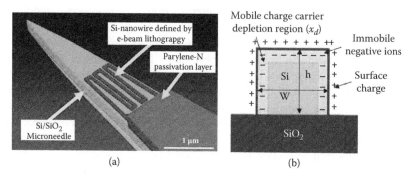

(a)   (b)

FIGURE 12.11 (a) Design of microneedle tip with integrated serpentine silicon nanowire structure on top surface; (b) cross-sectional view of silicon nanowire fabricated by e-beam lithography on silicon-on-insulator (SOI) substrate. (From Park, I., Z. Li, X. Li, A.P. Pisano, and R.S. Williams, 2007, *Biosensors and Bioelectronics* 22:2065–2070. With permission.)

sensing within cells, with an example quoted by the authors as neurotransmitter activities during the synaptic communication between neuron cells. Other workers have also developed complementary metal-oxide semiconductor (CMOS)-based nanosensors and have shown their ability to be used to detect cell signaling (Stern et al. 2007). For example, these workers studied T-lymphocyte activation where the activation of T cells induces intracellular signaling, one consequence of which is a release of acid that is detected by the sensor. The same sensor, when suitably derivatized, was capable of specific label-free detection of antibodies with limits of <100 fmol l^{-1}.

Silicon is one of the most common materials used to fabricate nanobiosensors; however, the use of carbon nanotubes and nanofibers is becoming more popular. We have already mentioned in this work how carbon nanostructures have enhanced some biosensing processes, but there are other ways in which these systems are unique (Wang and Lin 2008). Much of the early work on enzyme biosensors utilized oxidase enzymes; however, there has also been much interest in the use of dehydrogenase enzymes. However, these usually require a cofactor such as NAD/NADH. To oxidize the NADH often requires a high overpotential; however, carbon nanotubes offer a much-reduced overpotential (Musameh et al. 2002), indicating great promise for these materials in the development of sensitive, low-potential amperometric biosensors utilizing dehydrogenase enzymes.

Besides simply immobilizing these nanostructures on an electrode in random fashion, techniques for growing carbon nanotubes vertically from a surface also exist. These can be either as a tightly packed "forest" of nanotubes or more widely spaced. If the spacing between each nanotube is far enough, then each nanotube will act as an isolated nanoelectrode and the system as a whole as a nanoelectrode array, with the advantages noted earlier for microelectrode arrays. For example, a low-density nanoelectrode array could be grown from a surface coated with a patterned catalyst (Lin et al. 2004). Arrays of vertical carbon nanotubes could be vapor deposited onto a chromium electrode and the tops of the nanotubes selectively modified to allow covalent immobilization of glucose oxidase just on the ends of the nanotubes (Lin et al. 2004). This array of enzyme-modified nanoelectrodes could detect the reduction of enzymatically produced hydrogen peroxide in the presence of glucose at –0.2 V versus Ag/AgCl with no interference being observed from ascorbate, acetaminophen, or uric acid, thereby eliminating the need for permselective membrane barriers or artificial electron mediators and greatly simplifying the sensor design and fabrication.

Similar arrays were also used as the substrate for DNA biosensors where oligonucleotides are selectively grafted chemically onto the open ends of the nanotube array. Sensitivity could dramatically improve by lowering the nanotube density, the detection method being the oxidation of guanine residues mediated by a ruthenium species (Li et al. 2003). The hybridization of subattomole DNA targets can be detected. Other workers pyrolyzed iron phthalocyanine on gold electrodes to give a carbon nanotube "forest" that was then plasma treated to functionalize the ends of the tubes with acid groups. These could be substituted with single-stranded DNA and then used to detect hybridization of ferrocene-labeled target DNA (He and Dai 2004). The aligned arrays gave a 20-fold higher response than similar electrodes modified with "flat" nanotubes.

Immunosensors have also been constructed using these vertical nanotubes (Wang and Lin 2008). Arrays of nanotubes on pyrolytic graphite could be substituted with antibodies and used to detect species such as HSA (Yu et al. 2005). Combinations of these arrays with use of horseradish peroxidise–labeled secondary antibodies allowed detection of the target down to 75 pmol ml^{-1}. Other workers (Okuno et al. 2007) modified the arrays with antibodies to PSA and utilized differential pulse voltammetry to detect binding of the antigen at levels down to 0.25 ng ml^{-1}.

Carbon nanofibers have also been used in biosensors as reviewed here (Wang and Lin 2008). Workers have grafted oligonucleotides onto vertical arrays of nanofibers and onto glassy carbon and showed that 2.3×10^{14} DNA molecules cm^{-2} will hybridize to the nanofiber samples approximately eight times higher than for a glassy carbon sample (Baker et al. 2005). It has also proven possible using combined photolithographic and chemical or electrochemical techniques to spatially control functionalization across both the surface of an array and along the length of the vertical nanofibers (McKnight et al. 2006). This control of structure (shown schematically in Figure 12.12) could be visualized using fluorescent and electron microscopy. A wide

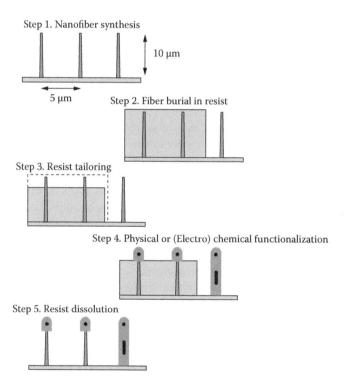

FIGURE 12.12 General scheme for photoresist-based blocking of chemical or electrochemical functionalization of arrays of vertically aligned carbon nanofibers. Resist layers may be used to block functionalization sites specifically along the nanofiber height (two fibers depicted at the left of each drawing) or site specifically at different regions of an array (single fiber depicted at the right of each drawing). (From McKnight, T.E., C. Peeraphatdit, S.W. Jones et al., 2006, *Chemistry of Materials* 18:3203–3211. With permission from the American Chemical Society.)

variety of metal oxide nanowires and nanotubes are also being investigated for use in biosensors, and much of this work is reviewed in Liu (2008).

12.8 APTAMERS

Aptamers are synthetic oligonucleic acids, or alternatively, peptides, that bind to a specific target molecule, although natural aptamers also exist in riboswitches. A riboswitch is a part of an mRNA molecule capable of binding a small target molecule and whose binding of the target affects and regulates the gene's activity. The usual method for the synthesis of aptamers is to utilize combinatorial methods to synthesize a wide range of random sequences (typically 10^{14}–10^{15} molecules) flanked on both sides with primers for PCR. These are then all screened for their ability to bind to a chosen target, usually immobilized on a column or on particles. Typical targets can be small molecules, proteins, nucleic acids, and even cells, tissues, and organisms. Unbound sequences are separated and the bound sequences isolated. The bound sequences are amplified using PCR, and then the cycle is repeated. As the number of cycles increases, the aptamer pool "evolves" so that only the sequences with the strongest binding remain. This process is known as systematic evolution of ligands by exponential enrichment (SELEX) (Ellington and Szostak 1990). The final aptamers are then isolated, their sequences determined, and then they can be synthesized in a pure form in sizable quantities.

Aptamers have been utilized for basic research and proposed to have clinical applications as macromolecular drugs. The main focus of interest in these materials is due to the fact that aptamers offer molecular recognition properties that rival those of antibodies. However, in many ways they are superior in that they can be chemically synthesized without the need for immunization of animals, they can display much better storage properties than antibodies, and they usually do not provoke any immune response—all of which are important if they are to be used in therapeutic applications.

Aptamers are not simple linear chains of nucleotides or amino acids but instead fold in a wide variety of three-dimensional shapes, which means that in a large pool of aptamers, some will bind to almost any target. Once the strongest binders are selected by SELEX and the structures of these deduced, they can be synthesized and used in sensing applications in the same manner as antibodies or DNA. A detailed discussion of this is outside the scope of this chapter but the subject has been recently and extensively reviewed (Mairal et al. 2008; De-los-Santos-Alvarez et al. 2008; Cheng et al. 2009; Mascini 2009).

The technologies of aptamer-based sensors are the same as described above for enzyme-, DNA-, and antibody-based sensors—it is just the recognition moiety that has changed, and therefore we will just give brief examples of aptamer-based sensors. An aptamer containing a tethered redox-active group could be immobilized onto a gold electrode via a thiol unit on the other end of the strand. Binding of thrombin from blood serum hindered electron transfer between the redox probe and the electrode and caused a drop in the recorded signal (Xiao et al. 2005). Aptamers to lysozyme could also be immobilized on electrodes and ruthenium moieties incorporated in the film. Binding of the cationic lysozyme displaced many of these ruthenium sensors and modified the electrochemical behavior of the sensor (Cheng et al. 2007). A variety of other electrochemical aptamer-based biosensors have also been reported (De-los-Santos-Alvarez et al. 2008). Optical methods such as SPR have also been utilized, with, for example, a number of aptamers to IgG-E being able to be immobilized on an SPR chip and SPR imaging used to assess which aptamer bound most strongly (Wang et al. 2007).

12.9 APPLICATIONS OF BIOSENSORS AND CONCLUSIONS

At this time the commercial biosensor market is dominated by glucose monitoring. The expected increase in diabetes due to greater levels of obesity and an aging population is predicted to increase the demand for these sensors (Luong et al. 2008). However, this is not the only market for biosensors, and we predict more will follow. One will almost certainly be the detection of sensors for biomarkers of various physiological conditions. We have within this chapter described construction of immunosensors and DNA-based sensors for markers for cancer, stroke, and various genetic conditions. It is inevitable that these will find their way into surgery and the marketplace.

As this chapter has shown, there are a wide variety of novel, exciting techniques becoming ever more widely available for the detection of specific nucleic acid sequences. The combination of this, with the miniaturization afforded by the use of microfluidics and also arraying techniques (which can lay thousands of different biological samples on a microscope slide), should greatly increase the throughput and reduce the time and cost of these measurements. We have already considered the possibility of the $1000 genome (Mardis 2006), and this opens up the possibility of having medical treatments personalized to give optimal efficacy for each individual patient.

Food pathogen detection is also a major potential market. On-site testing would reduce the possibility of contaminated food being allowed for public consumption, minimizing the dangers of widespread food poisoning that causes sickness and potentially death and the possible criminal charges, bad publicity, and litigation that can follow. The potential market for this has been estimated as $150 million per year (Nugen and Baeumner 2008). Current tests take time and require skilled personnel, and in this context a fast, inexpensive test for common pathogens would be of great interest to food manufacturers. Other tests such as for antibiotics or genetic modification could also find a market.

Finally, there is the issue of homeland security. Rapid detection of chemical or biological weapons would greatly minimize casualties and also the resultant cleanup operations in case of deliberate attacks using these materials. Biosensors have the potential to fill all the above applications, and it is clear that the demand for these systems can only increase.

REFERENCES

Ahmed, M.U., M.M. Hossain, and E. Tamiya. 2008. Electrochemical biosensors for medical and food applications. *Electroanalysis* 20:616–626.
Ahmed, S., C. Dack, G. Farace, G. Rigby, and P. Vadgama. 2005. Tissue-mplanted glucose needle electrodes: Early sensor stabilization and achievement of tissue-blood correlation during the run period. *Analytica Chimica Acta* 537:153–161.

Albers, J., T. Grunwald, E. Nebling, G. Piechotta, and R. Hintsche. 2003. Electrical bio-chip technology—A tool for microarrays and continuous monitoring. *Analytical and Bioanalytical Chemistry* 377:521–527.

Andreescu, S., and O.A. Sadik. 2005. Advanced electrochemical sensors for cell cancer monitoring. *Methods* 37:84–93.

Applegate, R.W., D.N. Schafer, W. Amir et al. 2007. Optically integrated microfluidic systems for cellular characterization and manipulation. *Journal of Optics A: Pure and Applied Optics* 9:S122–S128.

Arya, S.K., P.R. Solanki, M. Datta, and B.D. Malhotra. 2009. Recent advances in self-assembled monolayers based biomolecular electronic devices. *Biosensors and Bioelectronics* 24:2810–2817.

Aung, K.M.M., X.N. Ho and X.D. Su. 2008. DNA assembly on streptavidin modified surface: A study using quartz crystal microbalance with dissipation or resistance measurements. *Sensors and Actuators B: Chemical* 131:371–378.

Ayela, C., F. Roquet, L. Valera, C. Granier, L. Nicu, and M. Pugniere. 2007. Antibody–antigenic peptide interactions monitored by SPR and QCM-D: A model for SPR detection of IA-2 autoantibodies in human serum. *Biosensors and Bioelectronics* 22:3113–3119.

Bain, C.D., E.B. Troughton, Y.T. Tao, J. Evall, G.M. Whitesides, and R.G. Nuzzo. 1989. Formation of monolayer films by the spontaneous assembly of organic thiols from solution onto gold. *Journal of the American Chemical Society* 111:321–335.

Baker, S.E., K.Y. Tse, E. Hindin, B.M. Nichols, T.L. Clare, and R.J. Hamers. 2005. Covalent functionalization for biomolecular recognition on vertically aligned carbon nanofibers. *Chemistry of Materials* 17:4971–4978.

Bardea, A., F. Patolsky, A. Dagan, and I. Willner. 1999. Sensing and amplification of oligonucleotide-DNA interactions by means of impedance spectroscopy: A route to a Tay-Sachs detector. *Chemical Communications* 21–22.

Barton, A.C., F. Davis, and S.P.J. Higson. 2008b. Labeless immunosensor assay for the stroke marker protein neuron specific enolase based upon an AC impedance protocol. *Analytical Chemistry* 80:9411–9416.

Barton, A.C., F. Davis, and S.P.J. Higson. 2009b. Labeless immunosensor assay for the stroke marker protein S-100[β] based upon an AC impedance protocol. *Analytical Letters* 43: 2160–2170.

Barton, A.C., F. Davis, S.P.J. Higson. 2008a. Labeless immunosensor assay for prostate specific antigen with pg ml^{-1} limits of detection based upon an AC impedance protocol. *Analytical Chemistry* 80:6198–6205.

Barton, A.C., S.D. Collyer, F. Davis et al. 2004. Sonochemically fabricated microelectrode arrays for biosensors offering widespread applicability: Part I. *Biosensors and Bioelectronics* 20:328–337.

Barton, A.C., S.D. Collyer, F. Davis et al. 2009a. Labeless AC impedimetric antibody based sensors with pg ml^{-1} sensitivities for point-of-care biomedical applications. *Biosensors and Bioelectronics* 24:1090–1095.

Baselt, D.R., G.U. Lee, M. Natesan et al. 1998. A biosensor based on magnetoresistance technology. *Biosensors and Bioelectronics* 13:731–739.

Bentzen, E., D.W. Wright, and J.E. Crowe. 2006. Nanoscale tools for rapid and sensitive diagnosis of viruses. *Future Virology* 1:769–781.

Berkenpas, E., P. Millard, and C.M. Pereira da. 2006. Detection of *Escherichia coli* O157:H7 with langasite pure shear horizontal surface acoustic wave sensors. *Biosensors and Bioelectronics* 21:2255–2262.

Bidan, G., M. Billon, M.L. Calvo-Muñoz, and A. Dupont-Fillard. 2004. Bio-assemblies onto conducting polymer support: Implementation of DNA chips. *Molecular Crystals and Liquid Crystals* 418:255–270.

Bisoffi, M., B. Hjelle, D.C. Brown et al. 2008. Detection of viral bioagents using a shear horizontal surface acoustic wave biosensor. *Biosensors and Bioelectronics* 23:1397–1403.

Boduroglu, S., M. Cetinkaya, W.J. Dressick, A. Singh, and M.C. Demirel. 2007. Controlling the wettability and adhesion of nanostructured poly-(p-xylylene) films. *Langmuir* 23:11391–11395.

Briones, C., E.M. Marti, C.G. Navarro, V. Parro, E. Roman, and J.A.M. Gago. 2005. Structural and functional characterization of self-assembled monolayers of peptide nucleic acids and its interaction with complementary DNA. *Journal of Molecular Catalysis A-Chemical* 228:131–136.

Brockmann, J.M., A.G. Frutos, and R.C. Corn. 1999. A multistep chemical modification procedure to create DNA arrays on gold surfaces for the study or Protein-DNA interactions with surface plasmon resonance imaging. *Journal of the American Chemical Society* 121:8044–8051.

Byon, H.R., S. Kim, and H.C. Choi. 2008. Label-free biomolecular detection using carbon nanotube field effect transistors. *Nano* 3:415–31.

Cai, H., Y. Xu, P.G. He, and Y.Z. Fang. 2003. Indicator free DNA hybridization detection by impedance measurement based on the DNA-doped conducting polymer film formed on the carbon nanotube modified electrode. *Electroanalysis* 15:1864–1870.

Campas, M., and I. Katakis. 2004. DNA biochip arraying, detection and amplification strategies. *TrAC Trends in Analytical Chemistry* 23:49–62.

Caruso, F., E. Rodda, D.N. Furlong, and V. Haring. 1997. DNA binding and hybridisation on gold and derivatised surfaces. *Sensors and Actuators B: Chemical* 41:189–197.

Chen, S.-J., F.C. Chien, G.Y. Lin, and K.C. Lee. 2004. Enhancement of the resolution of surface plasmon resonance biosensors by control of the size and distribution of nanoparticles. *Optics Letters* 29:1390–1392.

Cheng, A.K.H., B. Ge, and H.-Z. Yu. 2007. Aptamer-based biosensors for label-free voltammetric detection of lysozyme. *Analytical Chemistry* 79:5158–5164.

Cheng, A.K.H., D. Sen, and H.-Z. Yu. 2009. Design and testing of aptamer-based electrochemical biosensors for proteins and small molecules. *Bioelectrochemistry* 77:1–12.

Chunder, A., K. Etcheverry, G. Londe, H.J. Cho, and L. Zhai. 2009. Conformal switchable superrhydrophobic/hydrophilic surfaces for microscale flow control. *Colloids and Surfaces A: Physicochemical and Engineering Aspects* 333:198–193.

Clark, L.C., and I.R. Lyons. 1962. Electrode systems for continuous monitoring in cardiovascular surgery. *Annals of the New York Academy of Sciences* 102:29–45.

Cosnier, S. 2005. Affinity biosensors based on electropolymerized films. *Electroanalysis* 17:1701–15.

Darder, M., E. Casero, F. Pariente, and E. Lorenzo. 2000. Biosensors based on membranebound enzymes immobilized in a 5-(octyldithio)-2-nitrobenzoic acid layer on gold electrodes. *Analytical Chemistry* 72:3784–3792.

Davis, F., A.V. Nabok, and S.P.J. Higson. 2004. Species differentiation by DNA-modified carbon electrodes using an AC impedimetric approach. *Biosensors and Bioelectronics* 20:1531–1538.

Davis, F., and S.P.J. Higson. 2005. Structured thin films as functional components within biosensors. *Biosensors and Bioelectronics* 21:1–20.

De-los-Santos-Alvarez, N., M.J. Lobo-Castañón, A.J. Miranda-Ordieres, and P. Tuñón-Blanco. 2008. Aptamers as recognition elements for label-free analytical devices. *TrAC-Trends in Analytical Chemistry* 27:437–446.

Delvaux, M., A. Walcaarius, and S.D. Champagne. 2005. Bienzyme HRP–GOx-modified gold nanoelectrodes for the sensitive amperometric detection of glucose at low overpotentials. *Biosensors and Bioelectronics* 20:1587–1594.

Delvaux, M., and S.D. Champagne 2003. Immobilisation of glucose oxidase within metallic nanotubes arrays for application to enzyme biosensors. *Biosensors and Bioelectronics* 18:943–951.

Deng, L, L.L. Bao, W.Z. Wei, H. Nie, and S.Z. Yao. 1996. Continuous measurement of bacterial populations on the surface of a solid medium with a thickness shear mode acoustic resonator in series. *Enzyme and Microbial Technology* 19:525–528.

Deobagkar, D.D., V. Limaye, S. Shinha, and R.D.S. Yadava. 2005. Acoustic wave immunosensing of *Escherichia coli* in water. *Sensors and Actuators B: Chemical* 104:85–89.

Dickert, F.L., P. Lieberzeit, and O. Hayden. 2003. Sensor strategies for microorganism detection—From physical principles to imprinting procedures. *Analytical and Bioanalytical Chemistry* 377:540–549.

Ellington, A.D., and J.W. Szostak. 1990. *In vitro* selection of RNA molecules that bind specific ligands. *Nature* 346:818–22.

Epstein, J.R., and D.R. Walt. 2003. Fluorescence-based fibre optic arrays: A universal platform for sensing. *Chemical Society Reviews* 32:203–214.

Eun, A.J.-C, L. Huang, F.-T. Chew, S.F.-Y. Li, and S.-M. Wong. 2002. Detection of two orchid viruses using quartz crystal microbalance (QCM) immunosensors. *Journal of Virological Methods* 99:71–79.

Fan, C., K.W. Plaxco, and A.J. Heeger. 2003. Electrochemical interrogation of conformational changes as a reagentless method for the sequence-specific detection of DNA. *Proceedings of the National Academy of Sciences of the United States of America* 100:9134–9137.

Fodor, S.P., J.L. Read, M.C. Pirrung, L. Stryer, A.T. Lu, and D. Solas. 1991. Light-directed, spatially addressable parallel chemical synthesis. *Science* 251:767–773.

Freeman, J., and L. Lyons. 2008. Clarifying details on the latest continuous glucose monitoring device. *Diabetes Spectrum* 21: 224.

Fritz, J., M.K. Baller, H.P. Lang et al. 2000. Translating biomolecular recognition into nanomechanics. *Science* 288:316–318.

Gajendragad, M.R., K.N.Y. Kamath, P.Y. Anil, K. Prabhudas, and C. Natarajan. 2001. Development and standardization of a piezo electric immunobiosensor for foot and mouth disease virus typing. *Veterinary Microbiology* 78:319–330.

Gajovic, N., K. Habermuller, A. Warsinke, W. Schuhmann, and F.W. Scheller. 1999. A pyruvate oxidase electrode based on an electrochemically deposited redox polymer. *Electroanalysis* 11:1377–1383.

Garifallou, G.-Z., G. Tsekenis, F. Davis et al. 2007. Labeless immunosensor assay for fluoroquinolone antibiotics based upon an AC impedance protocol. *Analytical Letters* 40:1412–1442.

Gautier, C., C. Cougnon, J.-F. Pilard, and N. Casse. 2005. Label-free detection of DNA hybridisation based on EIS investigation of conducting properties of functionalised polythiophene matrix. *Journal of Electroanalytical Chemistry* 587:276–283.

Gautier, C., C. Cougnon, J.-F. Pilard, N. Casse, B. Chenais, and M. Laulier. 2007. Detection and modelling of DNA hybridisation by EIS measurements. *Biosensors and Bioelectronics* 22:2025–2031.

Geetha, S., C.R.K. Rao, M. Vijayan, and D.C. Trivedi. 2006. Biosensing and drug delivery by polypyrrole. *Analytica Chimica Acta* 568:119–125.

Gerard, M., A. Chaubey, and B.D. Malhotra. 2002. Application of conducting polymers to biosensors. *Biosensors and Bioelectronics* 17:345–359.

Ghindilis, A.L., M.W. Smith, K.R. Scharztkopf et al. 2007. CombiMatrix oligonucleotide arrays: Genotyping and gene expression assays employing electrochemical detection. *Biosensors and Bioelectronics* 22: 1853–1860.

Gomez-Sjoberg, R., D.T. Morisette, and R. Bashir. 2005. Impedance microbiology-on-a-chip: Microfluidic bioprocessor for rapid detection of bacterial metabolism. *Journal of Microelectromechanical Systems* 14:829–838.

Gooding, J.J., V.G. Praig, and E.A.H. Hall. 1998. Platinum-catalysed enzyme electrodes immobilised on gold using self-assembled layers. *Analytical Chemistry* 70:2396–2402.

Grogan, C., R. Raiteri, and G.M.O'Connor. 2001. Characterisation of an antibody coated microcantilever as a potential immuno-based biosensor. *Biosensors and Bioelectronics* 17:201–207.

Guiseppi-Elie, A., S. Brahim, G. Slaughter, and K.R. Ward. 2005. Design of a subcutaneous implantable biochip for monitoring of glucose and lactate. *IEEE Sensors Journal* 5:345–355.

Hahm, J., and C.M. Lieber. 2004. Direct ultrasensitive electrical detection of DNA and DNA sequence variations using nanowire nanosensors. *Nano Letters* 4:51–54.

Hahn, C.D., C. Leitner, T. Weinbrenner et al. 2007. Self-assembled monolayers with latent aldehydes for protein immobilization. *Bioconjugate Chemistry* 18:247–253.

Hanazato, Y., M. Nakako, S. Satorus, and M. Mitsuo. 1989. Integrated multi-biosensors based on ion-sensitive field-effect transistor using photolithographic techniques. *IEEE Transactions on Electron Devices* 36:1303–1310.

Hansen, K.M., H.-F. Ji, and G. Wu. 2001. Cantilever-based optical deflection assay for discrimination of DNA single-nucleotide mismatches. *Analytical Chemistry* 73:1567–1571.

Hasunuma, T., S. Kuwabata, E. Fukusaki, and A. Kobayashi. 2004. Real-time quantification of methanol in plants using a hybrid alcohol oxidase–peroxidase biosensor. *Analytical Chemistry* 76:1500–1506.

He, P., and L. Dai. 2004. Aligned carbon nanotube–DNA electrochemical sensors. *Chemical Communications* 348–349.

Heller, A. 1990. Electrical wiring of redox enzymes. *Accounts of Chemical Research* 23:128–134.

Herne, T.M., and M.J. Tarlov. 1997. Characterisation of DNA probes immobilised on gold surfaces. *Journal of the American Chemical Society* 119:8916–8920.

Hiratsuka, A., K. Kojima, H. Muguruma, K.H. Lee, H. Suzuki, and I. Karube. 2005. Electron transfer mediator micro-biosensor fabrication by organic plasma process. *Biosensors and Bioelectronics* 21:957–964.

Hofmann, O., P. Miller, P. Sullivan et al. 2005. Thin-film organic photodiodes as integrated detectors for microscale chemiluminescence assays. *Sensors and Actuators B: Chemical* 106:878–884.

Hölzl, M., A. Tinazli, C. Leitner et al. 2007. Protein-resistant self-assembled monolayers on gold with latent aldehyde functions. *Langmuir* 23:5571–5577.

Howe, E., and G.A. Harding. 2000. Comparison of protocols for the optimisation of detection of bacteria using a surface acoustic wave (SAW) biosensor. *Biosensors and Bioelectronics* 15:641–649.

Hrapovic, S., Y.L. Liu, K.B. Male, and J.H.T. Luong. 2004. Electrochemical biosensing platforms using platinum nanoparticles and carbon nanotubes. *Analytical Chemistry* 76:1083–1088.

Huber, F., M. Hegner, C.H. Gerber, H.J. Guntherodt, and H.P. Lang. 2005. Label free analysis of transcription factors using microcantilever arrays. *Biosensors and Bioelectronics* 21:1599–1605.

Ivarsson, B., and M. Malmqvist. 2002. Surface plasmon resonance: Development and use of Biacore instruments for biomolecular interaction analysis. In *Biomolecular sensors*, ed. E. Gizeli, and C.R. Lowe, 241–268. London: Taylor & Francis.

Joos, T., M. Shrenk, P. Höpfl et al. 2000. Microarray ELISA for autoimmune diagnostics. *Electrophoresis* 21:2641–2650.

Juan, Z., and T.M. Swager. 2005. Poly(arylene ethynylene)s in chemosensing and biosensing. *Advances in Polymer Science* 177:151–179.

Karlsson, R. 2004. SPR for molecular interaction analysis: A review of emerging application areas. *Journal of Molecular Recognition* 17:151–161.

Keating, C.D., and M.J. Natan. 2003. Striped metal wires as building blocks and optical tags. *Advanced Materials* 15:451–454.

Kim, D.S., S.H. Lee, C.H. Ahn, J.Y. Lee, and T.H. Kwon. 2006a. Disposable integrated microfluidic biochip for blood typing by plastic microinjection moulding. *Lab on a Chip* 6:794–802.

Kim, J., N. Singh, and L.A. Lyon. 2006b. Label-free biosensing with hydrogel microlenses. *Angewandte Chemie-International Edition* 45:1446–1449.

Klein, E., P. Kerth, and L. Lebeau. 2008. Enhanced selective immobilization of biomolecules onto solid supports coated with semifluorinated self-assembled monolayers. *Biomaterials* 29:204–214.

Kudo, H., T. Sawada, E. Kazawa, H. Yoshida, Y. Iwasaki, and K. Mitsubayashi. 2006. A flexible and wearable glucose sensor based on functional polymers with Soft-MEMS techniques. *Biosensors and Bioelectronics* 22:558–562.

Lange, K., G. Blaess, A. Voigt, R. Gotzen, and M. Rapp. 2006. Integration of a surface acoustic wave biosensor in a microfluidic polymer chip. *Biosensors and Bioelectronics* 22:227–232.

Law, K.A., and S.P.J. Higson. 2005. Sonochemically fabricated acetylcholinesterase microelectrode arrays within a flow injection analyser for the determination of organophosphate pesticides. *Biosensors and Bioelectronics* 20:1914–1924.

Li, J., H.T. Ng, A. Cassell et al. 2003. Carbon nanotube nanoelectrode array for ultrasensitive DNA detection. *Nano Letters* 3:597–602.

Li, Z.H., R.B. Hayman, and D.R. Walt. 2008. Detection of single-molecule DNA hybridization using enzymatic amplification in an array of femtoliter-sized reaction vessels *Journal of the American Chemical Society* 130:12622–12623.

Ligler, F.S. 2009. Perspective on optical biosensors and integrated sensor systems. *Analytical Chemistry* 81:519–26.

Ligler, F.S., M. Breimer, J.P. Golden et al. 2002. Integrating waveguide biosensor. *Analytical Chemistry* 74:713–719.

Lin, J.H., and H.X. Ju. 2005. Electrochemical and chemiluminescent immunosensors for tumor markers. *Biosensors and Bioelectronics* 20:1461–1470.

Lin, Y., F. Lu, Y. Tu, and Z. Ren. 2004. Glucose biosensors based on carbon nanotube nanoelectrode ensembles. *Nano Letters* 4:191–195.

Liu, A. 2008. Towards development of chemosensors and biosensors with metal-oxide based nanowires or nanotubes. *Biosensors and Bioelectronics* 24:167–177.

Liu, Y., M. Wang, F. Zhao, Z. Xu, and S. Dong 2005. The direct electron transfer of glucose oxidase and glucose biosensor based on carbon nanotubes/chitosan matrix. *Biosensors and Bioelectronics* 21:984–988.

Livache, T., B. Fouque, A. Roget et al. 1998. Polypyrrole DNA chip on a silicon device: Example of Hepatitis C virus genotyping. *Analytical Biochemistry* 255:188–194.

Lueking, A., D.J. Cahill, and S. Mullner 2005. Protein biochips: A new and versatile platform technology for molecular medicine. *Drug Discovery Today* 10:789–794.

Lueking, A., M. Horn, H. Eickhoff, K. Büssow, H. Lehrach, and G. Walter. 1999. Protein microarrays for gene expression and antibody screening. *Analytical Biochemistry* 270:103–111.

Luo, X.-L., J.-J. Xu, W. Zhao, and H.-Y. Chen. 2004. Glucose biosensor based on ENFET doped with SiO2 nanoparticles. *Sensors and Actuators B: Chemical* 97:249–255.

Luong, J.H.T., K. Male, and J.D. Glennon. 2008. Biosensor technology: Technology push versus market pull. *Biotechnology Advances* 26:492–500.

Mairal, T., V.C. Ozalp, P.L. Sanchez, M. Mir, I. Katakis, and C.K. O'Sullivan. 2008. Aptamers: Molecular tools for analytical applications. *Analytical and Bioanalytical Chemistry* 390:989–1007.

Maljaars, C.E.P., A.C. Souza, K.M. Halkes et al. 2008. The application of neoglycopeptides in the development of sensitive surface plasmon resonance-based biosensors. *Biosensors and Bioelectronics* 24:60–65.

Mano, N., and A. Heller. 2005. Detection of glucose at 2 fM concentration. *Analytical Chemistry* 77:729–732.

Mardis, E.R. 2006. Anticipating the 1,000 dollar genome. *Genome Biology* 7:112.

Mascini, M. 2009. *Aptamers in bioanalysis.* Wiley-Blackwell.

McKnight, T.E., C. Peeraphatdit, S.W. Jones et al. 2006. Site-specific biochemical function-alization along the height of vertically aligned carbon nanofiber arrays. *Chemistry of Materials* 18:3203–3211.

McRipley, M.A., and R.A. Linsenmeier. 1996. Fabrication of a mediated glucose oxidase recessed microelectrode for the amperometric determination of glucose. *Journal of Electroanalytical Chemistry* 414:235–246.

Mena, M.L., P. Yanez-Sedeno, and J.M. Pingarron 2005. A comparison of different strate-gies for the construction of amperometric enzyme biosensors using gold nanoparticle-modified electrodes. *Analytical Biochemistry* 336:20–27.

Moll, N., E. Pascal, D.H. Dinh, et al. 2008. Multipurpose love acoustic wave immunosensor for bacteria, virus or proteins detection. *ITBM-RBM* 29:155–161.

Moussy, F., D.J. Harrison, D.W. O'Brien, and R.V. Rajotte. 1993. Performance of subcu-taneously implanted needle-type glucose sensors employing a novel trilayer coating. *Analytical Chemistry* 65:2072–2077.

Muramatsu, H., K. Kajiwara, E. Tamiya, and I. Karube. 1986. Piezoelectric immuno sensor for the detection of candida albicans microbes. *Analytical and Bioanalytical Chemistry* 188:257–261.

Musameh, M., J. Wang, A. Merkoci, and Y.H. Lin. 2002. Low-potential stable NADH detection at carbon-nanotube-modified glassy carbon electrodes. *Electrochemistry Communications* 4:743–746.

Nebling, E., T. Grunwald, J. Albers, P. Schafer, and R. Hintsche. 2004. Electrical detection of viral DNA using ultramicroelectrode arrays. *Analytical Chemistry* 76:689–696.

Nelson, B.P., T.E. Grimsrud, M.R. Liles, R.M. Goodman, and R.M. Corn. 2001. Surface plas-mon resonance imaging measurements of DNA and RNA hybridisation adsorption onto DNA microarrays. *Analytical Chemistry* 73:1–7.

Newman, J.D., L.J. Tigwell, A.P.F. Turner, and P.J. Warner. 2004. Biosensors: A clearer view. In *Biosensors 2004—The 8th world congress on biosensors.* New York: Elsevier.

Nugen, S.R., and A.J. Baeumner. 2008. Trends and opportunities in food pathogen detection. *Analytical and Bioanalytical Chemistry* 391:451–454.

O'Sullivan, C.K., and G.G. Guilbault. 1999. Commercial quartz crystal microbalances—Theory and applications. *Biosensors and Bioelectronics* 14:663–670.

Ohara, T.J., R. Rajagopalan, and A. Heller. 1994. Wired enzyme electrodes for amperometric determination of glucose or lactate in the presence of interfering substances. *Analytical Chemistry* 66:2451–2457.

Okahata, Y., Y. Matsunobu, K. Ijiro, M. Mukae, A. Murakami, and K. Makino. 1992. Hybridisation of nucleic acids immobilised on a quartz crystal microbalance. *Journal of the American Chemical Society* 114:8299–8300.

Okuno, J., K. Maehashi, K. Kerman, Y. Takamura, K. Matsumoto, and E. Tamiya. 2007. Label-free immunosensor for prostate-specific antigen based on single-walled carbon nano-tube array-modified microelectrodes. *Biosensors and Bioelectronics* 22:2377–2381.

Ouerghi, O., A. Touhami, N. Jaffrezic-Renault et al. 2002. Impedimetric immunosensor using avidin-biotin for antibody immobilization. *Bioelectrochemistry* 56:131–133.

Pang, P., W. Yang, S. Huang, Q. Cai, and S. Yao. 2007. Measurement of glucose concentra-tion in blood plasma based on a wireless magnetoelastic biosensor. *Analytical Letters* 40:897–904.

Park, I., Z. Li, X. Li, A.P. Pisano, and R.S. Williams. 2007. Towards the silicon nanowire-based sensor for intracellular biochemical detection. *Biosensors and Bioelectronics* 22:2065–2070.

Park, I.-S., and N. Kim. 1998. Thiolated *Salmonella* antibody immobilisation onto the gold surface of piezoelectric quartz crystal. *Biosensors and Bioelectronics* 13:1091–1097.

Patolsky, F., G.F. Zheng, and C.M. Lieber. 2006. Nanowire based biosensors. *Analytical Chemistry* 78:4260–4269.

Patolsky, F., G.F. Zheng, O. Hayden, M. Lakadamyali, X. Zhuang, and C.M. Lieber. 2004. Electrical detection of single viruses. *Proceedings of the National Academy of Sciences of the United States of America* 101:14017–14022.

Peng, H., C. Soeller, N. Vigar et al. 2005. Label-free electrochemical DNA sensor based on functionalised conducting copolymer. *Biosensors and Bioelectronics* 20:1821–1828.

Peng, H., L. Zhang, C. Soeller, and J. Travas-Sejdic. 2007. Synthesis of a functionalized polythiophene as an active substrate for a label-free electrochemical genosensor. *Polymer* 48:3413–3419.

Peng, H., L. Zhang, C. Soeller, and J. Travas-Sejdic. 2009. Conducting polymers for electrochemical DNA sensing. *Biomaterials* 30:2132–2148.

Pickup, J.C., F. Hussain, N.D. Evans, and N. Sachedina. 2005. *In vivo* glucose monitoring: The clinical reality and the promise. *Biosensors and Bioelectronics* 20:1897–1902.

Qu, X., L. Bao, X. Su, and W. Wei. 1998. A new method based on gelation of tachypleus amebocyte lysate for detection of *Escherichia* coliform using a series piezoelectric quartz crystal sensor. *Analytica Chimica Acta* 374:47–52.

Raiteri, R., M. Grattarola, H.-J. Butt, and P. Skládal. 1999. Micromechanical cantilever-based biosensors. *Sensors and Actuators B: Chemical* 79:115–126.

Ramanavicius, A., A. Kausaite, and A. Ramanaviciene. 2005. Polypyrrole-coated glucose oxidase nanoparticles for biosensor design. *Sensors and Actuators B: Chemical* 111:532–539.

Raz, S.R., M.G.E.G. Bremer, W. Haasnoot, and W. Norde. 2009. Label-free and multiplex detection of antibiotic residues in milk using imaging surface plasmon resonance-based immunosensor. *Analytical Chemistry* 81:7743–7749.

Ren, X.L., X.M. Meng, D. Chen, F.Q. Tang, and J. Jiao. 2005. Using silver nanoparticle to enhance current response of biosensor. *Biosensors and Bioelectronics* 21:433–37.

Rocha-Gaso, M.-I., C. March-Iborra, A. Montoya-Naides, and A. Arnan-Vives. 2009. Surface generated acoustic wave biosensors for the determination of pathogens. *Sensors* 9:5740–5769.

Sadik, O.A., A.O. Aluoch, and A. Zhou. 2009. Status of biomolecular recognition using electrochemical techniques. *Biosensors and Bioelectronics* 24:2749–65.

Sassolas, A., B.D. Leca-Bouvier, and L.J. Blum. 2008. DNA biosensors and microarrays. *Chemical Reviews* 108:109–113.

Sirkar, K., and M.V. Pishko. 1998. Amperometric biosensors based on oxidoreductases immobilized in photopolymerized poly(ethylene glycol) redox polymer hydrogels. *Analytical Chemistry* 70:2888–2894.

Spiller, E., A. Scholl, R. Alexy, K. Kummerer, and G.A. Urban. 2006. A microsystem for growth inhibition test of *Enterococcus faecalis* based on impedance measurement. *Sensors and Actuators B: Chemical* 118:182–191.

Steel, A.B., T.M. Herne, and M.J. Tarlov. 1998. Electrochemical quantification of DNA immobilised on gold. *Analytical Chemistry* 70:4670–4677.

Stern, E., J.F. Klemic, D.A. Routenberg et al. 2007. Label-free immunodetection with CMOS-compatible semiconducting nanowires. *Nature* 445:519–522.

Stoll, D., M.F. Templin, J. Bachmann, and T.O. Joos. 2005. Protein microarrays: Applications and future challenges. *Current Opinion in Drug Discovery and Development* 8:239–52.

Strong, Z.A., A.W. Wang, and C.F. McConaghy. 2002. Hydrogel-actuated capacitive transducer for wireless biosensors. *Biomedical Microdevices* 4:97–103.

Subramanian, A., J. Irudayaraj, and T. Ryan. 2006. A mixed self-assembled monolayer-based surface plasmon immunosensor for detection of *E. coli* O157:H7. *Biosensors and Bioelectronics* 21:998–1006.

Susmel, S., C.K. O'Sullivan, and G.G. Guibault. 2000. Human cytomegalovirus detection by a quartz crystal microbalance immunosensor. *Enzyme and Microbial Technology* 27:639–645.

Tang, H., J.H. Chen, L.H. Nie, Y.F. Kuang, and S.Z. Yao. 2007. A label-free electrochemical immunoassay for carcinoembryonic antigen (CEA) based on gold nanoparticles (AuNPs) and nonconductive polymer film. *Biosensors and Bioelectronics* 22:1061–1067.

Tansil, N.C., F. Xie, H. Xie, and Z.Q. Gao. 2005. An ultrasensitive nucleic acid biosensor based on the catalytic oxidation of guanine by a novel redox threading intercalator. *Chemical Communications* 28:1064–1066.

Tok, J.B.-H., F.Y.S. Chuang, M.C. Kao et al. 2006. Metallic striped nanowires as multiplexed immunoassay platforms for pathogen detection. *Angewandte Chemie–International Edition* 45:6900–6904.

Tombelli, S., M. Mascini, C. Sacco, and A.P.F. Turner. 2000. A DNA piezoelectric biosensor assay coupled with a polymerase chain reaction for bacterial toxicity determination in environmental samples. *Analytica Chimica Acta* 418:1–9.

Tsekenis, G., G.-Z. Garifallou, F. Davis et al. 2008a. Detection of fluoroquinolone antibiotics in milk via a labeless immunoassay based upon an alternating current impedance protocol. *Analytical Chemistry* 80:9233–9239.

Tsekenis, G., G.-Z. Garifallou, F. Davis et al. 2008b. Labeless immunosensor assay for myelin basic protein based upon an AC impedance protocol. *Analytical Chemistry* 80:2058–2062.

Ulman, A. 1991. *An introduction to ultrathin organic films: From Langmuir-Blodgett to self-assembly.* Boston: Academic Press.

Ulman, A. 1998. *Thin films: Self assembled monolayers of thiols (thin films).* New York: Academic Press.

Urban, G.A. 2009. Micro- and nanobiosensors—State of the art and trends. *Measurement Science and Technology* 20:012001.

Wang, J. 2006. Electrochemical biosensors: Towards point-of-care cancer diagnostics. *Biosensors and Bioelectronics* 21:1887–1892.

Wang, J., and M. Musameh. 2005. Carbon-nanotubes doped polypyrrole glucose biosensor. *Analytica Chimica Acta* 539:209–213.

Wang, J., and Y.H. Lin. 2008. Functionalized carbon nanotubes and nanofibres for biosensing applications. *Trends in Analytical Chemistry* 27:619–626.

Wang, J., M. Jiang, A. Fortes, and B. Mukherjee. 1999. New label-free DNA recognition based on doping nucleic-acid probes within conducting polymer films. *Analytica Chimica Acta* 402:7–12.

Wang, Z.Z., T. Wilkop, D.K. Xu, Y. Dong, G.Y. Ma, and Q. Cheng. 2007. Surface plasmon resonance imaging for affinity analysis of aptamer–protein interactions with PDMS microfluidic chips. *Analytical and Bioanalytical Chemistry* 389:819–825.

White, I.M., H. Oveys, X. Fan, T.L. Smith, and J. Zhang. 2006. Integrated multiplexed biosensors based on liquid core optical ring resonators and antiresonant reflecting optical waveguides. *Applied Physics Letters* 89:191106.

Wilson, M.S. 2005. Electrochemical immunosensors for the simultaneous detection of two tumor markers. *Analytical Chemistry* 77:1496–1502.

Wu, G., R.H. Datar, and K.M. Hansen. 2001. Bioassay of prostate-specific antigen (PSA) using microcantilevers. *Nature Biotechnology* 19:856–860.

Xia, Y., and G.M. Whitesides. 1998. Soft lithography. *Annual Review of Materials Science* 37:550–575.

Xiao, Y., A.A Luhin, A.J. Heeger, and K.W. Plaxco. 2005. Label-free electronic detection of thrombin in blood serum by using an aptamer-based sensor. *Angewandte Chemie-International Edition* 44:5456–5459.

Xiao, Y., F. Patolsky, E. Katz, J.F. Hainfeld, and I. Willner. 2003. "Plugging into Enzymes": Nanowiring of redox enzymes by a gold nanoparticle. *Science* 299:1877–1881.

Xiao, Y., H.-X. Ju, and H.-Y. Chen. 1999. Hydrogen peroxide sensor based on horseradish peroxidase-labelled Au colloids immobilised on gold electrode surface by cystamine monolayer. *Analytica Chimica Acta* 391: 73–82.

Xu, S., and X. Han. 2004. A novel method to construct a third-generation biosensor: Self-assembling gold nanoparticles on thiol-functionalised poly(styrene-co-acrylic acid) nanospheres. *Biosensors and Bioelectronics* 19:1117–1120.

Yu, X., S.N. Kim, F. Papadimitrakopoulos, and J.F. Rusling. 2005. Protein immunosensor using single-wall carbon nanotube forests with electrochemical detection of enzyme labels. *Molecular BioSystems* 1:70–78.

Zhang, J., H.P. Lang, A. Bietsch et al. 2006. Rapid and label-free nanomechanical detection of biomarker transcripts in human RNA. *Nature Nanotechnology* 1:214–220.

Zhang, S.X., N. Wang, H.J. Yu, Y.M. Niu, and C.Q. Sun. 2005. Covalent attachment of glucose oxidase to an Au electrode modified with gold nanoparticles for use as glucose biosensor. *Bioelectrochemistry* 67:15–22.

Zhang, Y.C., H.H. Kim, and A. Heller. 2003. Enzyme-amplified amperometric detection of 3000 copies of DNA in a 10μL droplet at 0.5 fM concentration. *Analytical Chemistry* 75:3267–3269.

Zheng, G., F. Patolsky, Y. Cui, W.U. Wang, and C.M. Lieber. 2005. Multiplexed electrical detection of cancer markers with nanowire sensor arrays. *Nature Biotechnology* 23:1294–1301.

Zhu, L.H., H.Y. Zhu, X.L. Yang, L.H. Xu, and C.Z. Li. 2007. Sensitive biosensors based on (dendrimer encapsulated Pt nanoparticles)/enzyme multilayers. *Electroanalysis* 19:698–703.

Ziegler, C. 2004. Cantilever-based biosensors. *Analytical and Bioanalytical Chemistry* 379:946–959.

Index

Printed and bound by CPI Group (UK) Ltd, Croydon, CR0 4YY

21/10/2024

01777107-0010